Stable Isotopes in Ecology and Environmental Science

METHODS IN ECOLOGY

Stable Isotopes in Ecology and Environmental Science

EDITED BY

KATE LAJTHA

Department of Biology, Boston University

AND

ROBERT H. MICHENER

Department of Biology, Boston University

OXFORD

BLACKWELL SCIENTIFIC PUBLICATIONS

LONDON EDINBURCH BOSTON

MELBOURNE PARIS BERLIN VIENNA

© 1994 by
Blackwell Scientific Publications
Editorial Offices:
Osney Mead, Oxford OX2 0EL
25 John Street, London WC1N 2BL
23 Ainslie Place, Edinburgh EH3 6AJ
238 Main Street, Cambridge
  Massachusetts 02142, USA
54 University Street, Carlton
  Victoria 3053, Australia

Other Editorial Offices:
Librairie Arnette SA
1, rue de Lille
75007 Paris
France

Blackwell Wissenschafts-Verlag GmbH
Düsseldorfer Str. 38
D-10707 Berlin
Germany

Blackwell MZV
Feldgasse 13
A-1238 Wien
Austria

First published 1994

Set by Excel Typesetters Co., Hong Kong
Printed and bound in Great Britain
at the Alden Press,
Oxford and Northampton

DISTRIBUTORS

  Marston Book Services Ltd
  PO Box 87
  Oxford OX2 0DT
  (*Orders*: Tel: 0865 791155
    Fax: 0865 791927
    Telex: 837515)

USA
  Blackwell Scientific Publications, Inc.
  238 Main Street
  Cambridge, MA 02142
  (*Orders*: Tel: 800 759-6102
    617 876-7000)

Canada
  Oxford University Press
  70 Wynford Drive
  Don Mills
  Ontario M3C 1J9
  (*Orders*: Tel: 416 441-2941)

Australia
  Blackwell Scientific Publications Pty Ltd.
  54 University Street
  Carlton, Victoria 3053
  (*Orders*: Tel: 03 347-5552)

A catalogue record for this title
is available from the British Library

ISBN 0-632-03154-9

Library of Congress
Cataloging in Publication Data

Lajtha, Kate.
    Stable isotopes in ecology
    and environmental science/
    Kate Lajtha, Robert Michener.
      p.   cm. – (Methods in ecology)
    Includes bibliographical references
    and index.
    ISBN 0-632-03154-9
    1. Stable isotopes in ecological research.
  I. Michener, Robert.
  II. Title. III. Series.
  QH541.15.S68L35 1994
  574.5′0724 – dc20

Contents

List of contributors

L.A. CIFUENTES *Department of Oceanography, Texas A&M University, College Station, Texas, USA*

R.B. COFFIN *US Environmental Protection Agency, Gulf Breeze Environmental Research Laboratory, Sabine Island, Gulf Breeze, Florida, USA*

N.M. CONWAY *515 Hastings Street, Pittsburgh, Pennsylvania, USA*

P.M. ELDERIDGE *Department Of Oceanography, Texas A&M University, College Station, Texas, USA*

M.L. FOGEL *Carnegie Institution of Washington, Geophysical Laboratory, Washington, District of Columbia, USA*

B. FRY *The Ecosystems Center, Marine Biological Laboratory, Woods Hole, Massachusetts, USA*

R. GOERICKE *Scripps Institution of Oceanography, University of California, San Diego, La Jolla, California, USA*

M.C. KENNICUTT II *Geochemical and Environmental Research Group, Texas A&M University, College Station, Texas, USA*

P.L. KOCH *Department of Geological and Geophysical Sciences, Princeton University, Princeton, New Jersey, USA*

I. KOIKE *Ocean Research Institute, University of Tokyo, Nakano, Tokyo, Japan*

K. LAJTHA *Department of Biology, Boston University, Boston, Massachusetts, USA*

S.A. MACKO *Department of Environmental Sciences, The University of Virginia, Charlottesville, Virginia, USA*

J.D. MARSHALL *Department of Forest Resources, University of Idaho, Moscow, Idaho, USA*

R.H. MICHENER *Department of Biology, Boston University, Boston, Massachusetts, USA*

J.P. MONTOYA *Department of Organismic and Evolutionary Biology, Harvard University, Cambridge, Massachusetts, USA*

K.J. NADELHOFFER *The Ecosystems Center, Marine Biological Laboratory, Woods Hole, Massachusetts, USA*

N.E. OSTROM *Department of Geological Sciences, Michigan State University, East Lansing, Michigan, USA*

D.M. SCHELL *Institute of Marine Science, University of Alaska, Fairbanks, Alaska, USA*

N. TUROSS *Smithsonian Institution, Conservation and Analytical Laboratory, Suitland, Maryland, USA*

C.L. VAN DOVER *Department of Marine Chemistry and Geochemistry, Woods Hole Oceanographic Institution, Woods Hole, Massachusetts, USA*

M. WAHLEN *Scripps Institution of Oceanography, University of California, San Diego, La Jolla, California, USA*

T. YOSHINARI *Wadsworth Center for Laboratories and Research, New York Department of Health and School of Public Health, State University of New York at Albany, New York, USA*

The Methods in Ecology Series

The explosion of new technologies has created the need for a set of concise and authoritative books to guide researchers through the wide range of methods and approaches that are available to ecologists. The aim of this series is to help graduate students and established scientists choose and employ a methodology suited to a particular problem. Each volume is not simply a recipe book, but takes a critical look at different approaches to the solution of a problem, whether in the laboratory or in the field, and whether involving the collection or the analysis of data.

Rather than reiterate established methods, authors have been encouraged to feature new technologies, often borrowed from other disciplines, that ecologists can apply to their work. Innovative techniques, properly used, can offer particularly exciting opportunities for the advancement of ecology.

Each book guides the reader through the range of methods available, letting ecologists know what they could, and could not, hope to learn by using particular methods or approaches. The underlying principles are discussed, as well as the assumptions made in using the methodology, and the potential pitfalls that could occur – the type of information usually passed on by word of mouth or learned by experience. The books also provide a source of reference to further detailed information in the literature. There can be no substitute for working in the laboratory of a real expert on a subject, but we envisage this Methods in Ecology Series as being the 'next best thing'. We hope that, by consulting these books, ecologists will learn what technologies and techniques are available, what their main advantages and disadvantages are, when and where not to use a particular method, and how to interpret the results.

Much is now expected of the science of ecology, as humankind struggles with a growing environmental crisis. Good methodology alone never solved any problem, but bad or inappropriate methodology can only make matters worse. Ecologists now have a powerful and rapidly growing set of methods and tools with which to confront fundamental problems of a theoretical and applied nature. We hope that this series will be a major contribution towards making these techniques known to a much wider audience.

John H. Lawton
Gene E. Likens

Introduction

The past decade has seen a rapid expansion in the use of natural abundance isotopes in ecological research. Geochemists and palaeo-oceanographers have developed a rigorous theoretical and empirical basis for the integration of isotopes into studies of global element cycles, past climatic conditions, hydrothermal vent systems and tracing rock sources. Similarly, plant biologists, ecologists and environmental chemists are currently developing the theoretical framework and the empirical database for the use of isotopes to study plants and animals. As a consequence, isotope analysis is rapidly becoming a standard tool for physiologists, ecologists and all scientists studying element or material cycles in the environment.

As access to isotope ratio mass spectrometers has increased and prices for sample analysis have decreased, ecologists from a broad range of disciplines, who are not necessarily trained as isotope chemists, have increasingly added stable isotope analysis to their research. This volume is intended as a review and assessment of the theory and practice of stable isotope analysis in a variety of ecological disciplines, with suggestions for both generalist ecologists who might be considering including such analyses to their studies, as well as for the more experienced isotope ecologist who is pioneering new uses and new directions in the field. This volume includes chapters on terrestrial, atmospheric, marine and estuarine systems, as well as a chapter on stable isotopes at the molecular level. The topics were chosen to both complement other existing volumes on stable isotopes in ecology and to highlight some of the fastest growing uses of stable isotopes.

Natural abundance isotopic signatures can be used to find patterns and mechanisms at the single organism plant and algae level as well as to trace food webs, understand palaeodiets and follow whole ecosystem cycling in both terrestrial and marine ecosystems. Technically, ecological research using stable isotopes includes the experimental addition of isotopically labelled compounds at tracer levels, as well as measurements of natural abundance signatures. Due to the shear volume of material written on uses of labelled isotopes, we have chosen to concentrate on natural abundance research. However, many chapters mention the use of tracer-level experiments and additions to, or verifications of, experiments using natural abundance signatures.

Theory

Stable isotope ratios are measured using an isotope ratio mass spectrometer, which measures the ratio of the heavy and light isotopes in a sample and compares this to a standard. In principle, one can measure the absolute abundance of the isotopes within a sample; however, these differences are typically quite small and are subject to problems, such as sample heterogeneity, day-to-day fluctuations within the mass spectrometer and sample preparation (Hayes, 1982). Therefore, in practice, the isotope ratio of a sample (R_{sa}) is compared to a standard (R_{std}), so that any fluctuations will be reflected equally in both standard and sample. R is expressed as the ratio of the heavy to light isotope. The differences in ratios are calculated in 'del' (δ) notation and have units of per mil (‰):

$$\delta \ (‰) = \frac{R_{sa} - R_{std}}{R_{std}} \times 1000 \quad \text{or} \quad (R_{sa}/R_{std} - 1) \times 1000.$$

Most ecologists are concerned with the stable isotopes of carbon ($^{13}C/^{12}C$), nitrogen ($^{15}N/^{14}N$), sulfur ($^{34}S/^{32}S$), oxygen ($^{18}O/^{16}O$) and hydrogen ($^{2}H/^{1}H$). All isotope values reported in the literature are referenced to primary standards. For carbon this is a marine limestone fossil, Pee Dee Belemnite (PDB) (Craig, 1953); for nitrogen, atmospheric air (Mariotti, 1983); for sulfur the troilite standard of the Canyon Diablo meteorite (CDT) (Krouse, 1980); and for hydrogen and oxygen, Vienna Standard Mean Ocean Water (V-SMOW) (Gonfiantini, 1978). Samples that contain more of the heavy isotope are referred to as 'enriched' and are 'heavier' than other samples; those that contain less of the heavy isotope are 'depleted' and are 'lighter' than other samples.

Many chemical and physical processes have a significant isotopic fractionation, which generally refers to an enrichment or depletion of the heavy isotope. For example, plants contain less ^{13}C than does atmospheric CO_2 because processes involved in both diffusion of CO_2 and enzymatic reactions discriminate against the heavier $^{13}CO_2$. These fractionations can occur during time-dependent or kinetic processes, as well as during equilibrium processes. An isotope effect, often referred to as the fractionation factor for the equilibrium reaction A $\leftrightharpoons$ B, is defined as:

$$\alpha = R_A/R_B.$$

Fractionation during unidirectional kinetic reactions, rather than in equilibrium reactions, is expressed as a ratio of reaction rates:

$$\alpha = k_1/k_2$$

where k_1 and k_2 are reaction rates of the light and heavy isotopes, respectively, by standard convention and the convention that we will follow. However, the reader should note that several isotope chemists place the heavy isotope in the numerator and the light isotope in the denominator.

In kinetic reactions the fractionation factor is an inherent or instantaneous measure of isotopic fractionation. However, the isotopic composition of the product can vary depending on the extent of the reaction. Thus authors have variously defined $\Delta\delta$, often referred to as Δ and as either isotopic discrimination or fractionation, as:

$$\Delta_{A/B} = \frac{\delta_A - \delta_B}{1 + \delta_B/1000}.$$

Note that the definition of this term as used here has units of per mil (‰), as does δ, yet refers to a difference between product and reactant rather than a value for a single compound. Also note that in most cases the denominator is very close to 1.0, and thus many authors have simplified the above definition of Δ to:

$$\Delta = \delta_A - \delta_B.$$

Because this term has received many numerical definitions, authors generally derive and define $\Delta\delta$ rather than relying on a standard convention, as do Goericke *et al.*, in Chapter 9.

Finally, isotopic segregation can also be expressed as the isotope enrichment factor, expressed as per mil:

$$\varepsilon = (\alpha - 1) \times 1000.$$

Note that when substrate is not limiting and the expression of the isotope factor is at a maximum,

$$\Delta \approx \varepsilon.$$

The usefulness of stable isotope ratios to ecologists arises from predictable physical and enzymatic-based discrimination between biological and non-biological materials which leads to different isotopic compositions (Ehleringer *et al.*, 1986). For instance, plants have been well characterized and divided into Calvin cycle (C_3), Slack–Hatch cycle (C_4) and Crassulacean acid metabolism (CAM) categories based on their method of photosynthesis (O'Leary, 1981; Osmond *et al.*, 1981), as discussed in Chapter 1. Further, several criteria must be met in order to use natural abundance stable isotope signatures in ecology. If isotopes are to be used as indicators of the origins of materials in the environment, then the potential sources must be isotopically distinct from each other and the signatures must either not change or change predictably, as materials are transported and transformed in the environment. If isotopes are to be used as indicators of ecological (chemical or biotic) processes, then the extent of isotopic fractionation, discrimination and segregation must be known for all reactions.

Overview of methods

Many ecologists using stable isotopes will, and perhaps should, choose to send their samples to outside laboratories that specialize in the analysis of stable isotopes. The costs for an individual to set up this type of laboratory are quite high, and typical startup budgets can be as high as $500 000; maintenance of the mass spectrometer and the costs of having a full-time, trained laboratory manager run the laboratory are also steep. One can contrast this to the analysis costs of a typical study, which could be about $5000–10 000 (with per sample charges averaging $10–100, depending on the sample matrix and isotope in question). For many scientists it is much more cost-effective to use an outside laboratory.

Methods of sample preparation vary for each isotope. The goal in stable isotope analysis is to convert a sample quantitatively to a suitable purified gas (typically CO_2, N_2 and H_2) which the mass spectrometer can then analyze. Sulfur can be analyzed as SO_2 or SF_6. Usually, organic samples are first dried (either in a 60°C oven or freeze-dried) and then ground to a fine powder using a mortar and pestle or a Wiley Mill with a fine mesh screen. Because of possible sample heterogeneity, the samples should be well ground and mixed. The samples can then be stored indefinitely in closed containers (such as scintillation vials or plastic bags), provided they are kept dry. If the investigator is interested in carbon isotopes for samples that may contain inorganic carbonates, the samples must first be acidified (usually with 1 N HCl, although some investigators are using dilute H_3PO_4; Showers & Angle, 1986), since carbonate isotopic values are heavy and will skew the results (Haines & Montague, 1979; Fry, 1988).

Carbon and nitrogen in organic matter

Many researchers then use an oxidation reaction either 'off-line' (sealed tubes in a muffle furnace, referred to as a Dumas combustion) or 'on-line' (sample preparation line connected directly to the mass spectrometer) to combust the organic sample to a gas. The off-line combustion involves mixing the sample (typically 5–20 mg, depending on the sample's organic content) with an oxidant, usually cupric oxide, in a vycor (quartz) tube. The sample must be in intimate contact with the CuO, which can be done several ways: shaking the sample vigorously with the CuO within the tube, grinding both in a mortar and pestle, or using a Wig-L-bug (Crescent Dental Manufacturing). Shaking or using a Wig-L-bug is preferred, since there is less chance of sample cross-contamination. Approximately 1 g of CuO is used, then about 0.5 g of Cu is placed on top of the sample mix within the vycor tube. Once all sample tubes are prepared, they are then sealed under vacuum. The tubes are placed in a muffle furnace and combusted at 900°C for 1–2 hours; the temperature is then reduced to 650°C and held at that temperature for 2 hours. This ensures that all CO is converted to CO_2, NO_x to N_2, and that halogens and sulfur are removed. This long cooldown also ensures that the excess oxygen is absorbed by the copper. The furnace is allowed to cool overnight to room temperature. The combined gases of CO_2, N_2 and H_2O can then be cryogenically separated and purified (Stump & Frazer, 1973; Boutton et al., 1983; Minigawa et al., 1984; Nevins et al., 1985). With manual samples, this is done on a vacuum line using liquid nitrogen and an ethanol/dry ice slush. The liquid nitrogen will trap CO_2 and water, and nitrogen is trapped in a sample bulb containing either molecular sieve or silica gel (Mariotti, 1984; Nevins et al., 1985) since these materials provide a large surface area to sorb nitrogen gas. After collection of N_2 gas, water and CO_2 are separated using a warmer dry ice/ethanol slush, which allows the CO_2 to sublimate and be transferred to another sample container. The remaining water can be transferred to a third container for later $\delta^{18}O$ and δD determinations, but one must be careful to collect all the water to prevent any isotopic fractionation. An important point to note for these and all following procedures is that the combustions and collections must be quantitative and close to 100% efficiency to prevent any fractionation. From here the gas samples are introduced into the isotope ratio mass spectrometer.

There are now semi-automated combustion systems available using elemental (CN) analyzers coupled to cryogenic purification systems that reduce sample preparation times and cost per analysis (Fry et al., 1992). A CN analyzer can be linked through a gas purification system to the mass spectrometer, which allows coupled analysis of carbon and nitrogen isotopic compositions. This type of system is appropriate for most organic tissue samples, sediment and soil samples containing sufficient organic matter, as well as materials such as collagen and some plankton samples. There are also encapsulators available that allow liquid samples to be analyzed. Some of the newer cryogenic systems can analyze samples containing as little as 1 µmol N and 1 µmol C. Depending on the type of system used and the type of sample being analyzed, 1–20 mg of material is loaded into a tin boat, folded, then placed in the sample carousel. Combustion and separation of the gases from the helium stream is, in principle, similar to the manual method. The sample is flash-combusted at

1600–1800°C in an oxygen stream, then the combustion gases are carried in the helium stream through a series of cryogenic traps, which are maintained at specific temperatures to collect H_2O, CO_2 and N_2. The gas of interest is then introduced into the mass spectrometer for analysis.

Another type of automated combustion system involves introducing the helium stream containing the combusted gases directly into the mass spectrometer. A series of chemical traps remove water and other contaminants. In general, a chromatographic column is used to separate CO_2 and N_2 before it enters the mass spectrometer, and both gases can be measured from the same combustion. The system is very rapid and can analyze around 100 samples per day. The results so far have been mixed for natural abundance measurements, however this type of system works well for enriched isotope work. Nevertheless, as more of these systems become available and are refined, the precision should improve for natural abundance samples.

Carbonates and dissolved inorganic carbon (DIC)

Inorganic carbonate samples (for example, foraminifera for palaeotemperature studies) are reacted under vacuum with 100% phosphoric acid and purified (Craig, 1953). This allows for the analysis of both $\delta^{13}C$ and $\delta^{18}O$ from the same sample, provided the H_3PO_4 is pure and contains no water (Coplen et al., 1983). In general, a system is used that allows the investigator to keep the acid and carbonate separate before the two are combined for the reaction. Once the system is under vacuum and free of water, the carbonate is introduced into the acid and allowed to react at a constant temperature. The time required for the reaction will vary with sample size and with the temperature of the acid (Coleman & Fry, 1991). Purification of the CO_2 gas is similar to that outlined above.

DIC in water samples is prepared by acidifying a water sample and stripping with CO_2-free gas under a partial vacuum (Kroopnick, 1974; McCorkle & Emerson, 1988), then isolating and purifying the gas. The same principle can be applied to samples of bicarbonate in blood for tracer studies (Moulton-Barrett et al., 1993).

Ammonia and nitrate $\delta^{15}N$ in water samples

Ammonium in water samples can be isolated using various steam distillation techniques (Velinsky et al., 1989) or using passive diffusion within a closed container (Brooks et al., 1989). Both procedures involve making the pH of the water sample basic, then trapping or collecting the NH_3 in an acid trap. Steam distillation techniques are good for large water samples containing low levels of NH_3, can be used on salt water solutions, and take about 30 minutes per sample (Velinsky et al., 1989). Another distillation procedure, often called passive distillation, also involves driving the ammonia into an acid trap by making the solution strongly basic and distilling the ammonia gas into a weak acid trap. However, this procedure takes longer and water samples must be concentrated to about 300 ml. Once the ammonia is collected in an acid trap using either procedure, zeolite is used to remove NH_3 from solution. The zeolite is then dried and can be analyzed using the sealed tube Dumas combustion. In both methods, the investigator must be careful to trap all of the NH_3 in all steps in order to avoid fractionation. Nitrate-N can be distilled using the same techniques after first reducing the nitrate to ammonia with Devarda's Alloy.

The passive diffusion techniques work well for samples such as soil solutions or Kjeldahl digests, and can be done in a batch fashion. Two different procedures are used, one involving suspending an acidified filter paper (usually with H_2SO_4) above the solution, the other wrapping the filter paper in PTFE tape and floating the packet in the solution. The solution is made basic and, using the same principle as the distillation technique, the ammonia diffuses onto the acidic filter paper. After the diffusion is complete (anywhere from 3–5 days), the filter paper is dried and can be combusted using the automated CN-mass spectrometer system. Nitrate-N is similarly isolated following the addition of Devarda's alloy to the solution. One of the disadvantages of this method is that low concentrations of NH_3 and NO_3 must be first concentrated into a volume of 100 ml in order for the diffusion technique to be effective. This could be done by boiling the solution (after first acidifying with H_2SO_4 to prevent NH_3 volitilization) or using ion exchange resins to separate and trap NH_3 and NO_3. Also, the method is not effective on certain salt solutions, such as seawater (M. Downs, personal communication 1993). There is also the possibility that boiling the water sample may break down any organics that are present, adding nitrogen to the solution. The passive diffusion procedures have not been well tested for natural abundance work, however the procedures do work well for ^{15}N-enriched samples. These are by no means the only techniques to measure NH_3 and NO_3 in water samples, and we refer the reader to the volume by Knowles & Blackburn (1993) for further details on these and other methods.

Oxygen in water

The measurement of [18]O in water samples can be accomplished using several different procedures. These techniques involve measuring $\delta^{18}O$ in H_2O, not dissolved oxygen. The most common procedure, applicable to larger water samples (such as ground water), uses 1–3 ml of water (Taylor, 1973; Wong *et al.*, 1987b; Socki *et al.*, 1992). The water sample is first placed in a suitable vessel such as a vacutainer or serum bottle. After removing the headspace atmosphere, a measured aliquot of CO_2 of known isotopic composition is then introduced into the headspace. The water is incubated at a controlled temperature for a period of time that allows the oxygen in the water to completely exchange with the oxygen in the CO_2, after which the headspace CO_2 is removed using cryogenic techniques, then analyzed. This technique has been automated and is suitable for larger volumes of water. For samples in the range of $25-100\,\mu l$, this procedure can be done manually using 6 mm pyrex tubes (Porter, personal communication 1992).

A different procedure is used for small volume samples in the range of $3-10\,\mu l$, which may be generated from samples such as small animal metabolic studies, plant water or combusted organic matter. This technique utilizes guanidine hydrochloride to release the oxygen. First, 100 mg of ultrapure guanidine hydrochloride is loaded into 9 mm pyrex tubes, attached to a vacuum line and evacuated. The tube and guanidine are gently heated to drive off excess water, then the water sample is transferred onto the guanidine hydrochloride and the sample tube is sealed. The tube is incubated at 260–290°C for 16 hours to form ammonium carbamate and ammonium chloride. The CO_2 is released from the ammonium carbamate by reacting the sample in a second vessel containing 100% phosphoric acid at 80°C for 1 hour, the acid also absorbs excess ammonia. From here the CO_2 is purified using a vacuum line and cryogenic distillation. For further details, see Wong *et al.* (1987b).

Deuterium

In order to measure ^2H or deuterium from organic tissue, the sample is combusted using an off-line, sealed tube procedure and the resulting water is collected quantitatively (Schiegl & Vogel, 1969). The procedure is also used for other types of water samples, such as plant water, ground water and water obtained for metabolic studies. The water is reduced to H_2 using either a vacuum line and uranium furnace, or using zinc in a sealed vessel (Krishnamurthy & DeNiro, 1982; Coleman *et al.*, 1982). Many investigators are using the zinc method, as it can be done in a batch fashion and avoids any problems associated with obtaining uranium for the furnace. The procedure involves using about 100 mg of zinc and placing it either in a 6 mm pyrex tube or in a special sample bulb. The tube/bulb is attached to a vacuum line, evacuated, and the zinc is heated to drive off any trapped water vapor. The water sample is transferred onto the zinc and the vessel is then closed off. The zinc and water are heated to about 500°C for 40 minutes, which reduces the water to H_2 gas. The hydrogen can then be introduced directly into the mass spectrometer. Although there is some speculation that the combusted samples can be stored for long periods, the H_2 should be analyzed within 24 hours.

Sulfur

The analysis of sulfur isotopes depends on the starting matrix, but in essence involves converting sulfur in the sample to SO_2 or SF_6. Sulfur hexafluoride has the advantage that fluorine has only one isotope, but the techniques involved are somewhat hazardous, therefore most laboratories use SO_2 gas. As yet this analysis is not automated, but ongoing research is working to at least semi-automate the procedure. Most of the procedures to isolate sulfur from its matrix (water, plant and animal tissue, soils, sulfides) generally involve oxidizing sulfur to sulfate in solution. The sulfate can then be precipitated as $BaSO_4$ using a 10% barium chloride solution. From here the sample is oxidized to SO_2 gas and introduced into the mass spectrometer. These procedures are generally not done in the laboratory of an ecologist, due to the labor, materials and time involved. For a more detailed description of sulfur preparation, see Krouse & Tabatabai (1986).

The reproducibility of isotope measurements will depend on the procedure and laboratory techniques of the investigator, but is typically ±0.2‰ for carbon, nitrogen and sulfur, and 0.3–2‰ for hydrogen. This methods section is a brief introduction to the procedures involved in preparing stable isotope samples. For more elaboration and further details on other methods, we refer the reader to volumes by Coleman & Fry (1991) and Knowles & Blackburn (1993).

In the preparation of this volume, we asked many colleagues to read and review each chapter. For their valuable advice and criticism we thank: Mark Altabet, Wood Hole Oceanographic Institution; Joe Berry, University of California; Colleen Cavanaugh, Harvard University; Mike Engel, University of Oklahoma; Brian Fry, Ecosystems Center, Marine Biological Laboratory; Patrica Glibert, Horn Point Environmental Laboratory; Meredith Hullar, Ecosystems Center, Marine Biological Laboratory; Ralph Keeling, National Center for Atmospheric Research; Bruno Marino, Harvard University; Marion O'Leary, University of Nebraska; and Patrick Parker, University of Texas at Austin. Lastly, we thank our colleagues at Boston University for their patience during the final stages in the preparation of this volume.

CHAPTER 1

Sources of variation in the stable isotopic composition of plants

K. LAJTHA & J.D. MARSHALL

1.1 Introduction

The use of stable isotopes of carbon, nitrogen, oxygen and hydrogen to elucidate plant photosynthetic pathways and to study other physiological processes has increased exponentially in the last two decades. Harmon Craig (1953, 1954), a geochemist and early pioneer of natural abundance stable isotopes, first measured isotopic values of plant materials, and he found that plants tended to have a fairly narrow $\delta^{13}C$ range of -25 to $-35‰$. However, he was not able to find large taxonomic or environmental effects on these values. Since that time ecologists have identified clear isotopic signatures based not only on taxonomic affiliation – due primarily to different photosynthetic pathways – but also on ecophysiological differences, such as photosynthetic water-use efficiency (WUE), source of water used and degree of symbiotic N_2 fixation. This chapter reviews the most common applications of stable isotope analysis in plant ecophysiology, comparing these techniques to more conventional ecophysiological tools. Now that this field is coming of age and large empirical databases are accumulating, scientists are discovering sources of error that sometimes cause a mismatch between theoretical and observed values, as well as confounding effects from sources and factors not previously considered. However, the discovery of confounding effects or unexpected results can, and has, led to important insights into physiological or ecological processes, as well as new uses of stable isotopes in plant ecophysiology.

1.2 Carbon isotopes

1.2.1 Photosynthetic pathways

Plants contain less ^{13}C than the atmosphere due to both enzymatic and physical processes that discriminate against the heavier isotope in favour of the lighter one. Differences in $\delta^{13}C$ among plants using the Calvin cycle (C_3), Hatch–Slack cycle (C_4) and Crassulacean acid metabolism (CAM) photosynthetic pathways are due to differences in fractionation at the diffusion, dissolution and carboxylation steps, and are discussed in detail by O'Leary (1981, 1988a). These differences have allowed ecologists to use isotopic signatures to identify photosynthetic pathways of many plant species.

In C_3 plants, atmospheric CO_2 diffuses through the boundary layer and the stomata into the internal gas space with an apparent fractionation ($\Delta\delta$) of ~4‰. The carboxylating enzyme ribulose bisphosphate carboxylase (rubisco) discriminates further against the heavier isotope, with a $\Delta\delta$ of about 29‰. If atmospheric diffusion were the sole limitation for CO_2 uptake, i.e., if all CO_2 entering the leaf was fixed, then we would expect to see only the fractionation of 4‰. This 4‰ would be subtracted from the $\delta^{13}C$ value for CO_2 in the atmosphere, which is about -8‰, yielding a $\delta^{13}C$ of -12‰. If, however, CO_2 uptake were purely enzyme limited, then only the rubisco fractionation would be observed, yielding a predicted leaf $\delta^{13}C$ value of about -37‰. In fact, median $\delta^{13}C$ values for C_3 plants lie between the two extremes, at about -27‰, slightly nearer the carboxylation-limited value. Variation in $\delta^{13}C$ values among C_3 plants depends on the balance between diffusive supply and enzymatic demand for CO_2; variation in this balance point has been exploited by ecologists (see Section 1.2.2).

Isotopic composition of C_4 plants is quite different from that of C_3 plants because the initial carboxylating enzyme is different. Phosphoenolpyruvate (PEP) carboxylase has a $\Delta\delta$ of -6‰ for the fixation of CO_2 (Farquhar, 1983). Enzyme-limited uptake would be expected to yield a leaf $\delta^{13}C$ value of about -2‰, but diffusion-limited uptake should be the same as that for C_3 plants, namely -12‰. Actual values cluster closer to about -14‰, slightly more negative than the diffusion-limited situation. This is attributed to 'leakage' of CO_2 from the bundle sheath, the site of CO_2 uptake by rubisco. Because rubisco fractionates against ^{13}C, the leaked CO_2 has a high $\delta^{13}C$, which will reduce the $\delta^{13}C$ of the carbon left behind (Ehleringer & Pearcy, 1983; Berry, 1989). Support for this hypothesis comes from evidence that C_4 plants with high bundle sheath permeability tend to be most negative in $\delta^{13}C$ (Hattersley, 1982). Other important considerations might include respiratory losses of fixed CO_2 or fractionation during development (O'Leary, 1988a). C_4 plants do not show evidence of variation in the balance of enzymatic vs. diffusive limitations; therefore, environmental effects on $\delta^{13}C$ cannot be interpreted as they can be for C_3 plants.

Plants using the CAM pathway have the same two carboxylating enzymes as do C_4 plants, but segregate the activities of the enzymes between night and day rather than between tissues. CO_2 is initially fixed by PEP carboxylase into C_4 acids at night. The carbon is then released from the acids and refixed by rubisco during the day. Some species are facultatively CAM rather than obligately CAM, meaning they may switch to daytime C_3 photosynthesis when conditions are favourable. When obligate-CAM plants fix CO_2 during the night using PEP carboxylase, they should fractionate like C_4 plants, with an associated leakage of CO_2 due to the transfer of C_4 acids to the daytime C_3 cycle (Farquhar, 1983). These species tend to have $\delta^{13}C$ values that cluster around $-11‰$. Facultative-CAM plants have shown $\delta^{13}C$ values intermediate between this value and that of obligate-C_3 plants. Thus carbon isotope signatures can separate obligate-CAM from C_3 plants, and can be used to estimate the proportion of CAM vs. C_3 photosynthesis in facultative-CAM plants (Osmond *et al.*, 1976; Teeri & Gurevitch, 1984; Kalisz & Teeri, 1986; Smith *et al.*, 1986; Mooney *et al.*, 1989; Kluge *et al.*, 1991).

Because carbon isotopes cannot effectively distinguish CAM from C_4 plants, this separation has traditionally been made by other methods, such as anatomy (succulence) or physiology (diurnal malic acid cycling). However, there is substantial empirical evidence that CAM plants are considerably enriched in deuterium relative to source water (Fig. 1.1; Ziegler *et al.*, 1976; Sternberg & DeNiro, 1983; Sternberg *et al.*, 1984a,b,c; Sternberg, 1989). The mechanism for this enrichment is not completely understood, but it is probably due to fractionations associated with

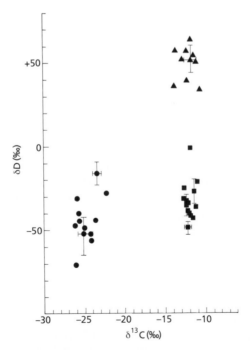

Fig. 1.1 Hydrogen vs. carbon isotope ratios of plant cellulose nitrate for plants having different photosynthetic pathways. ■ = C_4; ● = C_3; ▲ = Crassulacean acid metabolism; + = precision of isotopic analysis. (From Sternberg *et al.*, 1984a.)

carbohydrate metabolism unique to plants with CAM photosynthesis (Sternberg, 1989). Sternberg *et al.* (1985) used a combination of δD and ^{14}C analysis to confirm that *Stylites andicola*, a terrestrial vascular fern ally that lacks stomata, obtains its carbon from decomposing peat and that the CO_2 is processed via CAM. Similarly, δD values have been used to confirm the presence of CAM in aquatic plants when variations in $\delta^{13}C$ of source CO_2 precluded use of $\delta^{13}C$ from plant tissues (Sternberg *et al.*, 1984c).

1.2.2 Plant water-use efficiency

Perhaps the most common ecological use of carbon isotope ratios at natural abundance levels is as an indication of photosynthetic WUE in C_3 species. WUE is traditionally defined as A/E, or the ratio of net photosynthesis to transpiration. Farquhar *et al.* (1982) demonstrated that carbon isotope ratios in C_3 plants are a reliable long-term measure of intercellular CO_2 levels according to the equation:

$$\delta^{13}C_{leaf} = \delta^{13}C_{atm} - a - (b - a)c_i/c_a$$

where $\delta^{13}C_{atm}$ is $-7.8‰$, a is the fractionation caused by diffusion (4.4‰), b is the fractionation associated with carbon fixation (27‰), and c_i/c_a is the ratio of intercellular to ambient atmospheric concentrations of CO_2. Intercellular concentrations of CO_2 are of interest to ecologists because they can easily be estimated from gas exchange measurements, and c_i can be shown to be negatively related to WUE as:

$$A = (c_a - c_i)g/1.6$$

$$E = g(\text{LAVD})$$

$$\text{WUE} = A/E = (c_a - c_i)/(1.6(\text{LAVD}))$$

where A is net photosynthesis, E is transpiration, g is conductance to water vapour, 1.6 is the ratio of diffusivities of water vapour and CO_2 in air, and LAVD is the leaf to air vapour deficit (Farquhar & Richards, 1984). Because c_a is essentially constant, WUE varies with c_i and LAVD; if LAVD is assumed constant over the species or plants being considered, then $\delta^{13}C_{leaf}$ should be correlated with WUE.

This relationship can be understood intuitively if one considers the relative behaviour of ^{12}C and ^{13}C as CO_2 enters the leaf. When stomata are nearly closed and stomatal conductance is reduced, c_i falls and almost all of the intercellular CO_2 reacts with rubisco, whether it is $^{12}CO_2$ or $^{13}CO_2$. Reduced stomatal conductance means a reduction in water loss (E), yet because c_i is lowered, the CO_2 concentration gradient into the leaf increases, and thus there is a smaller reduction in CO_2 uptake (A)

and an overall increase in A/E. All other things being equal, less fractionation (a more positive $\delta^{13}C$ value) would thus be correlated with increased WUE. However, the above assumptions are not always met. In particular, LAVD might be reduced if high transpiration rates cooled the leaf or if gas exchange took place only during cool portions of the day. These would reduce leaf temperature (and therefore LAVD) during periods of carbon gain, increasing WUE. Such a pattern has been observed on comparing coastal to inland genotypes of Douglas-fir (Fig. 1.2; Zhang et al., 1993). In this study, inland genotypes maintained high conductance, and therefore high transpiration despite high LAVD in the mid-afternoon; coastal genotypes did not. Average LAVD, weighted by photosynthetic rates, was therefore different, as was WUE. However, these differences would not be reflected in c_i or in $\delta^{13}C$.

Traditional gas-exchange measurements have relied on infrared gas analysis (IRGA) and humidity sensors to measure A and E, respectively. Closed-system IRGA techniques can measure A/E on an instantaneous (15–30 seconds) basis, and either many measurements on one plant or many plants can be measured each day. Open-system, computer-controlled IRGA techniques can be used to get more integrated, day-long measurements of WUE on a single leaf or a single plant (Field et al., 1989). Although IRGA techniques can be used to understand short-term gas exchange, long-term estimation of WUE is more difficult because stomatal conductance and carboxylation efficiencies change as environmental conditions change over the life of a leaf. In contrast, carbon isotope analysis provides a technique for examining long-term WUE of plant tissue; this estimate integrates across the time during which carbon in the leaf was fixed. The results are not directly analogous to gas-exchange measurements, however, as $\delta^{13}C$ provides no information on absolute rates of gas exchange. Constraints on comparisons of isotope ratios among different species, or among plants grown in different environments, are still being clarified, as will be discussed below.

The use of $\delta^{13}C$ analysis to estimate long-term WUE has been the subject of detailed reviews and is becoming almost routine (Ehleringer, 1989, 1991; Ehleringer & Osmond, 1989; Farquhar et al., 1989a,b). A common application of isotope ratios has been for screening cultivars of dryland crop and rangeland grass species. The isotope data has compared well with various other methods of estimating WUE and the results have been remarkably consistent across environments (Farquhar & Richards, 1984; Condon et al., 1987; Hubick et al., 1988; Ehleringer et al., 1990; Johnson et al., 1990; Ehdaie et al., 1991; Read et al., 1991). Because these measurements are made with one species in a constant environment, environmental effects of temperature and LAVD are minimized.

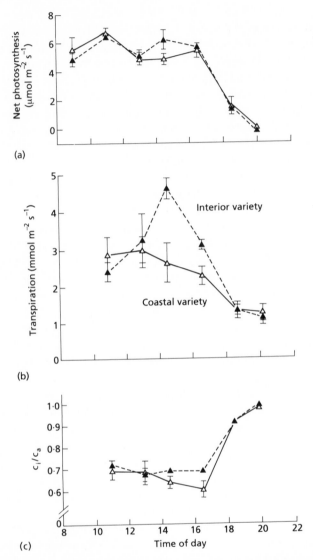

Fig. 1.2 Diurnal patterns of (a) photosynthesis (A), (b) transpiration (E) and (c) c_i/c_a from a common garden experiment including genotypes from both the coastal and interior races of Douglas fir. (From Zhang *et al.*, 1993.)

Reported heritabilities are often high (Hubick *et al.*, 1988; Farquhar *et al.*, 1989b; Hall *et al.*, 1990; Ehdaie *et al.*, 1991), suggesting that WUE is a trait for which breeding programmes can be readily designed.

Other insights have been gained by ecologists conducting surveys of carbon isotope ratios in natural ecosystems. Much of this research has

focused on WUE in desert ecosystems or across gradients of water availability in mesic systems. Within individual plants, Ehleringer *et al.*
(1987a) and Comstock & Ehleringer (1988) compared WUE in desert species that
had photosynthetic stems as well as leaves, and found that $\delta^{13}C$ values
were more positive, and thus WUE higher, in stem than in leaf tissue.
Instantaneous gas-exchange data confirmed isotope results. Comparing
different species within a habitat, Smedley *et al.* (1991) found that short-
lived annual or herbaceous species had significantly lower $\delta^{13}C$ values
than long-lived perennial species. Species active during the wetter, more
favourable months discriminated against ^{13}C more than species that
persisted over dry seasons, reflecting a higher WUE in the more drought-
tolerant species.

Ehleringer & Cooper (1988) compared desert plants that spanned a
gradient from wash (an intermittent streambed) to drier upland slope,
and found that leaf carbon isotope ratios were higher in plants from the
wash compared to those from the slope. This suggests that plants from
the drier habitat exhibited a higher WUE than plants from the wetter
habitat, although LAVD and temperature were not necessarily held
constant across this study. Garten & Taylor (1992) also compared plants
over a large environmental gradient, using trees in the Walker Branch
Watershed on ridges, coves and riparian zones, and found that plants in
drier microhabitats had more positive $\delta^{13}C$ values than in mesic areas.
Similarly, foliar $\delta^{13}C$ was more positive during a dry year than a wet year.
However, by factoring in differences in relative humidity (and thus
LAVD) and temperature, these authors calculated that A/E was higher
during the wet year than the dry year due to higher transpiration in the
dry year. Such a calculation demonstrates the importance of including
micrometeorological differences when comparing species over large environmental gradients.

In contrast to these studies, DeLucia & Schlesinger (1991) found that
drought tolerance of desert plants, as measured by habitat distribution
and the ability to maintain high rates of net photosynthesis at low pre-
dawn water potentials, was not correlated with high WUE. Foliar $\delta^{13}C$
was more negative (by inference, WUE was lower) in Great Basin shrubs
than in co-occurring trees, although the shrubs were more drought
tolerant. Note, however, that in both of these last two studies different
species were compared and environmental conditions, such as LAVD,
might have changed.

Few studies have attempted to correlate instantaneous, IRGA-derived
measurements of WUE with WUE measured using stable isotopes, although Toft *et al.* (1989) found excellent agreement between instantaneous
WUE and $\delta^{13}C$ values in experimental manipulations of a sagebrush
steppe ecosystem. Similarly, Lajtha & Barnes (1991) measured foliar

δ^{13}C as well as instantaneous WUE at A_{max} of Colorado pinyon pine and one-seed juniper in northern New Mexico. In general, juniper is more drought tolerant than pinyon, extending into drier habitats and maintaining positive carbon gain at lower water potentials. Pinyon, however, had higher foliar δ^{13}C than juniper, implying higher WUE. This inverse relationship agreed with the observations of DeLucia & Schlesinger (1991) for Great Basin vegetation. However, instantaneously measured WUE at A_{max} was significantly higher in juniper than pinyon during dry seasons. This discrepancy is probably due to the fact that isotopes reflect WUE over the entire lifetime of the leaf, and critical seasons and times of day can easily be missed when relying on IRGA measurements. Both measurements of WUE significantly increased with nitrogen fertilization in pinyon; δ^{13}C values suggested that WUE similarly increased with fertilization in juniper, but this was not reflected in instantaneous A or WUE measurements.

In a separate study, Lajtha & Getz (1993) found that neither A_{max} nor WUE measured at A_{max} varied significantly in either pinyon or juniper over a narrow water-availability gradient. However, δ^{13}C analysis suggested that while isotope-determined estimates of WUE did not vary significantly over the wetter end of the gradient, in the driest sites where pinyon was either scarce or absent, WUE in juniper increased significantly, agreeing with results of Ehleringer & Cooper (1988) for desert microhabitats (Fig. 1.3).

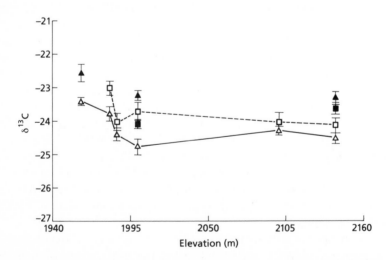

Fig. 1.3 Pinyon ($\square$) and juniper ($\triangle$) δ^{13}C over an elevational, and thus water availability, gradient in northern New Mexico. $\blacksquare$ and $\blacktriangle$ represent nitrogen-fertilized plants. δ^{13}C increases in the two lowest elevation sites; at the lowest elevation, only juniper was present. (From Lajtha & Getz, 1993.)

Altitude has been shown to affect carbon isotope ratios strongly and consistently. Körner *et al.* (1988), in a global survey of plants across altitudinal gradients, found consistent increases in $\delta^{13}C$ with altitude, although there was considerable variation among species at a given altitude and local trends along short gradients could vary (e.g., Friend *et al.*, 1989). Vitousek *et al.* (1990) found that foliar $\delta^{13}C$ of *Metrosideros polymorpha* increased with elevation on wet lava flows but not dry lava flows on the Mauna Loa volcano, and that values did not reflect patterns of precipitation or presumed water availability. Specific leaf weight, or leaf mass per area, increased significantly with elevation, and was strongly correlated with foliar $\delta^{13}C$. CO_2 gas-exchange data suggested that c_i did not change significantly with elevation and was not related to foliar $\delta^{13}C$. The authors hypothesized that there was an increased internal resistance to CO_2 diffusion from substomatal cavities, where c_i is estimated, to sites of carboxylation, where $\delta^{13}C$ is determined, in the thick *Metrosideros* leaves at the high elevation sites, thus decoupling WUE and $\delta^{13}C$. The authors did not rule out other possible explanations, such as differences in stomatal behaviour due to differences in cloudiness or soil drought, and did not monitor diurnal c_i/c_a throughout the photosynthetic period. However, changes in leaf internal resistance may contribute to variation in foliar $\delta^{13}C$, as was recognized by Farquhar *et al.* (1982).

In global surveys of $\delta^{13}C$ over altitudinal gradients, Körner *et al.* (1988, 1991) similarly found that $\delta^{13}C$ increased with altitude. Körner & Diemer (1987) noted that high-elevation plants from the Tyrollean Alps had thicker leaves than low-elevation plants. However, they attributed the presumed decrease in c_i/c_a to an increased carboxylation efficiency rather than to CO_2 concentration gradients within leaves. Körner *et al.* (1991) suggested that decreased oxygen partial pressures accounted for increased carboxylation efficiencies and, therefore, the decrease in discrimination with altitude. Finally, Morecroft & Woodward (1990) attributed the attitudinal gradient primarily to temperature effects on gas exchange, based on extrapolations from controlled-environment studies. Clearly, comparisons of $\delta^{13}C$ in plants over large environmental gradients are not simple to interpret.

Using foliar $\delta^{13}C$ values as an indicator of WUE is, of course, one step removed from a direct physiological measurement and thus is open to errors. This may particularly be true for studies comparing different species, or even single species over large environmental or elevational gradients. Isotope-derived estimates of WUE have advantages over instantaneously measured WUE, because isotope estimates integrate over all seasons of carbon gain and tissues from all canopy positions can be pooled into a single sample. WUE measured with gas-exchange techniques may give unambiguous and exact measures of CO_2 and water vapour flux at one moment, yet do not integrate over times of day or seasons, and there are known problems with estimating transpiration in closed chambers (Smith & Hollinger, 1988). Because single-moment measurements are easier than life-integrated measurements with gas-exchange equipment, ecologists have often expressed IRGA-derived estimates of WUE measured at A_{max}, the time of day when maximum photosynthetic rates are achieved (Field & Mooney, 1983; Field et al., 1983; Lajtha & Whitford, 1989; DeLucia & Schlesinger, 1991; Lajtha & Barnes, 1991). As it would be expected that WUE would change greatly with time of day due to changes in temperature, LAVD and stomatal opening, such measurements are insufficient to describe plant performance.

The correlation between IRGA-derived gas-exchange measurements and carbon isotope ratios was in part the basis upon which Farquhar et al. (1982) constructed their theory. The simplified form of the equation relating the two that has been presented above is mechanistic, except that the enzymatic discrimination, or b-value, of 27‰ is a result of curve-fitting, and was intended to account for numerous minor influences on carbon isotope ratios that were otherwise unaccounted for (Farquhar et al., 1982). Actual discrimination due to dissolution of CO_2 in water, diffusion of CO_2 through water, and rubisco is 29–30‰ (Roeske & O'Leary, 1984; Guy et al., 1987). Not surprisingly, some variation around this value has been observed. Brugnoli et al. (1988), who analysed the sugars and starch extracted from leaves on the assumption that these labile carbohydrates would most accurately reflect the previous day's gas exchange, attributed variation in b to differences in the proportion of CO_2 fixed by PEP carboxylase. Evans et al. (1986), and more recently von Caemmerer & Evans (1991), analysed discrimination against $^{13}CO_2$ by collecting the CO_2 remaining in the air stream after passage over a leaf in a cuvette. They found evidence of a significant CO_2 concentration gradient from the base of the stomata, where c_i is normally estimated, to the chloroplast. Perhaps more importantly, they found that the gradient was correlated with photosynthetic rate, which would have the effect of decreasing b as the photosynthetic rate is increased (Farquhar et al., 1989b; von Caemmerer & Evans, 1991). Because of these phenomena that are otherwise not accounted for, some variation in the relationship between gas exchange and carbon isotope ratio should clearly be expected.

A final change in carbon isotope ratios might occur in biosynthesis. Lipids can be as much as 10‰ lighter than the remainder of the tissue after synthesis. Other compounds are similarly depleted or enriched, but the differences are far less pronounced (O'Leary, 1981); for example, cellulose is typically 1–2‰ lighter than whole tissue (e.g., Leavitt & Long, 1986). In detailed work, sample variation can be reduced by extracting a single compound or class of compounds. Because cellulose is the most abundant single compound in plant tissue, it is frequently used where such control of variation is necessary (Park & Epstein, 1961; Freyer, 1979). It is worth noting that the b-value in the theory of Farquhar *et al.* (1982) is probably different for cellulose than for whole tissue, as has previously been demonstrated for starch (Brugnoli *et al.*, 1988).

Still, important insights into plant carbon budgets have been gained where carbon isotope ratios and gas-exchange data did not match. Perhaps the best example of this was observed in xylem-tapping mistletoes. Studies found discrepancies of up to 4‰ between carbon isotope ratios and c_i/c_a (Ehleringer *et al.*, 1985; Marshall & Ehleringer, 1990). These authors suggested that mistletoes obtained more than 60% of the carbon in their tissues from the host, a theory that was confirmed by subsequent work with a C_3 mistletoe on a CAM host (Schulze *et al.*, 1991). In these studies, explanations of apparent discrepancies in the carbon isotope data led to significant changes in our view of the biology of these plants.

1.2.3 Source of carbon

In closed-canopy or relatively dense forests, it has been shown that foliar $\delta^{13}C$ often increases with canopy height (Vogel, 1978a; Medina & Minchin, 1980; Medina *et al.*, 1986, 1991; Sternberg *et al.*, 1989). This pattern has been attributed to variation in the contribution of CO_2 derived from soil or plant respiration (c. -25 to -30‰) vs. atmospheric CO_2. The contribution of respired CO_2 to understory foliar $\delta^{13}C$ is still a debate, as lower light intensity in the understory could also lead to a lower c_i/c_a in plants and thus to a similar decrease in $\delta^{13}C$ (Ehleringer *et al.*, 1986; Ehleringer *et al.*, 1987b; Zimmerman & Ehleringer, 1990). Several authors did not find significant gradients in the isotopic composition of air or plants from the canopy to the forest floor, and thus attributed all foliar $\delta^{13}C$ differences, even within closed-canopy forests, to effects of changing light intensity on c_i/c_a (Francey *et al.*, 1985; Ehleringer *et al.*, 1987b). However, Sternberg *et al.* (1989) calculated that respired CO_2 accounted for 18% of total CO_2 0.5 m above the forest floor and 45–70% of the difference in $\delta^{13}C$ values between understory and canopy leaves in a Neotropical moist forest. Pearcy & Pfitsch (1991) found similar, though much less pronounced source effects 0.1 m above the forest floor in a Sequoia forest. This suggests that in closed-canopy forests, estimates of WUE from foliar $\delta^{13}C$ may need to be corrected for $\delta^{13}C_{atm}$, particularly near the ground surface.

In a study of floodplain forests along the Amazon River, Martinelli *et al.* (1991) also noted unexpected effects of recycled carbon on plant isotopic composition. Foliar $\delta^{13}C$ of C_3 understory plants increased downriver, whereas foliar $\delta^{13}C$ of C_4 semi-aquatic grasses decreased downriver. The authors suggested that inland C_3 plants, further from the Atlantic Ocean, were more exposed than downstream C_3 plants for more of the morning photoperiod to the ^{13}C-depleted biogenic CO_2 that is produced during the night, although more intense shading in the inland forests could produce a similar result. They similarly suggested that the change in δ^{13}C of C_4 semi-aquatic grasses downriver was more closely related to the content and isotopic composition of CO_2 outgassing from the river; downriver water was 1.5–2.0 times as supersaturated with CO_2 than upstream water, and the δ^{13}C of the CO_2 from the river decreased downstream. Significant carbon isotope gradients can thus exist across ecosystems and can affect the isotopic composition of plant species. This could be a confounding effect for attempts to relate WUE to foliar δ^{13}C, as well as for attempts to relate plant isotope composition to animal diet (see Chapter 4) or isotopic composition of the ancient atmosphere and past climate through tree ring or foliar analysis (Stuiver & Braziunas, 1987; Marino & McElroy, 1991). However, this effect might be significant only in dense closed-canopy forests and may be relatively insignificant in more open ecosystems. Clearly δ^{13}C of source CO_2 needs to be considered before interpretations of foliar δ^{13}C can be made.

1.3 Nitrogen isotopes

1.3.1 Nitrogen fixation
The use of stable isotopes to estimate the contribution of symbiotic nitrogen fixation to the total nitrogen economy of a plant has been demonstrated in crop, forest and desert ecosystems, and has been reviewed elsewhere (Shearer & Kohl, 1986, 1989; Virginia *et al.*, 1989).

Quantitative measurements of biological N_2 fixation are easily, cheaply and accurately made with small plants under controlled laboratory conditions using the acetylene reduction assay (Hardy *et al.*, 1973). Because nodules, or soil cores containing nodules, must be isolated for this assay, field measurements and measurements of deep-rooted shrubs or trees are difficult. The δ^{15}N method does not require the collection of nodules, does not disturb the soil ecosystem or plants in the field and requires only the collection of leaf material. Much like the δ^{13}C method to estimate WUE, δ^{15}N estimates of N_2 fixation do not give instantaneous rates, but rather integrate over the entire lifetime of the leaf and seasons of activity. However, this isotope method requires that several key assumptions are met, and also requires that conspecific N_2-fixing plants be grown hydroponically in a greenhouse.

The δ^{15}N method of estimating N_2 fixation is based on the fact that soil nitrogen is usually more abundant in ^{15}N than is atmospheric N_2 (Shearer *et al.*, 1978), and thus plants that rely on soil nitrogen should be more enriched in ^{15}N than plants that obtain nitrogen from the atmosphere through symbiotic fixation. To make this generalization quantitative, three measurements must be made:

1 the $\delta^{15}N$ of tissues from the fixing plant grown hydroponically in nitrogen-free solution, so that an isotope abundance can be measured under conditions of 100% fixed nitrogen including all fractionations within the plant ($\delta^{15}N_{hydro}$);

2 the $\delta^{15}N$ of tissues from the fixing plant grown in the field ($\delta^{15}N_{field}$);

3 the $\delta^{15}N$ of tissues from reference, non-fixing plants growing in the same environment as the fixing plant ($\delta^{15}N_{ref}$).

The proportional contribution of biologically fixed nitrogen to the N_2-fixing plant is then:

$$\%N \text{ fixed} = (\delta^{15}N_{ref} - \delta^{15}N_{field})/(\delta^{15}N_{ref} - \delta^{15}N_{hydro}).$$

The value of $\delta^{15}N_{hydro}$ is close to, but significantly different from, 0 for plants that have been grown hydroponically; lupine and varieties of soybean varied between -0.8 and $-1.6‰$, and the desert woody legume *Prosopis glandulosa* averaged $-1.3 \pm 0.5‰$ (Shearer *et al.*, 1983). Presumably, then, if the reference plant values are very different from this range, the value $-1.3‰$ could be assumed rather than conducting the difficult hydroponic experiment. However, differences between reference and fixing plants have tended to be less than 10‰, and thus an inaccurate value of $\delta^{15}N_{hydro}$ could be an important source of error.

Other sources of error include fractionation resulting from nitrogen metabolism before transport to the leaf within a plant, and inherent differences in nitrogen fractionation during uptake between different species of plants. The variation among plant parts for $\delta^{15}N$ is about 2‰, and thus it must be assumed either that the tissues chosen represent $\delta^{15}N$ of the entire plant or that similar plant parts will be equivalent between the reference and the test species (Shearer & Kohl, 1989). Ecologists generally use leaf tissue composited from various parts of the canopy from both reference and test species, and assume that leaf tissue is generally equivalent.

A more questionable assumption is that isotopic fractionation involved in plant uptake of soil nitrogen is the same between different species, i.e., that the reference plant accurately reflects the $\delta^{15}N$ value of the test species in the absence of N_2 fixation. In the study of Shearer *et al.* (1983) of several Sonoran Desert ecosystems, $\delta^{15}N$ of reference plants varied from 4.4 to 7.1‰ at one site (mean $5.7 \pm 0.3‰$) and from 7.2 to 12.1‰ (mean $9.3 \pm 0.5‰$) at another. A similar variation was seen by Schulze *et al.* (1991) in Namibia. This variation could be due either to microscale variations in the soil pools or differences in plant metabolism and/or rooting depth. Several studies showed great variability in soil $\delta^{15}N$. Broadbent *et al.* (1980) found that the $\delta^{15}N$ of various cultivated and virgin soils in California ranged from -8.4 to 10.2‰, and Cheng *et al.* (1964) found $\delta^{15}N$ values ranging from -1 to 16‰ in Iowa. Indeed, variations in soil $\delta^{15}N$ within a site are often greater than regional variations (Shearer *et al.*, 1978), although Shearer & Kohl (1989) correctly pointed out that only variations in the $\delta^{15}N$ of mineralized soil nitrogen is relevant to the isotopic composition of plants. Deep-rooted species might have access to older pools of soil nitrogen compared to shallow-rooted species, and both soil age and depth can translate into differences in $\delta^{15}N$ (Riga *et al.*, 1971; Mariotti *et al.*, 1980b; Ledgard *et al.*, 1984; Tiessen *et al.*, 1984; Gebauer & Schulze, 1991). In a chaparral ecosystem of California, Virginia *et al.* (1989) noted that differences between shallow-rooted and deep-rooted controls were as great as differences between the putative N_2-fixers and control plants. In the study of Shearer *et al.* (1983), the authors estimated the percentage of nitrogen fixed even when the differences between reference and control plants was less than 2‰, as well as when the difference between *Prosopis* and the control plant was within the variation seen among control species at the site.

With the exception of leguminous crop species, it is difficult to test this method quantitatively. Bremer & van Kessel (1990) compared natural abundance $\delta^{15}N$ and $\delta^{15}N$ isotope-dilution estimates of fixation of field-grown peas and lentils, and found that estimates were not significantly different in 18 of 21 comparisons, in part due to the variability within each estimate. However, these authors noted significant effects of the timing of nitrogen uptake from the soil, a problem that could be more severe in uncultivated plant communities.

Even when differences between test and control $\delta^{15}N$ values are small, however, natural abundance nitrogen isotopes could certainly be used as a qualitative indicator of the presence of N_2 fixation by a species in an ecosystem. Less successful, however, have been attempts to trace fixed nitrogen from a known N_2-fixer through the ecosystem, either through soil compartments or to other plants in the ecosystem (Binkley *et al.*, 1985; Lajtha & Schlesinger, 1986), probably due to multiple fractionations in the soil through repeated cycles of mineralization, nitrification, immobilization, plant uptake and denitrification. For instance, *Larrea* growing near fixing *Prosopis* had a higher tissue nitrogen content than *Larrea* not growing near *Prosopis*, suggesting an influence of *Prosopis*-fixed nitrogen on nearby plants, but tissue $\delta^{15}N$ values of *Larrea* near or away from *Prosopis* could not be distinguished (Lajtha & Schlesinger, 1986).

1.4 Hydrogen and oxygen isotopes

1.4.1 Isotopic fractionation in water
Kinetic fractionation of hydrogen and oxygen isotopes is exactly analogous to that of carbon, and can be attributed to the faster diffusion of mol-

ecules containing the lighter atoms. Kinetic fractionation factors for H_2O/DHO and $H_2^{16}O/H_2^{18}O$ are 1.025 and 1.0285, respectively (Merlivat, 1978); these factors describe the 2–3% faster diffusion of light water relative to heavy water. If one is primarily interested in the fate of water, enzymatic discrimination can be neglected because only a minor proportion of water is consumed as a substrate in plant metabolism.

Discrimination due to change in phase strongly affects the isotopic composition of water. Fractionation upon evaporation (or condensation) can be visualized in a simple way: the heavier isotope has more difficulty escaping from the liquid into the vapour phase and, conversely, is more readily condensed out, i.e., it is harder to lift and more inclined to fall. The equilibrium fractionation factor describes this difference quantitatively for a two-phase system at equilibrium. Equilibrium fractionation factors change with temperature, but values at 25°C are 1.079 and 1.0094 for H_2O/DHO and $H_2^{16}O/H_2^{18}O$, respectively (Majoube, 1971). Thus we would expect the isotope ratios of water vapour to be 7.9 and 0.9% more negative for δD and $\delta^{18}O$, respectively, than those of liquid water at equilibrium.

1.4.2 Isotopic composition of precipitation water

The fractionations described above underlie pronounced geographical and seasonal variation in the isotope ratios of precipitation. The following generalizations are made (Schiegl, 1970):

1 isotope ratios of precipitation become more negative with latitude;
2 become more negative with altitude;
3 become more negative as an air mass moves inland;
4 are more negative in winter, less negative in summer;
5 are generally more positive when more precipitation falls.

These patterns can be intuitively understood if one imagines the isotopic consequences of the hydrologic cycle. Water evaporates into the atmosphere with an isotopic signature that we will consider a constant, though it is in fact determined by the isotopic composition of the water, water temperature and relative humidity of the atmosphere (Merlivat & Jouzel, 1979). As the water vapour rises in convection cells, condensation occurs, clouds form and the heavy isotopes are concentrated in liquid water droplets by an amount described primarily by the equilibrium fractionation factor. Rain falls, carrying disproportionate amounts of the heavy isotopes, and the water vapour remaining in the cloud is consequently depleted. This process is repeated as the water vapour moves to higher latitudes, higher altitudes and as the water vapour moves inland over successive mountain ranges, resulting in more and more negative isotopic signatures. The depletion of heavy isotopes is especially pronounced in the cold temperatures of winter, but is partially offset when a single precipitation event is very intense (Lawrence & White, 1984).

Isotopic ratios of hydrogen and oxygen in precipitation are closely related, lying on a standard curve known as the meteoric water line (MWL; Craig, 1961). Near one end of the MWL lies the isotopic value for mean ocean water. Most of the line lies below this point, reflecting the depletion of heavy isotopes upon evaporation and subsequent condensation. The most depleted precipitation is observed in Antarctic snow. If water evaporates after it falls, it tends to deviate from the MWL, providing a means of determining whether, and sometimes to what extent, the water has been isotopically modified since it fell (e.g., Walker & Lance, 1991).

Outside the tropics, variation in the isotope ratio of precipitation on a given site is predominantly seasonal (Libby *et al.*, 1976; Dawson & Ehleringer, 1991). For example, values for δD in one study of precipitation in Austria range from −180‰ in December to −5‰ in May (Fig. 1.4). Similar data from Utah range from −220‰ for April snow to −15‰ from a summer thundershower. Ground water, or stream water, is typically at some intermediate value, the exact value depending on the proportion of groundwater recharge that occurs in each season (White, 1989). In this chapter the term ground water is used loosely (see White *et al.*, 1985), including perched water tables as well as water infiltrating toward the water table that has an isotopic signature indistinguishable from ground water.

1.4.3 Water uptake by plants

There is no fractionation of water upon uptake into the plant (Gonfiantini *et al.*, 1965; Dawson & Ehleringer, 1991), except perhaps under exceedingly unusual conditions. Furthermore, there is no evaporation, and therefore no fractionation, of water until it reaches the leaves. Therefore, the isotope ratio of xylem water can be used as a measure of the isotopic signature of the water being utilized. This technique is especially potent where one can assume that only two pools of water are available: surface water, comprised predominantly of recent precipitation, and ground water, for which isotopic values are frequently derived by measurement of spring water or stream water. Under these conditions one can use a two-ended mixing curve with the value for spring water at one end and the value for precipitation at the other; percentages of each water source can then be assigned by moving to the measured xylem water value on the curve. White (1989) measured the time course of xylem water δD following a rainfall event (Fig. 1.5). The data are presented on a two-ended mixing curve, showing the gradual return to groundwater values as the rain water was slowly depleted. Similar analyses have shown species

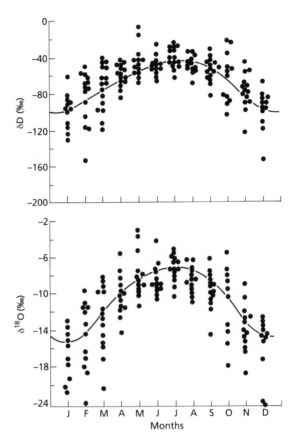

Fig. 1.4 Seasonal patterns of δD and δ^{18}O in precipitation in Austria. (From Libby *et al.*, 1976.)

differences in groundwater dependence (Ehleringer *et al.*, 1991; Flanagan *et al.*, 1992) and differences in uptake of ocean water or fresh water among species growing near the ocean shore (Sternberg & Swart, 1987).

Analysis based on the two-ended mixing curve has several short-comings, however. First, it is necessary to assume that no evaporation takes place between the precipitation event and uptake of the water by the plant. Enrichment of approximately 3‰ for δD and 0.1‰ for δ^{18}O has been observed due to interception losses in canopies (Pearce *et al.*, 1986) and further enrichment is likely at the soil surface, particularly at low humidity (Dinçer *et al.*, 1974; Allison & Hughes, 1983). Second, one must assume that all water is being drawn from two, and only two, homogeneous pools. This assumption is likely invalid if soils are deep; several authors have described considerable variation in δ^{18}O or δD of

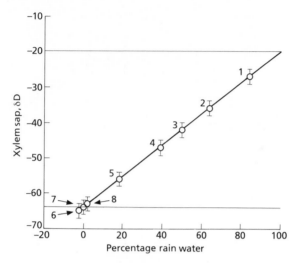

Fig. 1.5 A two-ended mixing curve showing relative proportions of rain water and ground water in the xylem of white pine trees over the course of 8 days following a rain event. (From White *et al.*, 1985.)

soil water with depth (Dinçer *et al.*, 1974; Allison & Hughes, 1983; Allison *et al.*, 1983). Still, if ground water and surface water are sufficiently different, and especially if they are separated by an intermediate zone of low moisture content, the method would seem to be extremely useful. For example, in similar studies conducted in different years, Flanagan *et al.* (1992) and J.D. Marshall (unpublished data) found xylem water in Utah juniper to range from the groundwater value, −95‰, in a dry year, to −65‰, approximately 40% rain water, in a wet year. Such large differences certainly provide useful information even if the detailed mechanisms underlying the differences are not well understood.

1.4.4 Isotope ratios of leaf water

As one might expect based on the foregoing discussion, transpiration from plant leaves leads to fractionation of the xylem water. As a result, leaf water is often considerably enriched in the heavier isotopes. If a record of leaf water isotope ratios were maintained in plant tissue, it might be possible to reconstruct climate or leaf physiology from the past by analysing δD or $\delta^{18}O$ of the plant tissue (Libby *et al.*, 1976; Yapp & Epstein, 1982a; Lawrence & White, 1984; Krishnamurthy & Epstein, 1985; Edwards & Fritz, 1986).

The problem of describing leaf water enrichment is an elaboration of the more general problem of water evaporation, upon which the seminal work was again conducted by geophysicists (Craig & Gordon, 1964). The isotopic composition of water in a transpiring leaf at steady state can be quantitatively described as follows:

$$R_L = \alpha^*[\alpha_k \cdot R_x((e_i - e_a)/e_i) + R_a(e_a/e_i)]$$

where R is the molar ratio of the heavy to light isotope, and the subscripts 'L', 'x', and 'a' refer to leaf, xylem and air, respectively; α^* is the equilibrium fractionation factor; α_k is the kinetic fractionation factor; e is partial pressure of water vapour; and the subscripts 'i' and 'a' refer to the intercellular spaces and ambient air, respectively. If R_x, R_a and leaf temperature are assumed constant, which would hold α^* and e_i constant, then R_L would be monotonically related to e_a/e_i, i.e., if leaf and air temperatures were equal, isotope ratios would be linearly related to relative humidity.

If leaf water comprised a single homogeneous, equilibrated pool, one would expect the isotope ratio of leaf water to match predictions based on the equations above. However, leaf water does not match this simple model (White, 1989; Yakir et al., 1990; Flanagan & Ehleringer, 1991a,b), for several reasons. Early attempts to account for inhomogeneity of leaf water divided the water into two pools; vein water (or xylem water) and evaporating water. Measured isotopic data were compared to steady-state predictions and the vein water contribution estimated by sliding down the two-ended mixing curve toward the vein-water value. However, vein-water contributions by this method were much greater than could be explained by anatomical measurements of vein volume (White, 1989; Flanagan & Ehleringer, 1991a). Thus, Yakir et al. (1990) proposed the existence of a third pool, probably symplastic water, with a slower turnover rate. In addition, enriched water may diffuse back into the vein water. One would expect high transpiration rates to hinder back-diffusion and therefore to reduce enrichment in the leaf water as a whole (White, 1989).

The picture emerging from these studies is complex, perhaps currently too complex, to make hydrogen or oxygen isotope ratios of leaf water commonly used measures of environmental conditions. However, where the rigorous data requirements can be met, this technique is likely to be an important tool in future ecophysiological research.

1.4.5 Hydrogen and oxygen isotopes in plant tissues

Hydrogen and oxygen isotope ratios could potentially be even more valuable tools if measurements of plant tissue could reliably be correlated with climate. Several authors have noted correlations between isotope data and temperature (Libby et al., 1976; Yapp & Epstein, 1982a), relative humidity (Yapp & Epstein, 1982b; Edwards & Fritz, 1986), or precipitation amount (Lawrence & White, 1984; Krishnamurthy & Epstein, 1985). However, as with carbon, significant fractionation occurs in biosynthesis (Epstein *et al.*, 1976; Northfelt *et al.*, 1981). Sample variation can be reduced by analysing only cellulose extracted from the tissue (Epstein *et al.*, 1976). Variation in δD can be further reduced if the hydroxyl hydrogens, which exchange with environmental hydrogen after biosynthesis, are replaced by nitrate prior to analysis, leaving only carbon-bound hydrogen (Epstein *et al.*, 1977).

The mechanisms controlling isotopic composition of the carbon-bound hydrogen or oxygen of nitrated cellulose are not well understood. It is clear that there are consistent fractionations of 27‰ for ^{18}O (Sternberg, 1989) and 155‰ for D (Yakir & DeNiro, 1990; Luo & Sternberg, 1992), both of which are associated with hydrogen exchange taking place during the synthesis of cellulose. It is also clear that the proportion of hydrogen exchanged on synthesis of cellulose depends on the substrate being consumed; cellulose synthesis from lipids replaces about two-thirds of the hydrogen, synthesis from starch replaces about one-third. The same proportions of oxygen and hydrogen are replaced (Luo & Sternberg, 1992). It would therefore seem that nitrated cellulose from tree stems would carry isotopic ratios intermediate between leaf water, where the photosynthate is first labelled, and stem water, where one- to two-thirds of the hydrogen and oxygen are replaced, offset by the fractionation factors above. Thus, cellulose isotopic composition should be only loosely correlated with xylem water – and substantially more enriched. Yet several authors have noted approximately equal hydrogen isotope ratios in stem cellulose and xylem water (Lawrence & White, 1984) or local surface water (Yapp & Epstein, 1982b). Clearly, more work is needed to identify the physiological determinants of the isotopic composition of cellulose.

1.5 Perspective and summary

The use of natural abundance stable isotopes to elucidate physiological processes in plants is one of the most common, and one of the oldest, applications of isotope analysis in ecology. Isotope analysis is now a well-established tool used to determine carbon fixation pathways of plant species, plant WUE, degree of nitrogen fixation and source of water used; new uses are rapidly being developed. In addition, understanding the physiological processes behind stable isotope signatures of primary producers has given researchers new tools to analyse animal palaeodiets (see Chapter 4), to trace food webs in both marine (see Chapter 7) and terrestrial ecosystems, and to trace carbon sources in estuarine systems (see Chapter 10), among other applications.

As with any other research tool, the use of stable isotopes has its limitations and constraints. In this chapter we have described many of the theoretical assumptions that underlie the interpretation of stable isotope signatures in plant ecophysiology, the types of empirical data that are commonly collected and cases where essential theoretical assumptions are violated in common practice. However, we fully expect that the potential of these powerful new isotope techniques will continue to expand in the future. It is not our intention to discourage the use of these techniques; to the contrary, there are many research applications to which they are ideally suited. As work continues to identify the detailed mechanisms underlying isotopic composition, the range of applications is sure to expand.

CHAPTER 2

Nitrogen isotope studies in forest ecosystems

K.J. NADELHOFFER & B. FRY

2.1 Introduction

The relative amounts of the two stable isotopes of nitrogen, ^{15}N and ^{14}N, vary predictably in soils and plant tissues of forests and other non-cultivated ecosystems. Slight fractionations, or discriminations against the heavier nitrogen isotope that can occur as nitrogen cycles through vegetation, soils and microbial biomass act over the long term to deplete vegetation and litter of ^{15}N and to enrich humus in ^{15}N. In this chapter, we discuss ways in which ^{15}N natural abundances can be used to make inferences about (i) rates and patterns of nitrogen cycling; (ii) the relative importances of nitrogen inputs vs. soil processes in supplying nitrogen for plant uptake; (iii) retention of nitrogen by forests; and (iv) forest ecosystem health. Although the primary focus is on the use of natural abundance nitrogen isotope signatures in forest ecosystem research, the uses of ^{15}N-labels in ecosystem-scale experiments on whole forests are also discussed. Such experiments offer possibilities for checking and improving estimates of nitrogen fluxes within and nitrogen losses from forests, and offer a powerful additional tool to the non-manipulative approach of natural abundance studies.

2.2 ^{15}N natural abundances in forest ecosystems

2.2.1 Patterns of variation in ^{15}N natural abundances

In forests, as in most terrestrial ecosystems, $\delta^{15}N$ values for plant tissues and bulk soils are generally between -10 and $+15‰$ (e.g., Hauck, 1973; Mariotti *et al.*, 1980b; Ledgard *et al.*, 1984). In absolute abundances, this represents a range of only 0.3626 to 0.3718 atom% ^{15}N. Although ^{15}N natural abundances in forests are generally close to atmospheric ^{15}N abundance, studies of individual forests typically show systematic variations in $\delta^{15}N$ values among vegetation, litter and soil samples. In general, tree tissues and fresh litter are slightly depleted in ^{15}N relative to soils and $\delta^{15}N$ values increase with depth in soil profiles. Nitrogen in plant tissues typically has $\delta^{15}N$ values ranging from about -5 to $+2‰$ (Fry, 1991), although values outside this range are occasionally observed. For example, Vitousek *et al.* (1989) reported $\delta^{15}N$ values as low as $-10‰$ for foliage on

very young soils (<200 years old) in Hawaiian forests and Virginia *et al.*
(1988) found foliar $\delta^{15}N$ values as high as +10‰ in desert woodlands. As
discussed in the previous chapter, tissues of tree species associated with
symbiotic nitrogen fixers generally have $\delta^{15}N$ values that lie within the
range of values reported for non-fixers, yet within individual sites, values
for nitrogen-fixing species sometimes differ from values of non-fixers,
opening the possibility of using natural abundances of nitrogen isotopes
to estimate the contributions of nitrogen fixation to meeting demands for
plant nitrogen uptake (Virginia *et al.*, 1988; Shearer & Kohl, 1989;
Vitousek *et al.*, 1989).

At the soil surface, organic matter $\delta^{15}N$ values are generally similar
to, or slightly greater than, values for plant litter, and increase to about
+8 ± 2‰ at depths of 20–40 cm (Riga *et al.*, 1971; Rennie *et al.*, 1976;
Létolle, 1980; Mariotti *et al.*, 1980b; Wada *et al.*, 1984; Shearer & Kohl,
1986). Where organic horizons exist, $\delta^{15}N$ values increase with depth in
Oi, Oe and Oa (or 'L', 'F' and 'H') horizons and mineral horizons have
greater $\delta^{15}N$ values than overlying organic horizons (Mariotti *et al.*, 1980b;
Gebauer & Schulze, 1991). A survey by Fry (1991) of forests and other
non-cultivated ecosystems located at a number of Long Term Ecological
Research (LTER) sites across North America illustrates the overall
pattern of progressive ^{15}N enrichment (increasing $\delta^{15}N$) of plant, litter,
organic soil and mineral soil in forests (Fig. 2.1).

The existence of a general pattern of ^{15}N distribution in forest ecosystems leads to questions such as: Why are plant tissues and litter
depleted in ^{15}N relative to soils? Why do $\delta^{15}N$ values increase with soil
depth in forest ecosystems? Can variations in the overall pattern of [15]N
distributions among different forests lead to insights about rates or
patterns of nitrogen cycling?

2.2.2 Causes of variation in ^{15}N contents

Isotopic compositions of forest nitrogen pools are determined by [15]N
contents of nitrogen inputs to and nitrogen outputs from forest ecosystems, and by isotopic fractionation that can occur during nitrogen
transformations within ecosystems. A schematic representation of major
nitrogen fluxes and pools (Fig. 2.2) is useful for considering the possible
implications of nitrogen inputs, outputs and isotopic fractionations for
[15]N abundances in forest ecosystems. Nitrogen bound in soil organic
matter (litter and humus) is generally the largest nitrogen pool. Organic
nitrogen in soils exists in a variety of forms, some of which are more
readily mineralizable than others (Stanford & Smith, 1972; Tiessen &
Stewart, 1983; Tate, 1987). Only a small fraction of total soil organic
nitrogen (N_t), generally less than 2% per year, is mineralized via
microbially-mediated decomposition. Nitrogen mineralized from soil

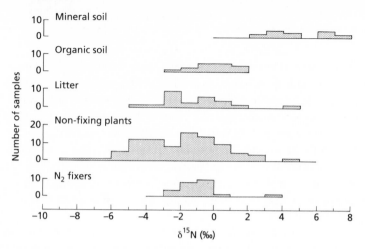

Fig. 2.1 $\delta^{15}N$ values of plants, litter and soils from forest and other non-cultivated terrestrial ecosystems in the US Long Term Ecological Research (LTER) network. Materials were collected from LTER sites at the Bonanza Creek Experimental Forest (Alaska), the H.J. Andrews Experimental Forest (Oregon), Niwot Ridge (Colorado), Central Plains Experimental Range (Colorado), Konza Prairie (Kansas), Cedar Creek Natural History Area (Minnesota), North Temperate Lakes (Wisconsin), W.K. Kellogg Biological Station (Michigan), Coweeta Hydrological Laboratory (North Carolina), North Inlet (South Carolina), Virginia Coastal Reserve (Virginia), Harvard Forest (Massachusetts), Hubbard Brook Experimental Forest (New Hampshire) and Luquillo Experimental Forest (Puerto Rico). (From Fry, 1991.)

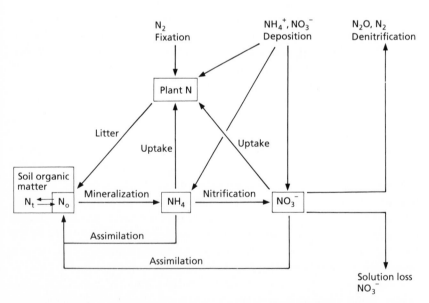

Fig. 2.2 Nitrogen transformations and processes affecting ^{15}N abundances in forest ecosystems. N_t, total organic nitrogen in soil; N_o, microbial nitrogen; NH_4^+, extractable soil ammonium; NO_3^-, extractable soil nitrate. (Partially adapted from Focht, 1973, and Shearer *et al.*, 1974a.)

organic matter is approximately matched by organic nitrogen inputs from plant detritus and microbial products. Ammonium produced by nitrogen mineralization has several potential fates including assimilation into microbial biomass, uptake by plant roots and nitrification. Nitrification occurs to varying degrees in forest soils and tends to be a more important process on relatively fertile sites where soils are not extremely acidic (pH > 4.0; see Aber *et al.*, 1991). Nitrate is more mobile in soils than ammonium. Nitrate that is not assimilated by microbial biomass or taken up by vegetation can be exported as NO_3 dissolved in solution or as N_2O or N_2 gas produced by denitrifying bacteria.

Contents and distributions of ^{15}N in forests are ultimately determined by the ^{15}N contents of nitrogen inputs and outputs. Nitrogen inputs that are depleted in ^{15}N, together with isotopic fractionation during nitrogen mineralization in deeper soils, are likely the ultimate causes of low $\delta^{15}N$ values in vegetation and litter at the soil surface. Enrichment of deeper soils in ^{15}N probably results from the relative isolation of soil nitrogen from atmospheric inputs, coupled with discrimination against ^{15}N during mineralization and loss of ^{15}N-depleted nitrogen from soils due to root uptake, nitrate leaching and denitrification (Fig. 2.2).

Nitrogen inputs and outputs

Dominant nitrogen inputs include direct deposition of nitrogen from the atmosphere and, in some forests, biological nitrogen fixation. Long-term and seasonal measurements of precipitation nitrogen in Europe (Freyer, 1978a), Africa (Heaton, 1987a) and North America (Garten, 1991) suggest that both ammonium and nitrate inputs to forests are relatively depleted in ^{15}N with values for these inputs ranging from roughly -10 to $0‰$ and with ammonium typically more depleted in ^{15}N than nitrate. Where nitrogen fixation occurs, $\delta^{15}N$ inputs of fixed nitrogen to plants are about $0 \pm 2‰$ because nitrogen fractionation during this process is very slight (Kohl & Shearer, 1980; Shearer & Kohl, 1989). Nitrogen can be exported from forests in dissolved forms (mostly nitrate) and in gaseous forms such as N_2O or N_2. We know of no time-integrated measures of ^{15}N contents of solution or gaseous nitrogen losses from forests. However, we expect that most nitrogen exported from forest ecosystems is depleted in ^{15}N relative to soils. Loss of ^{15}N-depleted nitrogen would explain why soils typically have positive $\delta^{15}N$ values when inputs have $\delta^{15}N$ values less than or equal to $0‰$.

Isotopic fractionation

Each of the various processes in forests that either chemically transforms or physically transports nitrogen offers an opportunity for isotopic fractionation (either differential conversion of the two stable nitrogen isotopes from substrate to product or differential transport of ^{15}N and ^{14}N). As described earlier, isotopic fractionation, α, occurring during chemical reactions can be described as the ratio of instantaneous reaction rates between the two stable nitrogen isotope species ($\alpha = {}^{14}k/{}^{15}k$, where ^{14}k and ^{15}k are reaction constants for ^{14}N and ^{15}N, respectively); instantaneous discrimination or enrichment factors are calculated as $\varepsilon = (\alpha - 1) \times 1000$. If the product is depleted in ^{15}N (lower $\delta^{15}N$) relative to the substrate, then $\varepsilon > 0$. The extent of isotopic fractionation that actually occurs during a specific nitrogen transformation depends upon a number of factors, including the substrate/product ratio, temperature, reaction rate and the specific enzymes or micro-organisms mediating the reaction (Mariotti *et al.*, 1981; Hübner, 1986).

Isotopic fractionation factors have been measured under laboratory conditions for both chemical equilibrium reactions (e.g., $NH_4^+ \rightleftharpoons NH_3$) and a number of biologically-mediated kinetic nitrogen transformations (e.g., $NH_4^+ \rightarrow NO_3^-$) (Létolle, 1980; Hübner, 1986). Fractionation in most ecosystems results in reaction products that are depleted in ^{15}N relative to substrates ($\delta^{15}N_{product} < \delta^{15}N_{substrate}$) and $\varepsilon > 0$ (Shearer & Kohl, 1986; Herman & Rundel, 1989). For example, NO_3^- produced during microbially-mediated nitrification of ammonium has been shown to be depleted in ^{15}N, while residual ammonium is, by mass balance, enriched in ^{15}N. Isotopic differences between substrate and product, however, can sometimes be suppressed, especially when substrate pools become so small that nearly all available nitrogen is consumed in the reaction. In such cases, the isotopic composition of the product is identical to the initial composition of the substrate.

The relationship between fractionation and product/substrate ratios is illustrated by Feigin *et al.* (1974b) who measured ^{15}N contents of soil nitrogen pools in an agricultural soil before and after addition of anhydrous ammonia fertilizer (Fig. 2.3). Prior to fertilization, soil nitrate $\delta^{15}N$ was slightly positive (c. +2‰). Increasing the ammonium substrate pool by fertilization in early May stimulated nitrification and lowered the ^{15}N content of soil nitrate while increasing the ^{15}N content of ammonium from its initial value of $-2.1‰$. After the ammonium fertilizer was consumed, nitrification rates diminished and nitrate $\delta^{15}N$ values returned to pre-fertilization levels. In this experiment, fractionation depended upon a relatively large pool of ammonium substrate; as this pool decreased, substrate limitation ultimately led to a very small fractionation during nitrification (Feigin *et al.*, 1974a). The reader is referred to Focht (1973) and Mariotti *et al.* (1981) and to reviews by Létolle (1980), Heaton (1986) and Hübner (1986) for more detailed discussions of variation in, and controls over, nitrogen isotopic fractionation.

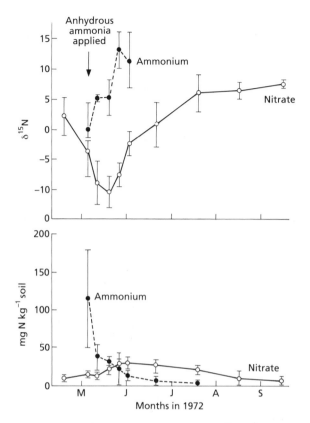

Fig. 2.3 The conversion of NH_4^+ to NO_3^- and changes in the $\delta^{15}N$ values of these to nitrogen forms following application of anhydrous NH_3 fertilizer to an agricultural field. (From Feigin *et al.*, 1974b.)

Nitrogen isotopic fractionation in forests

Isotopic fractionations relevant to forest nitrogen cycling have been measured under controlled conditions for a number of biological nitrogen transformations including N_2 fixation, nitrogen mineralization, nitrification, denitrification and microbial ammonium and nitrate assimilation (see Hübner, 1986 for a comprehensive summary). Processes that are important to consider are (i) those with high fluxes and slight fractionations, such as mineralization, nitrification and N_2 fixation, and (ii) those with small fluxes and potentially large fractionations such as ammonia volatilization and denitrification.

In forest soils, where amounts of substrate vary in both time and space and where microbial communities are dynamic and diverse, products of nitrogen transformations are often slightly depleted in ^{15}N relative to substrates. For example, Cheng *et al.* (1964) found that $\delta^{15}N$ values of nitrogen mineralized from five agricultural soils ranged from 1 to 10‰ lower than total soil nitrogen and that values for extractable ammonium were generally lower than likely ammonium precursors in soils including amino acids, hydroxy amino acids and hexosamine. Black & Waring (1977) reported $\delta^{15}N$ values for nitrate extracted from incubated soils collected from non-cultivated ecosystems in Queensland (Australia) that were 2.5–3.1‰ lower than values for bulk soil nitrogen. Their data suggest that fractionation occurred when organic nitrogen was converted to NH_4^+ (mineralization), when NH_4^+ was oxidized to NO_3^- (nitrification) or during both steps. In a study of ^{15}N natural abundances in western North American forests, Binkley *et al.* (1985) found that $\delta^{15}N$ values of extractable ammonium and nitrate were depleted by 3–10‰ relative to total soil nitrogen in pools in eight stands. Fractionations during nitrification (Delwiche & Steyn, 1970; Herman & Rundel, 1989) and microbial assimilation of ammonium and nitrate (Delwiche & Steyn, 1970; Wada & Hattori, 1978) have also been reported in soil incubations.

In a long-term field study, decreases in surface soil (0–10 cm) nitrogen content in forest plots deprived of leaf litter inputs for 28 years were accompanied by increases in $\delta^{15}N$ values of about 1.5‰ relative to soils in control plots (Nadelhoffer & Fry, 1988). Relative abundances of ^{15}N probably increased in the soils with no leaf litter inputs because isotopically light nitrogen was preferentially mineralized and removed from soil by root uptake faster than it was replaced by root sloughing and throughfall. Soils from these forested plots also showed increases in $\delta^{15}N$ values during laboratory incubations, further suggesting fractionation occurred during net nitrogen mineralization and nitrification.

Processes that lead to gaseous losses from ecosystems can also enrich residual soil nitrogen pools in ^{15}N. Isotopic fractionation during ammonia volatilization produces ^{15}N-depleted NH_3 gas and ^{15}N-enriched ammonium pools (Hübner, 1986). Isotopic fractionations during ammonia volatilization are probably unimportant in most forests, however, because forest soils (unlike many agricultural soils) are generally acidic enough to prevent ammonia losses. Denitrification, the reduction of NO_3^- to either N_2O or N_2, can also yield ^{15}N-depleted gasses and can enrich remaining NO_3^- pools in ^{15}N (Wellman *et al.*, 1968; Delwiche & Steyn, 1970; Blackmer & Bremner, 1977; Mariotti *et al.*, 1981). Effects of denitrification on nitrogen isotope distributions in most forests, however, are probably minor because denitrification losses are generally small relative to rates of net nitrogen mineralization, nitrification and nitrogen deposition (Bowden, 1986; Bowden *et al.*, 1990). Possible exceptions to this are forest soils subjected to elevated nitrogen inputs in throughfall or to periodic moisture saturation (Goodroad & Keeney, 1984; Schmidt *et al.*, 1988).

The consistent pattern of lower $\delta^{15}N$ values in plant tissues than in soils (Fig. 2.1) strongly suggests that isotopic fractionation occurs during mineralization of organic nitrogen in soils. However, isotopic fractionation occurring during nitrogen mineralization in forest soils is probably more complex than fractionations occurring during simple chemical conversions of substrates to products. Microbial enzymes responsible for degrading soil organic matter probably differentiate little, if at all, between ^{14}N and ^{15}N bound to the complex array of organic compounds in soils. Fractionations during microbial assimilation of ammonium are possible during periods when ammonium is present in excess of microbial demands (Hoch *et al.*, 1989). However, the most likely explanation for the gradual ^{15}N depletion of plant-available ammonium and ^{15}N enrichment of soil organic nitrogen involves microbial nitrogen metabolism. Heterotrophs generally become enriched in ^{15}N as a result of excreting ^{15}N-depleted nitrogen (Minagawa & Wada, 1984). This may occur because isotopic fractionations associated with catabolic processes such as peptide hydrolysis can increase the ^{15}N contents of residual peptides (Silfer *et al.*, 1992), leading to excretion of ^{15}N-depleted forms. Remineralization of ^{15}N-depleted ammonium by soil microbes probably contributes to lowering $\delta^{15}N$ values of exchangeable ammonium pools while raising $\delta^{15}N$ values of total soil nitrogen as microbial biomass turns over and becomes incorporated into soil organic matter. As such, isotopically light ammonium can become available for uptake by plant roots and, in some soils, oxidation to nitrate.

Fractionation can also occur during nitrification, rendering unreacted ammonium that is enriched in ^{15}N relative to the microbial or nitrate nitrogen produced. Most nitrate in forest soils is produced by autotrophic organisms that oxidize ammonium to derive metabolic energy. If the substrate for nitrification were entirely consumed, the nitrate would not differ in nitrogen isotopic composition from the ammonium substrate. If soil ammonium pools are large and are replenished while nitrification occurs, however, then some fractionation during nitrification is likely.

Nitrate, in contrast to ammonium, is highly mobile and generally does not accumulate in forest soils. It is normally either assimilated by microbes, denitrified, taken up by plant roots, or leached from soils as rapidly as it is produced. Fractionations during nitrate transformations are most likely when nitrate accumulates, such as when net nitrogen mineralization and nitrification rates are high due to elevated nitrogen inputs (Aber *et al.*, 1989, 1991) or when nitrification rates temporarily exceed nitrate consumption.

Uptake of either ammonium or nitrate by plant roots could discriminate against ^{15}N. However, when forest growth is nitrogen limited, isotopic fractionation is less likely because inorganic nitrogen is depleted close to active absorbing surfaces of roots. We emphasize that the extent of nitrogen fractionation in natural systems is likely to be quite variable; when it does occur, fractionations are probably slight because of small substrate pools and low rates of nitrogen supply relative to plant and microbial demands. The overall narrow range of δ^{15}N values among forest nitrogen pools and the consistent pattern of small differences in ^{15}N natural abundances between plants and soils are probably the cumulative result of many small fractionations occurring as nitrogen cycles among humus, inorganic, microbial and vegetation pools. Much improved, routine methods for measuring the isotopic compositions of nanomolar amounts of ammonium, nitrate and nitrogen gases are needed in order to further our understanding of isotopic fractionation in forest ecosystems.

2.2.3 Qualitative models

Controls on nitrogen isotopic compositions in forest ecosystems are considerably more complex than those controlling the isotopic compositions of simple chemical reactions, single nitrogen transformations mediated by specific micro-organisms in culture, and complex series of nitrogen transformations in soil incubations. Patterns of ^{15}N abundances in ecosystems can be influenced by (i) isotopic compositions of nitrogen inputs from precipitation, nitrogen fixation, fertilization or upslope ecosystems; (ii) isotopic fractionation during processes such as mineralization of organic nitrogen, nitrification, microbial assimilation of inorganic N and denitrification; (iii) differential sorption of ^{14}N and ^{15}N on soil exchange surfaces; and (iv) isotopic compositions of nitrogen removed from soil by leaching, runoff and plant roots (Bremner & Hauck, 1982). Therefore, it is probably impossible to construct quantitative models that would accurately predict the δ^{15}N values of forest nitrogen pools or that could allow for predicting rates of nitrogen cycling from patterns of ^{15}N natural abundances.

Although insufficient data on nitrogen isotope distributions in forests exist to allow for the development of quantitative, mechanistic models of nitrogen isotope dynamics, qualitative models (e.g., Shearer *et al.*, 1974a) can be constructed and used to examine relationships between patterns of nitrogen cycling and ^{15}N natural abundances. We propose such a model to explain the development and maintenance of the pattern of ^{15}N natural abundances that is characteristic of forest ecosystems (Fig. 2.4). The model is based on ^{15}N-depleted source terms and fractionations that are likely to occur during nitrogen transformations in forest soils. Precipitation and nitrogen-fixation inputs to forests are likely depleted in ^{15}N relative to bulk soil nitrogen as described above. Most of the isotopically light nitrogen inputs enter directly into plants and microbes or into

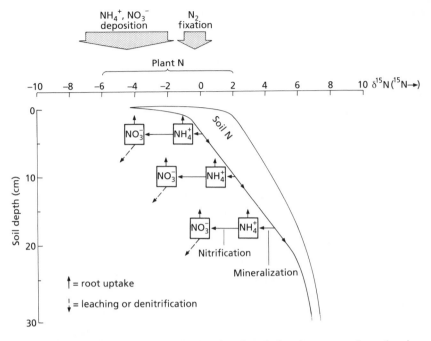

Fig. 2.4 A hypothetical model of isotope fractionations during nitrogen transformations in forest ecosystems. See text for discussion.

inorganic pools that are readily accessible to plants and microbes. Therefore, these inputs contribute to low $\delta^{15}N$ values of plants and litter. Fractionation or discrimination against ^{15}N during decomposition of litter (Turner *et al.*, 1983; Melillo *et al.*, 1989) and soil humus results in the production of isotopically light ammonium and the gradual ^{15}N enrichment of residual decomposing materials. Enrichment of humus in ^{15}N should continue as organic fragments continue to decompose and as humus particles move gradually downward in the soil profile and decrease in size, C:N ratio and total mass. When net nitrification occurs, fractionation can yield nitrate that is further depleted in ^{15}N and can contribute towards ^{15}N-enrichment of ammonium. Inputs to the ammonium pool from mineralization and precipitation, however, probably serve to maintain lower $\delta^{15}N$ values in ammonium than in soil organic matter. Although fractionations probably occur during nitrogen mineralization and nitrification throughout the profile, the products of these transformations probably become less depleted in ^{15}N (higher ^{15}N values) as substrate (humus nitrogen) becomes gradually enriched with depth. Uptake by plant roots of either ammonium or ammonium and nitrate from ^{15}N-depleted pools in soil serves to lower the ^{15}N abundance in biomass. In cases where plant demands for nitrate are exceeded, nitrate can accumulate in soil and increase the possibilities for fractionation during denitrification. When nitrate pools are large and denitrification occurs, we would expect $\delta^{15}N$ values of N_2 and N_2O exports to be low and residual nitrate pools to be enriched. We are, however, unaware of any direct field measures of ^{15}N abundances in denitrification products. Likewise, where nitrate leaching losses occur, dissolved nitrate exports (to locations downslope or deep in the profile) would likely have relatively low $\delta^{15}N$ values.

Combined measures of plant and soil ^{15}N natural abundances are consistent with our qualitative model (Fig. 2.4). For example, a study of non-cultivated pasture soils by Ledgard et al. (1984) showed that while nitrogen concentration decreased and $\delta^{15}N$ values increased with soil depth, the $\delta^{15}N$ values of mineralizable nitrogen and of nitrogen taken up by plants grown in soils collected from four depths (to 60 cm) were always at least 3‰ less than bulk soil nitrogen values. Increases in $\delta^{15}N$ values with soil depth were associated with decreases in the relative abundances of sand- and silt-sized humus particles and with increases in clay-sized humus particles. Similar patterns were noted by Tiessen et al. (1984) who reported increases in $\delta^{15}N$ values of 2–3‰ between sand- and silt-sized humus particles and between silt- and clay-sized particles in uncultivated grasslands. In addition, sand- and silt-sized humus particles in cultivated but otherwise similar soils had higher $\delta^{15}N$ values than equal-sized particles in the uncultivated soils. Results of these two studies suggest that $\delta^{15}N$ values increase with both organic matter age and extent of decomposition as well as with depth in soil. Isotopic analysis of nitrogen in soils and roots in two Norway spruce (Picea abies) forests in Germany by Gebauer & Schulze (1991) provides evidence that nitrogen mineralized and taken up by forest vegetation is depleted in ^{15}N relative to soil (Fig. 2.5). In each forest, the $\delta^{15}N$ values of fine (non-woody) roots increased with soil depth and were 1–4‰ lower than $\delta^{15}N$ values for soils sampled from the same depths as roots, suggesting that much of the structural nitrogen in roots is acquired locally from soils surrounding roots and that plant available nitrogen throughout the rooting zone is depleted in ^{15}N relative to organically bound nitrogen in soil.

Preferential sorption of $^{15}NH_4^+$ on clays and other cation-exchange surfaces leading to slight ^{15}N depletions in soil solution ammonium relative to adsorbed ammonium have been reported (Delwiche & Steyn, 1970; Karamanos & Rennie, 1978). These studies suggest that chromatographic processes in soil profiles could lead to slightly lower $\delta^{15}N$ values in more mobile nitrogen pools and thereby lower the ^{15}N content of ammonium available for uptake by roots or for oxidation by nitrifiers. Therefore, chromatographic effects could interact with biochemical frac-

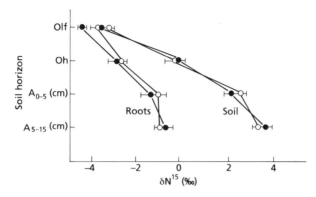

Fig. 2.5 $\delta^{15}N$ abundances in soils and plant roots in two forests in northeastern Bavaria (the Fichtelgebirge). $\bigcirc$ = declining; $\bullet$ = healthy. (From Gebauer & Schulze, 1991.)

tionations and contribute to lowering the ^{15}N abundances of nitrogen that is more readily cycled (see Section 2.2.1). Also, several investigators have reported that although $\delta^{15}N$ values typically increase with depth in surface soils (see Section 2.2.1), values can decrease again deeper in the profile (50–500 cm) where nitrogen concentrations are low and where nitrogen pools are derived largely from illuvial inputs (e.g., Delwiche & Steyn, 1970; Karamanos & Rennie, 1978). Low $\delta^{15}N$ values in deep soils are consistent with the idea that nitrate exported below the rooting zone is depleted in ^{15}N relative to soils in the rooting zone and is to some degree immobilized deeper in soil profiles.

2.3 ^{15}N natural abundance case studies

Several key studies suggest that surveys of ^{15}N natural abundances both within forests and across landscapes and regions can provide useful information on nitrogen cycling and ecosystem functioning. Here we briefly review four case studies that link patterns of nitrogen cycling to distributions of ^{15}N natural abundances.

2.3.1 Soil development and nitrogen inputs

Analyses of nitrogen isotopes in foliage and soils in forests along a gradient of primary succession in Hawaiian rainforests (Vitousek *et al.*, 1989) demonstrate how patterns of ^{15}N abundances can be used to draw inferences about relationships between nitrogen cycling in forests and soil development. Foliage of non-nitrogen-fixing tree species growing on soils developing on young parent material (<200-year-old lava) was the most depleted in ^{15}N, with $\delta^{15}N$ values for one tree fern (*Cibitobium*) as low as −10.1‰. The dominant non-fixing tree (*Metrosideros polymorpha*) averaged $-5.9‰$ $\delta^{15}N$ on a young soil (197 years old) and increased to values as high as $+0.7‰$ on much older soil ($>60\,000$ years old). Abundances of ^{15}N in both non-fixing plants and mineral soils increased with soil age, and foliar $\delta^{15}N$ values were generally about 4‰ lower than soils across sites. In contrast to non-fixers, foliar $\delta^{15}N$ values of exotic nitrogen fixers (especially *Myrica fava*) were generally close to $0 \pm 1‰$. Vitousek and co-workers interpreted the low $\delta^{15}N$ values in foliage and soils of their younger sites as reflecting the low ^{15}N abundances in precipitation, the dominant nitrogen input to the young soils, and small losses of isotopically light nitrogen. Nitrogen losses from the younger soils along their gradient are probably minimal because of low soil nitrogen content, slow nitrogen cycling rates and efficient uptake of mineralized nitrogen. They proposed that $\delta^{15}N$ values in these forests increase with age as soil nitrogen accumulates, nitrogen fixation inputs increase, rates of nitrogen cycling between plants and soils increase, and as losses of ^{15}N-depleted nitrogen forms increase. Along this age sequence, therefore, low $\delta^{15}N$ values in plants and soils are indicative of early stages of soil development, of precipitation as an important nitrogen input, of low nitrogen cycling rates and of small nitrogen losses relative to inputs. Higher $\delta^{15}N$ values suggest less reliance on precipitation in meeting plant nitrogen uptake demands, higher rates of nitrogen cycling, greater reliance by plants on mineralized and fixed nitrogen, and greater nitrogen losses.

2.3.2 Nitrogen cycling rates

In comparing nitrogen isotopic distributions in four non-cultivated brush and pasture ecosystems, Mariotti *et al.* (1981) found that while $\delta^{15}N$ values increased with soil depth in each ecosystem, differences in $\delta^{15}N$ values between the soil surface and B horizons were greatest in their least fertile site and smallest in their most fertile site. Moreover, infertile sites had the lowest $\delta^{15}N$ averaged across soil depths and fertile sites had the highest. They noted that their results were consistent with other evidence showing that highly disturbed ecosystems, such as cultivated fields, are characterized by high levels of biological activity and are typically enriched in ^{15}N relative to unploughed forests and pastures. Therefore, it might be expected that ecosystems with higher rates of biological activity and nitrogen cycling between plants and soils would tend toward higher soil $\delta^{15}N$ values.

Higher $\delta^{15}N$ values may especially result when pool sizes of ammonium and nitrate increase following large or chronic disturbances, such that fractionations during transformations of these compounds are not substrate limited and become more pronounced. Residual nitrogen pools in vegetation and soils then increase in ^{15}N as more highly fractionated, ^{15}N-

depleted compounds are lost from disturbed ecosystem. Increases in soil nitrogen turnover and nitrogen mineralization rates following disturbance also increase the relative amount of plant-available nitrogen supplied from the mineralization of older, more decomposed humus (Fig. 2.4) and probably also result in exports of ^{15}N-depleted nitrogen compounds.

2.3.3 Chronic nitrogen deposition

Recent measures of ^{15}N natural abundances conducted at sites of nitrogen deposition studies also suggest that losses of isotopically light nitrogen can increase in forests subjected to elevated nitrogen inputs from atmospheric and other sources. Isotopic analyses of grass (*Deschampsia flexulosa*) samples collected from Scots pine (*Pinus sylvestris*) plots in northern Sweden subjected to 18 years of nitrogen fertilization (treatment totals: 0, 660, 1320 and 1980 kg N over 18 years) showed that plant available nitrogen increased linearly with rate of nitrogen application from δ^{15}N = -0.7 to 11.0‰ during the treatment period (Högberg, 1990). A likely explanation for these increases is that nitrogen inputs increased both inorganic nitrogen pool sizes and nitrogen cycling rates, thereby increasing the extent of isotopic fractionation. Cumulative nitrogen losses of ^{15}N-depleted products (nitrate and possibly nitrogen gases) served to gradually increase the ^{15}N content of actively cycling nitrogen. On an annual basis, the isotopic shifts were slight (c. 0.5‰) and undetectable against background sample variability. On a decade timescale, however, effects of nitrogen additions (especially as urea) on ^{15}N contents were clear. Chronic additions probably lead to increased nitrogen cycling between soil and vegetation and to increased nitrogen losses. Nitrogen exported from the forest in soluble and gaseous forms was likely depleted in ^{15}N relative to residual substrate nitrogen pools thereby leading to gradual enrichment of plant available nitrogen in ^{15}N.

2.3.4 Forest decline

A study of stable isotope distributions in soil and plant components in declining and apparently healthy Norway spruce forests in the Fichtelgebirge of northeastern Bavaria (Gebauer, 1991; Gebauer & Schulze, 1991) demonstrates how patterns of ^{15}N natural abundances can provide critical insights into interactions between nitrogen cycles and biological processes. Nitrogen isotope distributions were characteristic of patterns for forests (see Section 2.2.1) with plant tissues depleted in ^{15}N relative to soils and with δ^{15}N values increasing and nitrogen content decreasing with soil depth. As already noted, δ^{15}N values of roots also increased with depth in soil profiles (Fig. 2.5). The overall positive correlation between δ^{15}N values in roots and soils suggested that structural nitrogen in fine roots is derived primarily from horizons where roots grow. However, differences in $\delta^{15}N$ values between roots and soils increased with depth. The increasing differences could have resulted from greater isotopic fractionation with depth, changes in the relative importance of ammonium and nitrate uptake by roots, or nitrogen mobility within root systems. All three factors could contribute, but the authors suggest that ammonium uptake is probably more important in the more acidic organic horizon while nitrification and nitrate uptake by plants is probably greater in mineral than in organic horizons. Therefore, isotopic fractionation associated with nitrification in mineral soils could have been largely responsible for the greater differences between root and soil $\delta^{15}N$ values in deeper horizons. Above-ground tissues (needles, twigs) had $\delta^{15}N$ values similar to roots in organic horizons and lower than roots in mineral horizons. They concluded, therefore, that most nitrogen in above-ground plant tissues was derived from sources near the soil surface.

Gebauer & Schulze (1991) also reported small but significant differences between the declining and healthy sites in both needle and twig ^{15}N abundances. Across age classes and crown position, $\delta^{15}N$ values of needles and twigs were 0.5–1.0‰ higher at the declining site. In addition, older needles were more enriched in ^{15}N relative to young needles on the declining site than on the healthy site. They speculated that this difference occurred because the onset of catabolic activity was accelerated in needles on the declining site and ^{15}N-depleted compounds were preferentially exported from senescent needles. As a result, needle litter produced on the declining site is more enriched in ^{15}N than litter produced on the healthy site. The authors suggest that these mechanisms will feedback over the long term to increase soil $\delta^{15}N$ values in the declining site.

2.3.5 *[15]N and other isotopes*
Results of ^{15}N natural abundance studies of forest ecosystems support Hauck's (1973) contention that examining the variation in ^{15}N abundances at large scales might be used to improve our understanding of relationships among various nitrogen-cycling processes under different environmental conditions. Additional studies of ^{15}N natural abundances in forests and other ecosystems developing under wider ranges of environmental conditions and subjected to various types and levels of disturbance would likely yield more useful information. A particularly productive approach would be to link measures of ^{15}N abundances to measures of ^{13}C natural abundances to allow for comparing ecosystem processes that fractionate nitrogen and carbon isotopes (e.g., photosynthesis and respiration). For example, Gebauer & Schulze (1991) concluded, based on patterns of ^{13}C and ^{15}N natural abundances in healthy

and declining forest stands, that reduced tree growth at the declining site resulted more from perturbations of nutrient cycles than from direct effects of atmospheric pollutants on photosynthesis. Also, measuring soil ^{15}N abundances and determining ages of soil organic matter particles using ^{14}C dating (Martel & Paul, 1974; O'Brien, 1984) would allow for linking patterns of nitrogen cycling to rates of organic matter turnover in soils. Use of accelerator mass spectrometry (AMS), a new technique useful for measuring ^{14}C contents in small samples, to determine ages of recently formed humus particles by exploiting the diminishing atmospheric ^{14}C signal (Trumbore et al., 1989) would likely increase our understanding of carbon and nitrogen interactions in ecosystems.

2.4 Large-scale ^{15}N labelling studies

Nitrogen isotopes are generally used for studying ecological processes in one of two ways. Either patterns of ^{15}N natural abundances are examined in order to make qualitative or semi-quantitative inferences about nitrogen cycling process over long timescales, as has been discussed above, or trace amounts of ^{15}N-enriched compounds (60 to >99 atom% ^{15}N) are applied in a single pulse and followed through soil cores, plants or small field plots at short timescales (e.g., Schimel et al., 1990; Davidson et al., 1990; McKane et al., 1990). In the latter approach, rates of cycling between mineral, microbial, organic, solution, gaseous and plant nitrogen pools are estimated by sampling over short periods (hours to days) and using isotope-dilution equations to calculate gross and net nitrogen fluxes. Both approaches have increased our basic understanding of nitrogen-cycling processes. Natural abundance studies can provide information on overall patterns of nitrogen cycling, sources of nitrogen inputs, relative importances of nitrogen inputs and losses, and perturbations in nutrient cycles at scales ranging from decades to centuries. Past use of ^{15}N tracers at small scales has provided important information about processes such as gross nitrogen mineralization, microbial ammonium and nitrate assimilation, denitrification and competition for nitrogen between plants and microbes over periods of days to several weeks.

A novel approach to the use of stable isotopes in ecology is to use ^{15}N labels at large scales. The use of ^{15}N labels at ecosystem scales, combines aspects of ^{15}N-tracer studies that traditionally have been done at small scales with ^{15}N natural abundance techniques that, as discussed in this volume, are generally applied at larger scales. The use of ^{15}N tracers at ecosystem scales, as discussed below, requires adding enough ^{15}N to induce small but detectable shifts in ^{15}N natural abundances in critical ecosystem nitrogen pools. Ecosystem-scale ^{15}N-labelling studies, there-fore, employ analytical techniques used in natural abundance studies.

2.4.1 *Rationale and utility*

Agronomists have used large-scale applications of ^{15}N since the 1970s to study fertilizer nitrogen use by crops and nitrogen retention in soils (Hauck, 1973; Hauck & Bremner, 1976). The relatively small variation in natural abundances of ^{15}N is important in these studies, where large ^{15}N additions have typically been made to produce strong signals over the natural background levels. Such studies show that application of ^{15}N-labelled materials to large plots ($>100\,m^2$) and even small catchments (c. 1–10 ha) can serve as a powerful tool in ecosystems research. Use of this technique could, for example, allow ecologists to (i) follow nitrogen additions as they enter and cycle through forest ecosystems; (ii) test and apply results of small-scale studies to larger scales; and (iii) test and refine nitrogen-cycling components of ecosystem models.

Ecosystem-scale additions of ^{15}N to high atom% ^{15}N targets, typically 10–50 atom% in microcosm and small plot NH_4^+ and NO_3^- pools, would be prohibited by the cost of ^{15}N (1994 prices were US $150–350\,g^{-1}$). However, the narrow range of ^{15}N isotopic abundances in forest and other ecosystems (see Section 2.2.1) and improvements in automated sample analysis (e.g., Fry *et al.*, 1992) and use of highly sensitive mass spectrometers now permit ecosystem-scale ^{15}N-labelling studies. Because δ^{15}N values of natural materials range between -10 and $+15$‰ (or 0.3626–0.3718 atom% ^{15}N) and because isotopic fractionations in ecosystems are generally small, compounds with δ^{15}N values between 100 and 1000‰ (or 0.4029 to 0.7326 atom% ^{15}N) can be used as tracers at large scales.

2.4.2 *Methods*

Measurements of ^{15}N abundances in ecosystems prior to and following large-scale additions of ^{15}N-labelled nitrogen to large plots (c. 0.05–1 ha), or even small catchments (c. 10 ha), can be used to derive information on nitrogen cycling and on interactions between nitrogen transformations and biological processes in ecosystems. Amounts of ^{15}N required and isotope purchase costs need not be prohibitive. For example, the amount of ^{15}N required to enrich 50 kg nitrogen fertilizer with a δ^{15}N value of 0‰ to a value of 100‰ can be estimated as follows. Increasing the fertilizer δ^{15}N value requires increasing the ^{15}N:^{14}N ratio by 100‰ (or 10%) from 0.0036765 to 0.0040442. The ^{15}N mass needed to enrich 50 kg nitrogen can be approximated as the product of mass of nitrogen fertilizer and the difference in ^{15}N:^{14}N ratios between nitrogen fertilizer labelled at 100‰ and fertilizer at atmospheric ^{15}N abundance or:

$$50\,kg\,N \times (0.0040442 - 0.0036765) = 0.0184\,kg\,^{15}N. \qquad (2.1)$$

Therefore, $18\,g\,^{15}$N would be enough to apply 50 kg nitrogen fertilizer with $\delta^{15}N = 100‰$ to a 1 ha forest plot. At current prices for ^{15}N (c. US \$150–350 g^{-1}) purchase cost for isotope would be about \$2000. This is a small expense compared to costs of conducting other aspects of field experiments and to costs of ^{15}N analysis at natural abundance levels (\$10–30 per sample using an isotope ratio mass spectrometer).

Tracing the movement of ^{15}N-labelled nitrogen additions into ecosystem nitrogen pools requires mass-balance techniques. Assuming conservation of mass where

$$(m_i \times \delta^{15}N_i) + (m_{lab} \times \delta^{15}N_{lab}) = (m_f \times \delta^{15}N_f) \tag{2.2}$$

and where m_i = initial mass of the ecosystem nitrogen pool; m_{lab} = mass of ^{15}N label incorporated into the nitrogen pool; m_f = final mass of the ecosystem nitrogen pool; $\delta^{15}N_i$ = initial ^{15}N abundance in the ecosystem nitrogen pool; $\delta^{15}N_{lab}$ = ^{15}N abundance of added nitrogen; and $\delta^{15}N_f$ = final ^{15}N abundance of the ecosystem nitrogen pool.

The following mass-balance equation can then be derived:

$$m_{lab} = m_f \times (\delta^{15}N_f - \delta^{15}N_i)/(\delta^{15}N_{lab} - \delta^{15}N_i). \tag{2.3}$$

Equation 2.3 can be used to estimate fluxes of labelled nitrogen into relatively small pools, such as decomposing litter, where changes in mass on an areal basis can be reliably detected. Implicit in Equation 2.3, however, is the assumption that no net fractionation is associated with nitrogen fluxes through the pool in question. This is probably a reasonable assumption given the narrow range of $\delta^{15}N$ values in ecosystems and the small fractionations that occur as nitrogen cycles through ecosystems. This assumption should be kept in mind, however, when interpreting data from large-scale tracer experiments.

The following equation (Equation 2.4) is required for estimating fluxes into large nitrogen pools where changes in mass can only be detected over extremely long intervals (decades or longer) and where $m_f \sim m_i$,

$$m_{lab} = m_i \times (\delta^{15}N_f - \delta^{15}N_i)/(\delta^{15}N_{lab} - \delta^{15}N_f). \tag{2.4}$$

For example, soil nitrogen pools can exceed 1000 kg N ha^{-1} in organic horizons alone and can range from 3000 to 4000 kg N ha^{-1} in the rooting zone of soil profiles in temperate hardwood forests (Bormann *et al.*, 1977). Spatial and sampling variability are high enough to prevent detecting changes as large as 10% of total mass (100–400 kg N). If, however, changes in soil $\delta^{15}N$ values can be detected following large-scale additions of ^{15}N-labelled nitrogen to a forest, then the importance of soil as a sink for the added nitrogen can be assessed (see Section 2.4.3).

It is critical that variability in ^{15}N abundances of ecosystem nitrogen pools be assessed prior to additions of ^{15}N-labelled materials. Variations among samples of tissue, soil and other forest components are likely to greatly exceed analytical variations. For example, our experience indicates that $\delta^{15}N$ values of Oa (or 'H') horizons on a 0.1 ha plot can occupy a range of 2–4‰ whereas current techniques used with modern isotope ratio mass spectrometers can produce highly repeatable measures of ^{15}N abundance (precision <0.2‰ $\delta^{15}N$).

2.4.3 Examples and results

As part of an ongoing, multi-investigator study of the effects of nitrogen and sulfur deposition on forest biogeochemistry (The Bear Brooks Watershed Study or 'BBWS'; Rustad *et al.*, 1993; Nadelhoffer *et al.*, unpublished data), we are tracing the movements of ^{15}N-labelled nitrate additions being applied to large plots in a northern hardwood forest in eastern Maine, USA. A limited number of ^{15}N analyses done on samples collected from nitrogen-amended plots illustrates how plot-level ^{15}N experiments can be used to measure movements of applied nitrate into forest ecosystem compartments such as live biomass, decomposing litter and soils.

Baseline analyses showed that litter and soil $\delta^{15}N$ values on the BBWS plots are within the normal range of values for forests with $\delta^{15}N$ values of -2.0 ± 2, $+3.0 \pm 1.5$ and $+6 \pm 1.5$‰ for tree foliage, organic soil (Oe + Oa horizons) and 0–10 cm mineral soil, respectively. A spray irrigation system was used to apply labelled nitrate-nitrogen to replicated ($n = 3$) plots (15 m × 15 m) in a beech–maple–red spruce forest ecosystem on a spodosol at rates of 0, 28 and 56 kg N ha^{-1} yr^{-1} in about 15 applications spread evenly across growing seasons (see David *et al.*, 1990, and Rustad *et al.*, 1993, for details of design). $\delta^{15}N$ values of the nitrate-nitrogen additions were increased from 0 to about 350‰ (analysed as 349 ± 2‰) by adding H^{15}NO$_3$ (98 atom% ^{15}N) to non-labelled nitric acid. Differences in nitrogen isotopic compositions of samples from control and treated plots are now being used to estimate fluxes of the added nitrate into ecosystem components on BBWS plots.

Composite samples from a control and treated plot were analysed for ^{15}N during the first year of treatment in an initial attempt to compare live foliage, fresh decomposing litter and soils as sinks for ecosystem-scale nitrate additions (Fig. 2.6). These results clearly show that the level of ^{15}N enrichment in the applied nitrate is adequate to allow tracing fluxes of the added nitrogen into forest ecosystem pools. After four applications of nitrate (when about 22 kg ha^{-1} of labelled nitrogen had been applied) the ^{15}N label was detectable in beech but not in red spruce. These results suggest that beech may function to retain and cycle nitrate in this forest whereas red spruce may be less important. Decomposing litter also appears to serve as a sink for nitrate additions. The $\delta^{15}N$ value of decomposing beech litter (<1 year old) on the control plot was slightly

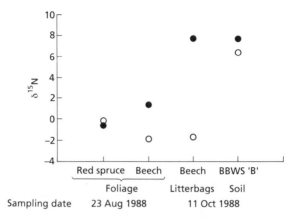

Fig. 2.6 Isotopic composition of forest nitrogen pools in two BBWS plots after 4 (foliage) and 11 weeks (litter and soil) of ^{15}N-labelled HNO_3 additions ($\delta^{15}N = 250\permil$). Treatments were started on plots on 25 July 1988. Foliage, beech litterbag (1 year old) and B horizon soil (incubated below O Horizons in Nytex 250 μm-mesh bags) samples were collected on the dates indicated. Each data point represents one pooled sample from a single control or treated plot. Greater $\delta^{15}N$ values indicate uptake of labelled NO_3^- additions. O = control; ● = high nitrogen.

negative after decomposing for one season, whereas similar litter on the treated plot showed a large shift in $\delta^{15}N$ (10‰).

The shifts in $\delta^{15}N$ values occurring as litter decays into humus are minor (≪2‰; see Section 2.2.1) compared to the 10‰ $\delta^{15}N$ shift in litter decomposing on this nitrogen-amended plot. Samples of homogenized mineral soils (BBWS 'B') incubated in fine-mesh bags below the oxygen horizon showed a smaller but detectable $\delta^{15}N$ increase of about 1.5‰ on the ^{15}N-labelled plot relative to mineral soil incubated on the control plot, suggesting that processes in mineral soils might also serve to retain nitrate in forest ecosystems.

Although shifts in nitrogen pool $\delta^{15}N$ values on plots subjected to isotopically labelled nitrogen additions can identify ecosystem components serving as sinks for nitrogen, mass balances must be calculated (see Equations 2.3 and 2.4, Section 2.4.2) to compare fluxes of labelled nitrogen into different ecosystem components. This is because changes in nitrogen pool $\delta^{15}N$ values for a given mass of ^{15}N label entering an ecosystem pool are inversely proportional to pool size.

Sequential changes in $\delta^{15}N$ values on treated plots across four years of isotopically labeled nitrogen additions (1988 to 1991) and mass-balance techniques were used to compare the potentials of different components in the BBWS for retaining nitrate (Table 2.1). Estimates of nitrogen pool sizes based on field measurements and on measured shifts in ^{15}N abundances in vegetation, decomposing litter and soil samples composited at the plot level were used to estimate fluxes into ecosystem compartments on plots treated with $56\,\mathrm{kg\,NO_3^- \text{-} N\,ha^{-1}\,yr^{-1}}$. For assimilation by foliage, the difference between mean $\delta^{15}N$ values of samples collected in August of 1988 and 1989 and estimated 1989 foliar nitrogen content were used in combination with Equation 2.3. For wood, growth estimates and differences in $\delta^{15}N$ values of sapwood collected from control and treated plots following one year of treatment were used in combination with Equation 2.3. Nitrate-nitrogen assimilation by decomposing litter was estimated using Equation 2.3 and shifts in $\delta^{15}N$ values of decomposing litter and litter nitrogen contents after 1 year. For organic and mineral soils, we compared $\delta^{15}N$ values of samples collected before ^{15}N-labelled nitrate was added and following four years of treatment. Annual $\delta^{15}N$ increases of 0.7‰ $\delta^{15}N$ were calculated as total 4-year $\delta^{15}N$ shifts (detectable c. 3.0‰ differences) divided by 4. Soil nitrogen stocks in year 4 and Equation 2.4 were used to estimate fluxes of labelled nitrogen additions in soils.

Table 2.1 Estimates of NO_3-N retention using ^{15}N-tracer additions to experimentally acidified plots in a beech–maple–spruce forest at the Bear Brooks Watershed Study site in Beddington, Maine, USA. Plots were treated with $56\,\mathrm{kg\,NO_3^- \text{-} N\,ha^{-1}\,yr^{-1}}$ with nitrate labelled at 350‰ $\delta^{15}N$. See text for details

Component	Mass ($\mathrm{kg\,ha^{-1}}$)	Percentage of nitrogen	Nitrogen pool ($\mathrm{kg\,ha^{-1}}$)	$\delta^{15}N_i$ (‰)	$\delta^{15}N_f$ (‰)	$\Delta\delta^{15}N$	Nitrogen retained ($\mathrm{kg\,ha^{-1}}$)
Foliage							
Beech	1 200	2.5	30.0	2.0	31.7	29.7	2.6
Maple	1 000	2.5	25.0	2.0*	31.7*	29.7	2.1
Birch	400	2.1	8.4	−0.1	16.1	16.2	0.4
Spruce	1 600	1.0	16.0	−1.0	−0.6	0.4	0.0
Wood produced							
Deciduous	4 000	0.1	4.0	0.0	50.0	50.0	0.6
Spruce	100	0.1	0.1	0.0	10.0	10.0	0.0
Decomposition							
Beech	1 100	0.9	9.9	−0.9	13.9	14.8	0.4
Birch	400	0.9	3.6	−1.9	6.4	8.3	0.1
Spruce	400	0.4	1.6	−1.7	11.5	13.2	0.1
Wood	1 000	0.1	1.0	−1.2	22.3	23.5	0.1
Soils							
Organic	26 250	2.0	525.0	2.9	3.6	0.7	1.1
Mineral	10^6	0.2	2000.0	6.1	6.8	0.7	4.1
Total							11.6

$$\text{Percentage retention} = \frac{\text{N retained (kg)}}{\text{N applied (kg)}} \times 100 = \frac{11.6}{56} = 21\%$$

* Maple values estimated based on beech.

Mass-balance estimates of nitrate-nitrogen assimilation by forest eco-system components showed that about 20% of the nitrate-nitrogen added to plots was retained (Table 2.1). This retention estimate was consistent with results from the same plots showing high nitrate concentrations in soil solutions collected from deep in the B horizon showing large amounts of nitrate being lost below the rooting zone. Mass-balance calculations also showed that vegetation components with high nitrogen concentrations and soils served as the most important biological sinks for the nitrate additions. It should be noted that the sensitivity for tracing movements of isotopically labelled nitrogen into pools is inversely proportional to pool size. This approach allows for detecting small net fluxes into small nitro-gen pools (e.g., current year wood production, species litter cohorts) which, in total, can be important at the ecosystem level. A corollary is that small $\delta^{15}N$ shifts in large pools result from large fluxes. Organic and mineral soils together could serve to retain up to about $7\,kg\,NO_3^-$-$N\,ha^{-1}\,yr^{-1}$ as could deciduous tree foliage. Nitrate assimilation in grow-ing wood and immobilization into recent litter together could account for about $2\,kg\,NO_3^-$-N retention. All measured components combined could serve to retain about $17\,kg\,NO_3^-$-$N\,ha^{-1}$ or about one-third of the total amount of nitrate-nitrogen added annually.

This mass-balance approach is subject to some uncertainties. For example, nitrate assimilation into soils could be matched by nitrogen losses of native soil nitrogen. Also estimates of nitrate-nitrogen assim-ilation into any ecosystem component can be affected by measurement errors for pool sizes. Nevertheless, this approach leads to important insights into processes controlling nitrate retention, and can be used to identify an upper limit for nitrate retention in ecosystems. The results suggest that soils and litter can serve to immobilize substantial amounts of nitrate even though ammonium may be the preferred form of inorganic nitrogen for soil microbes. The results also suggest that nitrate uptake differs among tree species and strongly suggest that forest species com-position can influence nitrate retention.

2.5 Conclusions

Studies comparing patterns of ^{15}N natural abundances among forests and among other non-cultivated ecosystems suggest that variations in the relative distributions of nitrogen isotopes can provide useful information on patterns of nitrogen cycling. For example, relatively low $\delta^{15}N$ values in vegetation, or in both vegetation and soil, could indicate that rates of nitrogen losses from an ecosystem are low and that vegetation and soil microflora assimilate most of the nitrogen that enters the system and almost all nitrogen mineralized in soil. In contrast, relatively high rates of nitrogen export from ecosystems via solution or gaseous losses could, over scales of decades or longer, lead to higher $\delta^{15}N$ values in vegetation and surface soils. Residual nitrogen pools in soils could become relatively enriched in ^{15}N if nitrogen losses occur when ammonium or nitrate concentrations in soils are high and when substantial fractionation of nitrogen isotopes during ammonia volatilization (in circumneutral or basic soils), nitrification or denitrification occurs. Also, variations in ^{15}N natural abundances among forest tree species may provide information about forms of nitrogen taken up by plants and about depths in soils from which plant nitrogen is derived. Such observations, however, are based on a small number of studies. Additional studies of ^{15}N natural abundances in ecosystems at various stages of development and across wide ranges of environmental conditions should be pursued to more thoroughly explore how distributions of nitrogen isotopes in ecosystems may be linked to nitrogen cycling and other biogeochemical processes. Direct measurements of ^{15}N in soil ammonium and nitrate pools and in N_2O and N_2 produced during denitrification have rarely been attempted, but are needed to improve our understanding of controls on ^{15}N distributions in forest ecosystems.

Large-scale ^{15}N tracer studies are possible because of the relatively narrow range of $\delta^{15}N$ values characteristic of terrestrial ecosystems and because modern mass spectrometers can be used to detect relatively small changes in ^{15}N abundances that can result from the movement of ^{15}N labels into ecosystem compartments. Tracing the movements of ^{15}N labels through large plots or small catchments can provide information about processes controlling nitrogen fluxes within ecosystems and can be used to estimate upper limits of nitrogen retention in forests. Further development of this approach in ecological research is encouraged.

Acknowledgements
This work was supported in part by the US National Science Foundation (BSR-9009190) and the US Environmental Protection Agency (EPA). Although some research described in this article has been funded by the USEPA, it has not been subjected to the agency's review and does not necessarily reflect the views of the EPA. Therefore, no official endorsement by the USEPA should be inferred.

CHAPTER 3

Pollution studies using stable isotopes

S.A. MACKO & N.E. OSTROM

3.1 Introduction

Stable isotope analysis of organic and inorganic materials at natural abundance levels is a very powerful tool in the resolution of the sources, history and pathways that a material can have as it enters an environment. Included in the potential for the delineation of materials through their natural abundance isotopic signatures is the potential to identify a pollutant by an isotopic characterization and quantification by isotopic mass balance of end members. The purpose of this chapter is to present aspects of the use of stable isotopes in pollution research. Specifically, applications are discussed for the determination of sources of groundwater nitrogen contamination from fertilizers or animal wastes, the tracing of sewage in sediments, identifying sulfur and nitrogen compounds in atmospheric pollution and their eventual precipitation into aquatic environments and finally, the estimation of oil contamination in sediments. These perspectives represent only a few of the many possible applications of isotope analyses to furthering our understanding of the fate and effects of pollutants on the environment.

The industrial and agricultural activities of man have resulted in the dispersal of pollutants into virtually all areas of the atmosphere, hydrosphere, pedosphere and biosphere. Many of these compounds have deleterious effects on humans and/or the environment. Owing to biological and physical alterations during transport, the fate of contaminating components is often difficult to trace. In addition, there are numerous undesirable chemicals found in the environment that are of natural origin that can be confused with anthropogenic contaminants. For this reason, it is often desirable to determine the origins of specific compounds or trace the flow of effluents in the natural environment. Once the source and fate of pollutants are understood, steps may be taken to minimize the release or hazards of effluents. Stable isotopes have been used extensively to assess the origins of geological, biological and atmospheric materials (Kaplan, 1975). Their application to pollution studies has shown great value and will continue to grow as the concern for the environment increases. In addition, it must be remembered that two criteria must be satisfied in order to use stable isotope abundances as indicators of the origins of materials in the environment: (i) the primary sources of interest must be isotopically distinct from each other; and (ii) the isotopic signature of the source must not change as the material is transported and transformed in the environment or must do so in a predictable manner.

3.2 Sources of inorganic nitrogen in surface and ground waters

Nitrogen is recognized as an essential and often limiting element in primary production. The natural cycling of nitrogen in the environment has been extensively altered by the activities of humans. One example of this disturbance has been the observation that nitrate concentrations have been rising in ground waters on a world-wide basis (Young, 1983; Madison & Brunett, 1985). High nitrate concentrations in drinking waters are linked to methaemoglobinaemia, a potentially fatal disease resulting from oxygenation inhibition, and the metabolic production of carcinogenic nitrosamines (Schuphan, 1974; Shuval & Gruener, 1974; Follet & Walker, 1989). For this reason the World Health Organization and US Environmental Protection Agency have set recommended maximum safe drinking water concentrations of nitrate and nitrite at 10 and $1\,mg\text{-}at\,l^{-1}$, respectively (World Health Organization, 1984; USEPA, 1976). In addition, high concentrations of nitrate may exacerbate eutrophication in surface waters.

It is commonly believed that application of fertilizers and/or the influence of animal or sewage wastes have contributed to the abundance of nitrate in ground water. The natural abundances of the stable isotopes of nitrogen offer a tracer whereby the origin(s) of nitrate in surface or ground waters may be determined. Once the origin of nitrate in drinking water is understood, corrective measures may be taken to prevent or minimize further contamination.

The isotopic composition of nitrate is a function of its source and any isotopic segregation that may have occurred during its generation or transport to ground waters. As nitrate is derived from a number of different sources and processes, its origin must be considered in the context of the entire nitrogen cycle. What follows is a brief description of the environmental nitrogen cycle as it applies to the origin and isotopic composition of nitrate. More detailed descriptions of the nitrogen cycle may be obtained elsewhere (Delwiche, 1970; Söderlund & Svensson, 1976; Clarke, 1981; Rosswall, 1983; Winteringham, 1984; Wada & Hattori, 1991).

3.2.1 Origin and isotopic composition of nitrate

Nitrogen in soils is ultimately derived from two sources, atmospheric nitrogen and geological materials. Contributions of nitrate from geological materials are not considered to be important except on a localized basis. The majority of nitrogen in the biosphere and hydrosphere is derived from atmospheric nitrogen gas via biological nitrogen fixation (Fig. 3.1, step 1), although in certain environments larger portions derive from nitrate or ammonium deposition from anthropogenic sources (Aber *et al.*, 1989). Atmospheric nitrogen is also used to produce ammonium for fertilizers. In recent years production of fertilizers by this process has increased and is expected to equal the total amount of nitrogen bound as organic matter by terrestrial nitrogen fixation by the year 2000 (Söderlund & Svensson, 1976).

Once incorporated in soils, organic matter is utilized as a substrate for microbial growth in many reactions. Mineralization is the process whereby ammonium is generated from organic matter during decomposition (Fig. 3.1, step 2). Once generated or applied as fertilizer, ammonium is susceptible to volatilization under conditions of high pH (Fig. 3.1, step 3) or oxidation to nitrite and nitrate (Fig. 3.1, steps 4 and 5), by microbial nitrification. Nitrate is readily soluble, is not as reactive with the soil organic matter complex as ammonium and is easily transported by the movement of water through soil (Fig. 3.1, step 6). Nitrate may be lost to soils or ground waters through denitrification (Fig. 3.1, step 7). This process requires anoxic or near anoxic conditions and results in the production of N_2, N_2O and other nitrogen oxide gases. Denitrification has been shown to occur in soils, phreatic waters and occasionally in aquifers

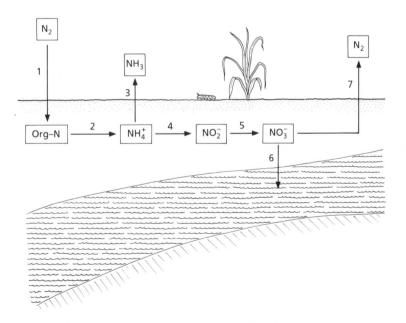

Fig. 3.1 A simplified diagram of the nitrogen cycle in soils and ground waters.

(Sexstone *et al.*, 1985; Myrold, 1988; Lowrance & Pionke, 1989) and is the mechanism that completes the nitrogen cycle by returning nitrogen to the atmosphere.

Each step in the nitrogen cycle may result in isotopic fractionation (Shearer *et al.*, 1974a). For each of the reactions illustrated in Fig. 3.1 (excluding nitrogen fixation and nutrient uptake), isotopic differences between substrate and reactant are thought to be 10–30‰ or more, with the product depleted in ^{15}N (Urey, 1947; Delwiche & Steyn, 1970; Freyer & Aly, 1975; Kohl & Shearer, 1980; Mariotti *et al.*, 1981, 1982; Yoshida, 1988; Böttcher *et al.*, 1990). Such fractionations are sometimes observed following fertilizer additions (Feigin *et al.*, 1974b). However, fractionations in more pristine natural systems are generally smaller. Differences between substrate and reactant in $\delta^{15}N$ during nitrogen fixation are rarely greater that 2‰ (Hoering & Ford, 1960; Delwiche & Steyn, 1970; Kohl & Shearer, 1980; Macko *et al.*, 1987). Similarly, fractionation during nitrogen mineralization in soils is small (e.g., Cheng *et al.*, 1964; Shearer *et al.*, 1974a). Fractionation during assimilation of inorganic nitrogen by plants in terrestrial systems has generally been shown to be small (Kohl *et al.*, 1973; Amarger *et al.*, 1977; Kohl & Shearer, 1980); however, isotopic fractionation factors for nitrate or ammonium uptake by microbes or aquatic algae when supplied with inorganic nitrogen are commonly in the range of −10 to −20‰ (Delwiche & Steyn, 1970; Estep & Vigg, 1985; Macko *et al.*, 1987). The extent and magnitude of nitrate fractionation is difficult to estimate in the natural environment, and therefore the use of $\delta^{15}N$ to trace origins of nitrate is considered to be a semi-quantitative or qualitative technique (Hauck *et al.*, 1972; Hauck, 1973; Heaton, 1986). Considering that the variation between sources of nitrate in ground waters are generally less than 10‰, the effects of fractionation are a serious concern to studies using this technique.

Despite the potential for significant alteration of isotopic signatures during transformation of nitrogenous compounds, fractionation may not always be observed. For example, if all of the available nitrate is converted to ammonium during denitrification no isotopic segregation would be observed (Cook *et al.*, 1973). When a series of reactions are involved, the step which is rate limiting to the overall process will control the expression of the net or observed fractionation. This occurs during mineralization and nitrification of ammonium in soils. Two isotope effects have been reported to dominate during these reactions (Feigin *et al.*, 1974a; Freyer & Aly, 1975; Mariotti *et al.*, 1980a; see Fig. 2.3). When a large reservoir of ammonium is present, as would occur after recent fertilizer application, nitrification (step 4 in Fig. 3.1) is limiting and nitrate produced by this series of reactions will be depleted in ^{15}N. Soil nitrate compositions that are depleted in ^{15}N by greater that 10‰ with respect to soil organic matter have been attributed to this isotope effect (Rennie *et al.*, 1976; Karanamos & Rennie, 1980a,b). However, differences in $\delta^{15}N$ values between nitrate and ammonium are difficult to predict because both the amount of ammonium and its ^{15}N composition can vary with changes in rates of nitrogen mineralization, assimilation and other processes. Some studies show a slight depletion of nitrate in ^{15}N (<2‰) relative to ammonium and its organic precursors (see Chapter 2). Small differences such as this may permit the use of isotopes for identifying the source(s) of nitrate in ground water. Three sources of nitrate to surface and ground waters have been recognized as major contributors to ground water systems: (i) fertilizer nitrogen; (ii) natural soil-derived nitrate; and (iii) animal waste- or sewage-derived nitrate. Other sources, including precipitation, nitrogen fixation and geological materials may be significant at specific locations but in general are not considered important.

3.2.2 Tracing soil-derived vs. fertilizer nitrate

The use of nitrogen-based fertilizers has increased dramatically in recent years. Between 1975 and 1985 fertilizer usage rose 160% in undeveloped countries and 55% on a world-wide basis (Follet & Walker, 1989). Surveys by the US Geological Survey (USGS) indicate significant increases in groundwater nitrate that were related to agricultural activity in 81% of sampling locations (Smith, 1987; Smith *et al.*, 1987). One-quarter of all fertilizer applied may be lost to ground waters and this may represent the largest source of nitrate delivered to ground and surface waters (Hallberg, 1986, 1989; Keeney, 1982). Clearly, nitrate contamination from fertilizer application is recognized.

Industrial fertilizers are produced using the Haber process which converts atmospheric nitrogen to ammonium. Most fertilizers retain the isotopic signature of atmospheric nitrogen, defined as 0‰, although further processing to create nitrate-, nitrite- or urea-based compounds may result in a wider spread of values (Fig. 3.2). The isotopic composition of fertilizers may change as nitrogen compounds interact with the soil organic matter complex or experience partial volatilization. In a survey of soils from Texas that were predominantly fertilizer influenced, Kreitler (1975, 1979) noted a difference of 2–3‰ between the fertilizers used and underlying groundwater nitrate. This difference was associated with volatilization of ammonia following fertilization, leaving the residual ammonium enriched in ^{15}N. Losses of ammonia are particularly a concern in alkaline soils and may exceed 50% of the fertilizer nitrogen applied (Rashid, 1977). The isotopic shift of such a loss could be quite dramatic. Naturally, this is not a concern if oxidized forms of nitrogen are the only

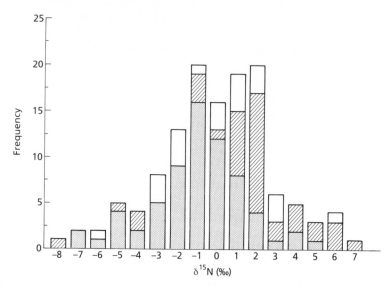

Fig. 3.2 Histogram of the $\delta^{15}N$ of various fertilizers separated into NO_3^- (▨), NH_4^+ (☐) and reduced nitrogen (☐). Values from: Kohl *et al.* (1971, 1973); Edwards (1973); Feigin *et al.* (1974b); Freyer & Aly (1974); Shearer *et al.* (1974b); Aly (1975); Rennie *et al.* (1976); Black & Waring (1977); Kreitler (1979); Aly *et al.* (1981, 1982); Flipse & Bonner (1985).

fertilizers used. Therefore, in isotope tracer studies a knowledge of soil pH, type of fertilizer applied and the $\delta^{15}N$ of the fertilizers used, are important for the success of the study.

Soils, particularly agricultural soils, generally have $\delta^{15}N$ values that range between 3 and 12‰ (Rennie *et al.*, 1976; Shearer *et al.*, 1978; Sweeney *et al.*, 1978; Tiessen *et al.*, 1984), although there are exceptions to this pattern, particularly in forested ecosystems (Nadelhoffer & Fry, 1988). If nitrate produced from mineralization and nitrification closely matched these values, as would happen if fractionation was minimal, then discriminating between fertilizer-derived and soil-derived nitrate would be relatively simple (Feigin *et al.*, 1974a,b). Nitrate produced from mineralization and nitrification in some systems has been shown to be within this range, thus facilitating identification of this source of nitrate (Feigin *et al.*, 1974a,b; but see Binkley *et al.*, 1985). For example, in a confined aquifer of the Western Kalahari a range of 4–8‰ was found for soil-derived nitrate (Fig. 3.3; Heaton *et al.*, 1983).

A distinction in $\delta^{15}N$ between soil- and fertilizer-derived nitrate prompted the first use of nitrogen isotopes to trace fertilizer contamination (Kohl *et al.*, 1971). In this and later studies, the observed decrease in the $\delta^{15}N$ of nitrate correlated with the increase in nitrate concentration suggested increased contributions of fertilizer to the background levels of

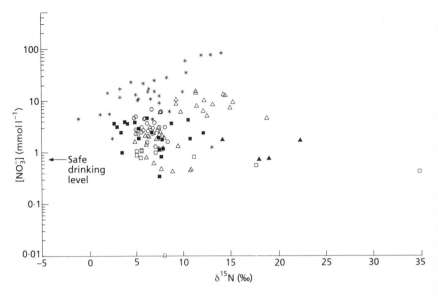

Fig. 3.3 The concentration and isotopic composition of NO_3^- in ground waters from selected locations. □ = West Kalahari, ground water; ■ = Long Island, New York; ○ = Springbok Flats, South Africa; △ West Kalahari, phreatic water; ▲ = Grand Cayman Islands; * = Runnels County, Texas. Values from: Kreitler & Jones (1975); Kreitler *et al.* (1978); Kreitler & Browning (1983); Heaton *et al.* (1983); Heaton (1984, 1985).

soil-derived nitrate (Kohl *et al.*, 1971; Freyer & Aly, 1974, 1975; Aly, 1975; Aly *et al.*, 1981, 1982). However, calculation of the relative contribution of nitrate-nitrogen from each source is not always possible owing to fractionations during the many reactions that fertilizer nitrogen experiences after application to soils (Hauck *et al.*, 1972).

Although soil-derived nitrate is considered 'natural' or associated with undisturbed ecosystems, its presence in ground waters may be enhanced by human activities. Simple clearing and cultivation of soil can dramatically increase the rate of organic matter mineralization. Losses of total soil organic matter as a consequence of initial ploughing or clearing vary between 8 and 61% (Winteringham, 1984). Mineralization following cultivation would thus release a great quantity of nitrogen into the soil that could readily be transported into ground waters.

In Runnels County, Texas, ground waters with very high concentrations of nitrate (average of $250\,\text{mg}\,NO_3\text{l}^{-1}$) were found to have $\delta^{15}N$ values predominantly in the range of 2–8‰ (Fig. 3.3; Kreitler, 1975; Kreitler & Jones, 1975). Appreciable quantities of fertilizer had not been used in this area, and thus these values indicate soil nitrogen as the source of nitrate contamination. Age-dating indicated that the ground waters were less than 20 years old, which substantiates the conclusion that contamination may have been associated with extensive terracing that occurred in that area in the 1950s. Similar isotope values and nitrate concentrations indicated soil-derived nitrate contamination in the ground waters of Springbok Flats, South Africa (Fig. 3.3; Heaton, 1985). Nitrate contamination in Springbok Flats was associated with cultivation practices that have approximately tripled in the last 20 years. However, high concentrations of nitrate from soil mineralization in ground waters are not always associated with cultivation. Initial concentrations of nitrate in confined aquifers of the western Kalahari were in the range of 0.5 to 1.6 mmol l^{-1} for waters greater that 3000 years BP (Heaton *et al.*, 1983). Values for $\delta^{15}N$ between 4 and 8‰ confirmed that this nitrate originated from soil mineralization and, considering the age of these waters, cultivation is not likely to have influenced this process.

Contamination from animal and/or sewage waste has been considered an additional cause of high nitrate concentrations in ground waters (Kreitler *et al.*, 1978). The isotopic composition of fresh animal excrement is not usually distinct from $\delta^{15}N$ values typical of soil- or fertilizer-derived nitrate. In one study average $\delta^{15}N$ values for cow and pig excrement ranged between 1.7 and 4.8‰ (Gormly & Spaulding, 1979). Nonetheless, nitrate collected from barnyard or feedlot soils was consistently more enriched in ^{15}N, with values as high as 44‰. More commonly, the range in $\delta^{15}N$ for barnyard soils is 10–22‰ (Kreitler, 1975; Kreitler & Jones, 1975; Gormly & Spaulding, 1979). This enrichment is considered to be a consequence of fractionation during volatilization of ammonia gas, which leaves the remaining ammonium enriched in ^{15}N. Hydrolysis of urea to form ammonium results in an increase in pH, which in turn favours volatilization of ammonia. This causes a decrease in pH and the remaining ammonium to be enriched in ^{15}N. Subsequent nitrification produces nitrate with $\delta^{15}N$ values typically greater than 10‰ (Kreitler & Jones, 1975; Kreitler, 1975). Observations of nitrate $\delta^{15}N$ values in the range 10–22‰ have indicated ground waters contaminated with animal waste as point sources in the vicinity of barnyards and cattle watering wells (Fig. 3.3; Kreitler, 1975; Kreitler & Jones, 1975; Kreitler, 1979; Gormly & Spaulding, 1979; Heaton, 1984) and as a non-point source associated with septic tank effluent (Kreitler & Browning, 1983).

The occurrence of denitrification in ground waters can compromise the success of the $\delta^{15}N$ technique to indicate sources of nitrate. Denitrification is associated with extensive isotopic fractionation and it has been estimated that as little as 20% total nitrate removal by this process will result in an increase of 8‰ in the $\delta^{15}N$ of the remaining nitrate (Heaton, 1984). Denitrification requires anoxic conditions, the presence of appropriate microbes and an available organic carbon substrate. There is little doubt that denitrification occurs in some soils, shallow ground waters and aquifers, although it is not a ubiquitous process (Hallberg, 1989). Therefore, it is important to assess whether or not denitrification may be occurring when evaluating isotope data on nitrate.

Several techniques have been used to evaluate the presence or absence of denitrification in aquatic environments. The most simple is determination of dissolved oxygen; if dissolved oxygen is present in detectable quantities then denitrification cannot occur. This approach has been used in the ground waters of Springbok Flats and the Kalahari, South Africa (Heaton, 1984, 1985). However, detection of oxygen in waters does not preclude the possibility that the waters were at some point anoxic and that oxygen diffused in later. For this reason determination of ratios of nitrogen to argon gases provides an additional constraint on the presence of denitrification. The N_2/Ar ratio should reflect equilibration with the atmosphere at the temperature and time of groundwater formation. If excess nitrogen entered by denitrification then the N_2/Ar ratio will increase and reflect the magnitude of this reaction. Ratios of nitrogen to argon have been useful in determining the extent of denitrification in oceanic and groundwater environments (Richards & Benson, 1961; Cline, 1975; Vogel et al., 1981; Heaton et al., 1983). As denitrification produces nitrogen depleted in ^{15}N, determination of the $\delta^{15}N$ of dissolved nitrogen gas can also be used to recognize this process (Richards & Benson, 1961; Vogel et al., 1981). Because the reservoir of nitrogen in dissolved N_2 may be much greater than that of nitrate, denitrification may alter the isotopic composition of nitrate without affecting the $\delta^{15}N$ of N_2 (Cline & Kaplan, 1975).

3.2.3 $\delta^{18}O$ of nitrate

A promising new technique, not only for recognizing denitrification but also assisting in delineation of sources, is the determination of the oxygen isotopic composition of nitrate. Oxygen, as well as nitrogen, is fractionated during denitrification, although the magnitude of fractionation for oxygen is approximately half of that for nitrogen (Böttcher et al., 1990). The degree to which fractionation of oxygen and nitrogen occurs is likely to vary with changes in environment and microbial population. Nonetheless, when an enrichment in the heavy isotopes of both oxygen and nitrogen is observed in the residual nitrate in ground waters, then it may be concluded that denitrification is occurring.

The oxygen isotopic composition of nitrate is also a reflection of its origin. Oxygen in industrial fertilizers originates from the atmosphere, whereas oxygen in nitrate produced during nitrification is derived two-thirds from water and one-third from oxygen (Andersson & Hooper,

1983; Kumar *et al.*, 1983; Hollocher, 1984). These two sources may differ by 30‰ or more and thus provides a clear distinction that may prove useful in tracing the fate of fertilizer nitrate in both terrestrial and aquatic environments (Amberger & Schmidt, 1987).

3.3 Tracing atmospheric nitrogen compounds

Considerable attention has been focused on nitrogen oxides in the atmosphere as a consequence of their contribution to acid rain and the destruction of stratospheric ozone. Tracing nitrogen compounds in the atmosphere by stable isotope data is complicated by both the number of compounds present and potential for many reactions and phase changes to occur (see Crutzen, 1981 for review). Each reaction or phase change may result in a segregation of the isotopes, making determination of sources difficult. Nonetheless, while there remains considerable research to be done, preliminary studies have indicated that stable isotopes may provide insight into the origins and cycling of nitrogen compounds in the atmosphere.

Atmospheric nitrogen gases may be derived from soils, fertilizers, combustion of fuels and animal waste. Natural processes such as ammonia volatilization, nitrification and denitrification release ammonia and nitrogen oxides. As mentioned previously, volatilization of ammonia involves a large fractionation of nearly 40‰ with ^{14}N preferentially incorporated in the vapour phase (Urey, 1947). This effect has been verified by observations of low $\delta^{15}N$ values for gaseous ammonia in barnyards and over high pH soils (Fig. 3.4; Freyer, 1978a; Kreitler, 1979). Unfortunately, other sources of atmospheric ammonia may also be depleted in ^{15}N and it may be for this reason that seasonal studies of ammonium in rain have shown no consistent isotopic trends (Fig. 3.4; Freyer, 1978a; Heaton & Collet, 1985; Heaton, 1987a).

Seasonal variation in the $\delta^{15}N$ of rain nitrate has been observed, with depleted values being associated with increased soil activity in early spring or summer (Fig. 3.4; Freyer, 1978a; Heaton & Collet, 1985). As temperatures rise in the spring and fertilizers are applied, nitrification may become more prominent in soils and favour the release of isotopically depleted nitrogen oxides. Release of ^{15}N-depleted nitrous oxide ($<-60‰$) has been observed in laboratory studies, but has not yet been confirmed in soil environments (Yoshida & Matsuo, 1983; Yoshida, 1988). The higher $\delta^{15}N$ values within the lower atmosphere prevalent at other times of the year may reflect release of nitrogen oxides from combustion of fossil fuels that have been characterized by $\delta^{15}N$ values between -1.8 to 3.9‰ (Fig. 3.4; Moore, 1977; Freyer, 1978a).

Despite these encouraging initial results, it must be realized that nitrogen compounds are generated in, and removed from, the atmos-

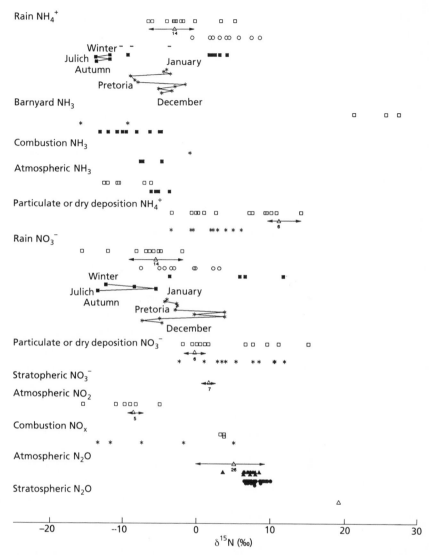

Fig. 3.4 The $\delta^{15}N$ of nitrogen compounds in rain and in the atmosphere. Arrows indicate a range in values and the number beneath arrows is the number of samples analysed. □ = Moore (1977), Boulder, Colorado; ■ = Freyer (1978a,b), Julich, Germany; ○ = Hoering (1957), Fayettesville, Arkansas; ● = Yoshida & Matsuo (1983), Eastern Tropical Pacific; △ = Moore (1974), Boulder Colorado; ▲ = Yoshida *et al.* (1984), Japan; ∗ = Heaton (1987a), Pretoria, South Africa; ▰▰ = Wada *et al.* (1975), Tokyo, Japan.

phere and may experience exchange reactions that have the potential to vary isotopic abundances. For example, nitrate may be generated during electrical storms from atmospheric N_2 and be depleted in ^{15}N by approximately 15‰ during equilibration of these two species (Ingerson, 1953). Whether or not equilibrium is reached between these species is uncertain, as no fractionation was observed in laboratory simulations and field data (Hoering, 1957). Particulate ammonium and nitrate in atmospheric samples has been suggested to be of soil origin based on similar $\delta^{15}N$ values (Moore, 1974). However, these results could also be explained by removal of condensable nitrogen compounds following isotopic equilibration in the atmosphere (Moore, 1977).

Although complicated by atmospheric processes, isotopic tracing of atmospheric gases may be successful within specific locales. For example, the removal of nitrous gases from chemical plant emissions is commonly done by the Spindel–Taylor process, that also entraps ^{15}N enriched materials (Hübner, 1986). NO_x gases associated with one such plant were clearly identified by $\delta^{15}N$ values as low as $-140‰$ (Maass & Weise, 1981). Oxygen isotope ratios have proved useful in distinguishing nitrous oxides derived from denitrification and nitrification (Wahlen & Yoshinari, 1985; Yoshinari & Wahlen, 1985). Combined oxygen and nitrogen isotope ratios in nitrogen oxides may provide a powerful technique in tracing sources and the fate of atmospheric compounds. For further details, see Chapter 6.

3.4 Tracing sewage in aquatic environments

Stable isotopes have frequently been used to determine the origins of organic matter in sediments. In addition, the stable isotopes of carbon, nitrogen, sulfur and hydrogen have provided a means to trace the flow and fate of sewage and other anthropogenic effluents in aquatic systems (Burnett & Schaeffer, 1980; Sweeney & Kaplan, 1980a; Sweeney et al., 1980; Rau et al., 1981; Macko, 1983; Ostrom & Macko, 1991; Coakley et al., 1991).

The flow of sewage organic matter in aquatic environments can only be traced if it has an isotopically distinct signature, with respect to other sources of organic matter, and is not altered during transport and diagenesis. Generally, it is assumed that changes in isotopic abundances are small or negligible during diagenesis (Sackett, 1964; Sweeney et al., 1980; Arthur et al., 1985; Dean et al., 1986; Schidlowski et al., 1987) and this has been verified in one sewage influenced environment (Myers, 1974; Sweeney et al., 1980). However, in some organic-rich sediments isotopic shifts as a consequence of diagenesis have been observed (Behrens & Frishman, 1971; Macko, 1981; Ostrom & Macko, 1992).

In several environments, sewage has successfully been distinguished from marine or freshwater productivity on the basis of stable isotope ratios. In a study of a dump site in the New York Bight, anthropogenic waste and marine organic matter were characterized by distinct $\delta^{13}C$ values of -26.2 and $-22.0‰$, respectively (Burnett & Schaeffer, 1980). Transects of sediment samples clearly demonstrated a predominance of carbon with a terrigenous source at the dump site with an increasing influence of marine organic matter with distance from the site (Fig. 3.5).

In a series of studies off the coast of Southern California the isotopic difference between sewage and marine sources in $\delta^{13}C$ was small (Myers, 1974; Sweeney $et\ al.$, 1980). However, the nitrogen isotopic composition of sewage effluent (2.5‰) was depleted in ^{15}N by over 7‰ with respect to marine organic matter (Sweeney & Kaplan, 1980a; Sweeney $et\ al.$, 1980). Sulfur isotopes have also been used to trace the flow of sewage into the marine environment in this area. Marine organic matter and effluent from a treatment plant were characterized by $\delta^{34}S$ values of $\geqslant 10$ and 0‰, respectively. Organic-rich sewage deposited in the marine environment may create anoxic conditions in the sediment. Sulfide resulting from sulfate reduction will add isotopically-light sulfur ($-20‰$) and shift the $\delta^{34}S$ to approximately $-10‰$ (Sweeney & Kaplan, 1980b; Sweeney $et\ al.$, 1980).

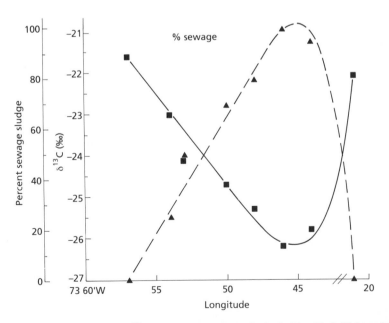

Fig. 3.5 Transect of sediment $\delta^{13}C$ values across a dump site in the New York Bight and the estimated abundance of sewage derived material comprising the sediment. ■ = $\delta^{13}C$; ▲ = percent sewage sludge. (From Burnett & Schaeffer, 1980.)

Additional information on the fate of sewage effluent in the marine environment may be gained by determination of hydrogen isotope ratios. In temperate regions, terrigenous organic matter is generally characterized by δD values greater than $-100‰$, whereas marine organic hydrogen has values that are commonly less than this value (Schiegl & Vogel, 1970; Nissenbaum, 1973; Stuermer et al., 1978; Estep & Hoering, 1980). This distinction permitted Rau et al. (1981) to demonstrate the assimilation of sewage-derived organic matter by organisms in the vicinity of outflows off southern California.

With a knowledge of the isotopic composition of the sources of organic matter influencing an environment it is possible to quantitatively assess their relative contributions. This calculation is based on the premise that the isotopic composition of a sediment is the sum for all sources of the fraction contributed from a source (F_i) multiplied by its isotope ratio (δ_i) (isotopic mass balance):

$$\delta_{sample} = \sum_{i=1}^{n} F_i \delta_i$$

Using a single isotope, this equation may be solved to estimate the relative contribution from two sources:

$$F_1 = (\delta_s - \delta_2)/(\delta_1 - \delta_2)$$

where δ_s = the isotopic composition of the sample mixture.

This model has been used to trace the influence of sewage-derived organic matter in the marine environment off the coast of southern California and in the New York Bight (Fig. 3.5; Burnett & Schaeffer, 1980; Sweeney et al., 1980). If two stable isotopes are used it is possible to determine the relative contributions to a sample from three sources based on a simultaneous solution of two isotope proportionation equations and the distribution equation for the three components (Harrigan et al., 1989).

3.5 Tracing atmospheric sulfur pollution in lakes
Industrial activities may contribute substantial quantities of sulfur to the environment. For example, the INCO nickel smelter plant at Copper Cliff, Ontario, releases an estimated 3500 tons of sulfur dioxide on a daily basis (Lusis & Wiebe, 1976). Concentrations of sulfur compounds in air, rain and dry precipitation have been used to track the fate of pollutant sulfur. However, stable isotope ratios can provide a more distinct determination of origin, particularly when multiple sources are involved (Nriagu et al., 1991). The use of sulfur isotope ratios to trace pollutants in the atmosphere, soils and vegetation has been the subject of recent review (Krouse 1980, 1988). In this section, the $\delta^{34}S$ of lake sediments

will be discussed as a means to understand historical as well as recent sources of atmospheric pollution.

Contributions of atmospheric pollutants to lakes has been evidenced by increased concentrations of dissolved sulfate and sedimentary sulfur in the vicinity of industrial plants. Lakes as far away as 90 km from smelters may have appreciably elevated sulfate concentrations (Nriagu & Harvey, 1978). The $\delta^{34}S$ of sulfate might provide a confirmation of the origins of excess lake sulfur. However, in North America industrial emissions are frequently similar in $\delta^{34}S$ to other natural sources of sulfur (Nriagu & Harvey, 1978; Saltzman et al., 1983; Dickman & Thode, 1985; Nriagu & Soon, 1985; Thode et al., 1987). Enigmatically, surficial sediments in lakes experiencing high inputs of anthropogenic sulfur have shown dramatic depletions in ^{34}S, as much as 15‰, toward the sediment–water interface (Nriagu & Coker, 1983; Nriagu & Soon, 1985; Fry, 1986; Thode et al., 1987). However, these shifts are not apparent in pristine lakes (Fig. 3.6).

The shifts in $\delta^{34}S$ and elemental sulfur abundances observed in sediments of some lakes clearly is a reflection of sulfur loading. The low C:S ratio for surficial sediments in pollutant-influenced lakes provides a strong indication that the source of additional sulfur is mineral and not organic. Dissimilatory sulfate reduction is accompanied by a large fractionation and has been shown to produce sulfides depleted in ^{34}S by as much as 50‰ with respect to the initial sulfate present (Thode et al., 1951; Kaplan & Rittenberg, 1964; Krouse et al., 1967; Chambers et al., 1975). Isotopically depleted sulfides may precipitate upon reacting with sedimentary iron and organic matter, thereby decreasing $\delta^{34}S$ and increasing total sulfur abundances (Nriagu & Soon, 1985; Fry, 1989). Nonetheless, in the vicinity of contaminated lakes atmospheric pollutants, terrestrial vegetation, soils, minerals and dissolved sulfate have all been shown to be significantly enriched in ^{34}S with respect to surficial sediments (Nriagu & Soon, 1985; Thode et al., 1987). A similar distinction is evident between surficial sediments and sources of sulfur in C:S values. Surficial sediments in contaminated lakes have been characterized by C:S values of 1.4–3.9 (Fig. 3.6), whereas ratios for soils and seston in lakes are generally 16–30 and 60–120, respectively (Stevenson, 1982; Nriagu & Soon, 1985).

The magnitude of the difference between lake sulfates and sediments has been shown to increase with increasing sulfate concentrations and presumed sulfur loading (Dickman & Thode, 1985; Fry, 1989). This effect is consistent with isotopic fractionation during kinetic reactions, where the full expression of the fractionation factor is not seen until the substrate, in this case sulfate, is not limiting. The lack of, or small difference between, sediments and dissolved sulfate $\delta^{34}S$ in pristine lakes is due to

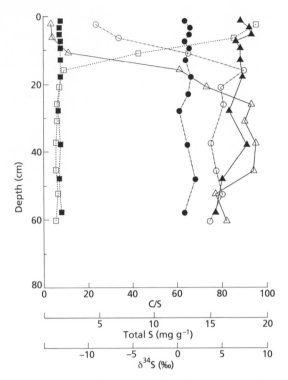

Fig. 3.6 Comparisons of sediment $\delta^{34}S$, carbon/sulfur (C/S) and total sulfur in a pristine (Turkey) and a polluted (McFarlane) lake. △ = McFarlane Lake C/S; ▲ = Turkey Lake C/S; □ = McFarlane Lake total sulfur; --■-- = Turkey Lake total sulfur; --○-- = McFarlane Lake $\delta^{34}S$; ● = Turkey Lake $\delta^{34}S$. (From Nriagu & Coker, 1983.)

the fact that sulfate reduction in lakes is limited by the availability of sulfate (Schindler, 1985; Fry, 1989). Experimental studies have shown that fractionation during sulfate reduction is not evident at concentrations less than $10\,\mu mol\,l^{-1}$ (Harrison & Thode, 1958). In addition, sulfate reduction may be inhibited by, and less energetically favourable than, methanogenesis at concentrations below $30\,\mu mol\,l^{-1}$ (Lovley & Klug, 1986). Therefore, the similarity in $\delta^{34}S$ between lake sulfate and deep sediments in sulfur-loaded lakes suggests that at the time of deposition sulfate concentrations were similar to those presently found in pristine lakes (Fig. 3.6). With this understanding, dating of the sediment interval where changes in $\delta^{34}S$ and sulfur concentrations occur may indicate when atmospheric pollution began. In lakes near Sudbury, Ontario, the shift in sedimentary $\delta^{34}S$ and percentage sulfur is first apparent near 1890, a date consistent with the beginning of metal smelting in that area (Nriagu & Coker, 1983).

3.6 Stable isotopes in studies of oil pollution

Because petroleum has unusual stable carbon isotopic compositions relative to most marine materials, isotope analysis has proven to be a good indicator of the contribution of petroleum carbon to the total organic matter in a sediment (Botello *et al.*, 1980; Botello & Macko, 1980, 1982; Farran *et al.*, 1987). Analyses of this sort yield the best evidence in regions of intense hydrocarbon addition. Calder & Parker (1968) found effluents in the Houston Ship Channel which were strongly depleted in ^{13}C and calculated that 50% of the carbon in the dissolved fraction was likely due to petrochemical pollution. At a more sensitive level, extracts of sediments can also be used to indicate hydrocarbon pollution. Macko *et al.* (1981) found isotopic depletions of carbon in the lipid extracts of sediments from the South Texas coast, indicating areas which saw a chronic influence of oil tanker traffic as well as in an area which was impacted by a small oil spill 2 years prior to the analyses. During the natural degradation of oil, the chemistry of a spilled oil changes quite drastically; the isotopic compositions remain essentially unaltered during those processes and can usually be distinguished from naturally occurring lipid carbon (Macko *et al.*, 1981; Fig. 3.7).

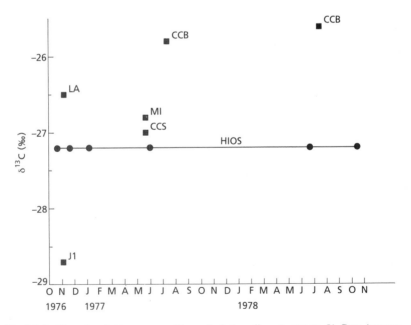

Fig. 3.7 Stable carbon isotope compositions of whole sediment extracts. J1, Port Aransas Jetty; CCS, Corpus Christi Ship Channel; MI, Mud Island; CCB, Corpus Christi Bay; LA, Lydia Anne Ship Channel; HIOS, Harbor Island Oil Spill. Unpolluted isotope signatures from this region generally range from −15 to −20‰. (From Macko *et al.*, 1981.)

Stable nitrogen isotope ratios also show promise in the tracking of spilled petroleum owing to the large range in isotopic signatures for petroleum reservoirs throughout the world. Even more powerful in the identification of the source of a spilled oil would be the use of a dual isotope tag of nitrogen and carbon for a specific nitrogen-rich fraction of an oil, such as the asphaltenes. Sources of tarballs, which are rich in the asphaltene component, avail themselves to such a duel tracer. Macko & Parker (1983) made use of the dual isotope label to indicate tarballs originating from the *Ixtoc I* spill and from a smaller tanker spill, the *Burmah Agate*, in the Gulf of Mexico.

3.7 Conclusions

The characterization of the isotopic compositions of pollutants can provide useful information with regard to its source and quantity in an environment. Groundwater contamination by nitrogenous organic wastes, or from fertilizers, can be assessed through distinct differences between their isotopic signatures. Atmospheric inputs of nitrogen or sulfur from agricultural or industrial processes can be identified in the isotopes of those elements, as well as the associated oxygen isotopes. Sewage transported into aquatic environments has characteristic isotopic signatures which make resolution and quantification possible. Petroleum also has an unusual carbon isotopic content with respect to much of the biosphere, and inputs from this source can be estimated using mass-balance equations. Nitrogen isotopic compositions of petroleum have a fairly wide range and may be useful in the delineation of different sources of contamination.

CHAPTER 4

Tracing the diets of fossil animals using stable isotopes

P.L. KOCH, M.L. FOGEL & N. TUROSS

4.1 Introduction

Diet influences the morphology, behaviour, ecological interactions and evolutionary history of animals. As a consequence, palaeobiologists and archaeologists have devoted substantial effort to the study of diet. Preserved gut contents and faecal matter provide the most direct source of palaeodietary information (Dreimanis, 1968; Mead *et al.*, 1986). However, these materials are quite rare in the fossil record, and only offer a glimpse of an animal's diet immediately prior to death. Historically, comparative and functional morphology of the teeth and jaws have served as important indirect sources of palaeodietary inferences (Crompton & Hiiemae, 1969; Rensberger, 1986). Study of abrasion and wear patterns on vertebrate teeth has allowed semi-quantitative analysis of palaeodiets (Kay, 1975; Walker *et al.*, 1978; Teaford, 1988). In recent years, analyses of the trace element and isotope chemistry of teeth and bones have revolutionized palaeodietary research, potentially allowing exploration of dietary preferences, food webs and microhabitat utilization (Toots & Voorhies, 1965; Brown, 1974; van der Merwe, 1982; Sillen & Kavanaugh, 1982; Schoeninger, 1985; DeNiro, 1987).

In this review, we explore the rapidly expanding use of stable carbon, nitrogen and oxygen isotopes in palaeodietary research. Isotopic palaeodietary information has been obtained both from organic remains, such as collagenous and non-collagenous bone proteins, shell proteins and individual amino acids, and from biominerals in bone, tooth enamel, dentin, bird eggshells and mollusc shells. The stable isotopic compositions of food and fluids ingested by animals have a strong influence on the isotopic compositions of the tissues they synthesize. The isotopic composition of animal tissues can serve as a natural tracer of different dietary inputs with distinct isotopic signatures. However, the precise relationship between the isotopic compositions of ingested materials and any particular tissue or molecular component is quite complex, responding to changes in nutritional status, turnover rate of the tissue and biosynthetic pathway. As a result, stable isotope analysis provides more than just a tracer of the materials that go into an animal; it offers a view of the biological processes within an organism.

This chapter is not an exhaustive review of the literature on stable isotopes and palaeodiet. Rather, we focus on recent applications that have attempted to expand the application of isotopic reconstruction in both time and source materials. We examine the carbon and nitrogen isotope records preserved in fossilized organic constituents, such as collagen, non-collagenous proteins and amino acids. We then discuss carbon and oxygen isotopes in biogenic minerals, including hydroxyapatite in bones and teeth and calcium carbonate in bird eggshells and land snail shells. For each major type of isotopic source material, we evaluate current methods and the isotopic dietary controls in modern organisms, then consider selected palaeodietary applications. The success of different methods is evaluated in light of the significant impact that post-mortem, diagenetic alteration may have on fossilized tissues. Finally, we stress studies of fossil vertebrates other than hominids, as the archaeological aspects of stable isotope analysis have been reviewed previously (van der Merwe, 1982; Schoeninger, 1985; DeNiro, 1987).

4.2 Isotope analysis of organic matter from the fossil record

After an animal's death, intact biochemicals in soft tissues can be quickly degraded by proteases and other degradative enzymes (Allison & Briggs, 1991). Cellular material is decomposed further by microbial processes and physical weathering, particularly if the carcass remains on the surface (Behrensmeyer, 1978) or is buried in wet, aerobic soils (Hare, 1980). Often only the organic matter contained in mineralized tissues, such as teeth and bones, survives in the fossil record. Bones and teeth, however, are not closed systems; even these mineralized tissues are subject to weathering, infiltration by exogenous organic matter, dissolution and decay (Fig. 4.1; Retallick, 1984; Piepenbrink, 1989). A fundamental step in any isotopic palaeodietary study is to determine if the organic material preserved in mineralized tissues is indigenous and, if so, whether these indigenous biomolecules retain their original isotopic compositions with high fidelity. Consequently, in many studies, an effort has been made to ensure that the material recovered from mineralized tissues is directly related to a characteristic biochemical component, such as collagen, the most abundant protein in most animals. We review palaeodietary studies using collagen and techniques for assessing collagen purity, as well as several studies that have attempted to explore palaeodiets with non-collagenous organic compounds.

4.2.1 Collagen

The major structural protein in all vertebrates is collagen, which is found in bone, tooth dentin, skin and muscle tissue (Kemp, 1984). This protein has an unique amino acid composition; one-third of its amino acid residues

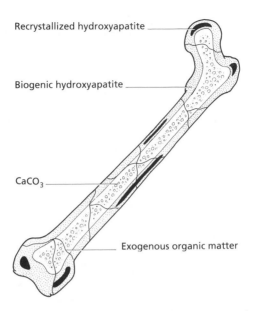

Fig. 4.1 Potential diagenetic alterations experienced by fossil bones and teeth. Calcium carbonate and exogenous organic matter may invade pore spaces. With time, poorly crystalline bone apatite will recrystallize to form larger more stable crystals.

are glycine. Other characteristic amino acids include hydroxyproline (10%) and proline (12.3%), with correspondingly lower percentages of acidic amino acids, such as glutamate (7.2%) and aspartate (4.7%) (Herring, 1972). Collagen in bones is deposited and reworked by growth and remodelling, albeit at a much slower rate in adults than in growing juveniles (Libby *et al.*, 1964). Thus the isotopic composition of bone collagen reflects that of the diet averaged over a substantial fraction of an animal's life (Neuberger & Richards, 1964; Libby *et al.*, 1964). Protein in tooth dentin and enamel, however, is not extensively reworked, and its isotopic composition reflects the diet during the period of tooth formation early in the organism's life. Accordingly, each tissue provides an exclusive piece of information concerning an organism's diet.

Methods
Insoluble collagen is obtained for isotope analysis by dissolving away biogenic and diagenetic minerals, non-collagenous proteins, contaminating organic material and preservatives added after sample collection. Preservatives introduce a vexing set of problems (Moore *et al.*, 1989), and whenever possible, it is wise to avoid analysis of treated bones. Different methods have been successfully utilized to isolate pure collagen when the material has not been extensively degraded or contaminated (Table 4.1).

Table 4.1 Techniques for isolating collagen from modern or fossil materials

Technique	Procedure	Product	Reference
Gelatin extraction	Dissolve collagen in weak acid; dry	Gelatin	Longin, 1971; Schoeninger & DeNiro, 1984
Demineralization with EDTA	Bone chucks demineralized in EDTA solution	Collagen	Tuross et al., 1988
Collagenase	Hydrolyse collagen with enzymatic reaction	Low molecular weight hydrolysis products	DeNiro & Weiner, 1988c
Resin purification	Hydrolyse residue and purify on ion-exchange (XAD) resin	Amino acid mixtures	Stafford et al., 1988
Individual amino acid analysis	Hydrolyse collagen; column chromatography or gas chromatography	Individual amino acids	Hare et al., 1991; Silfer et al., 1991

The simplest method is to demineralize the bone with dilute (0.1 to 1N) hydrochloric acid or organic chelating agents (e.g., EDTA), wash to neutrality with distilled water, and then freeze-dry the organic residue (Tuross et al., 1988).

Following simple demineralization, it is important to verify that the residual material contains only collagen. One definitive test is to determine whether the residue has an amino acid composition similar to modern collagen. Collagen has an amino acid distribution that is significantly different than other proteins and potential contaminants (e.g., glycine : aspartic acid = 6–7; the presence of hydroxyproline; Herring, 1972). Quantitative amino acid analysis of decalcified bone or dentin can conclusively demonstrate the chemical integrity of fossil collagen (DeNiro & Weiner, 1988a; Tuross et al., 1988).

Because many isotope laboratories do not have access to routine amino acid analysis, the integrity of fossil collagen has been assayed by a variety of less definitive techniques. First, if collagen is well preserved, a replica or 'ghost' will remain following demineralization of bones or teeth. The presence of a replica is diagnostic of intact collagen (Tuross et al., 1988). If a collagen replica is absent at this stage, even though the bone or tooth possesses morphological integrity, the organic residue is probably unsuitable for routine isotope analysis. A second criterion for purity is the colour of the collagen 'ghost'. Pure collagen is pale yellow to white in colour; if the residue has a brownish hue, it is most likely contaminated by humic material formed after burial and must be presumed to have altered isotopic values. Finally, the molar or atomic ratio of carbon to nitrogen (C/N) in the residue is an indicator of collagen purity. Fossil collagen with C/N values in the range of modern collagen (2.9–3.5) is thought to retain its original isotopic composition (DeNiro, 1985; Ambrose, 1990). During diagenesis, C/N values can rise (e.g., 4.0–15.0) or fall (e.g., 1.0–2.0) due to deamination of amino acids or invasion by soil humic acids and inorganic material (DeNiro, 1985; DeNiro & Weiner, 1988a).

More complicated methods are available to obtain isotopically faithful collagen for palaeodietary analysis. When a sample is contaminated with exogenous organic matter, collagen may be selectively solubilized with hot, dilute acid or collagenase, leaving the contaminant as an insoluble residue. The 'collagenous' material in solution is collected by freeze-drying (Longin, 1971; DeNiro & Weiner, 1988a,c). Alternatively, amino acids derived from collagen may be separated from contaminating humic substances on an ion-exchange column (Stafford et al., 1988). When using these isolation methods, it is still very important to assess purity using C/N and amino acid composition before isotope analysis. Finally, in situations where collagen peptide bonds have been substantially hydrolysed, and different amino acids have been removed or added to the fossil, the only method that may provide access to relatively unaltered organic material is isotope analysis of individual amino acids (Hare et al., 1991).

Following isolation and purification, 1–5 mg of collagen is combusted in an evacuated quartz tube with a mixture of cupric oxide and copper or silver metal at 850–900°C, then slowly cooled to ensure quantitative reaction of carbon to CO_2 and nitrogen to N_2 (Minagawa et al., 1984). Further details of the procedure can be found in the Introduction to this volume.

Carbon isotopes in collagen

An animal utilizes food resources to obtain energy and to supply raw materials for growth. The isotopic composition of an animal's tissues are offset from the mean isotopic composition of its diet by a consistent, usually small amount (see Chapter 7; Vogel, 1978b; Tieszen et al., 1983; Fry & Sherr, 1984). In laboratory-reared mammals and birds, the difference between the $\delta^{13}C$ value of collagen and diet is variable, but usually small (i.e., from 1 to 4‰; DeNiro & Epstein, 1978; Tieszen & Boutton, 1989; Hare et al., 1991; Hobson & Clark, 1992). In mammals and birds collected from field settings where the mean diet was known, the difference in $\delta^{13}C$ between collagen and diet was about 5‰ (Vogel, 1978b; von Schirnding et al., 1982; Schoeninger & DeNiro, 1984; Lee-Thorp et al., 1989a; Fig. 4.2). This difference between fractionations measured in the laboratory versus a natural setting is puzzling, and it may indicate that the isotopic composition of collagen is controlled more

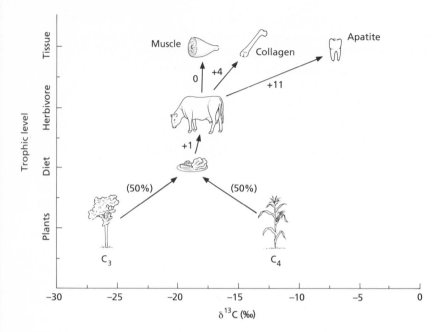

Fig. 4.2 Model of carbon isotope variation in diet and the fractionation of carbon between diet and different tissues in mammalian herbivores.

by particular dietary components, such as proteins, than by bulk diet (Krueger & Sullivan, 1984; Ambrose & Norr, 1993). Until the causes of this variability in isotopic spacing between collagen and diet are better known, collagen $\delta^{13}C$ values will provide only a qualitative measure of the abundance of particular food sources in animal diets.

The $\delta^{13}C$ values of animal collagen reflect the isotopic compositions of plants at the base of the food chain in an ecosystem. Plants vary in carbon isotopic composition in response to physiological and environmental factors (Fogel & Cifuentes, 1993; for a review see Chapter 1). As a consequence, carbon isotope studies of palaeodiet have been successful in determining: (i) aspects of feeding ecology; (ii) qualitative palaeoclimatic change; (iii) microhabitat utilization (e.g., 'the canopy effect'); and (iv) historical changes in the $\delta^{13}C$ value of atmospheric CO_2. Of primary interest in palaeodietary research has been determining the proportions of C_3 (Calvin cycle) versus C_4 (Hatch–Slack cycle) plants in the diets of hominids and herbivores from various times in the past (e.g., Ambrose & DeNiro, 1989; Stafford *et al.*, 1994). Plants using different photosynthetic pathways have different mean $\delta^{13}C$ values: $-26.5 \pm 2‰$ for C_3 vs. $-12.5 \pm 1‰$ for C_4 plants (see Chapter 1). Because C_4 photosynthesis occurs chiefly in grasses adapted for warm or dry habitats, whereas C_3 plants include all trees, most shrubs and herbs, and grasses adapted to cool/wet climates, carbon isotope analysis can sometimes discriminate between animals that eat grass (grazers) and animals that eat leaves, twigs and herbs (browsers). Isotopic studies are especially valuable for understanding the diets of extinct animals, such as proboscideans or large edentates (sloths and glyptodonts), that may have been able to eat very different types of food. For example, Krueger (1991) determined that late Pleistocene mastodonts from the east coast of the USA consumed a 100% C_3 diet, whereas mammoths from the same region had a substantial fraction of C_4 plants in their diets. In contrast, Koch (1991) found no carbon isotope difference between these two species in the late Pleistocene of Michigan and western New York. These studies highlight the flexibility in feeding habits of these extinct proboscideans.

Carnivores and omnivores have $\delta^{13}C$ values that are influenced by the isotopic composition of the animals that they eat. They incorporate carbon from different sources within their prey (fats, proteins, carbohydrates) that may be isotopically distinct. Despite this potential isotopic variability across dietary resources, the difference in $\delta^{13}C$ between collagen in prey and predator is quite small (Schoeninger & DeNiro, 1984). In order to understand human feeding ecology, the remains of potential human prey species have been analysed and compared with human remains from the same archaeological sites (Schwarcz *et al.*, 1985; Katzenberg, 1989; Tuross *et al.*, 1994). In addition, some investigators analysed modern organisms at the same location to unravel prehistoric feeding relationships (Sealy & van der Merwe, 1985; Keegan & DeNiro, 1988; Ambrose & DeNiro, 1989). Museum curators are often more willing to provide faunal remains than ancient human bones for destructive isotope analysis. Analysis of domestic dogs, which presumably consumed human refuse, has provided valuable dietary information about human populations (Burleigh & Brothwell, 1978; Noe-Nygaard, 1988).

Because the proportion of C_4 grasses relative to C_3 grasses in a region is sensitive to temperature and humidity (Teeri & Stowe, 1976; Teeri *et al.*, 1980; Hattersley, 1983), carbon isotope analysis of grazers can provide qualitative estimates of palaeoclimatic change. For example, Stafford *et al.* (1994) examined the $\delta^{13}C$ value of collagen in a suite of bison from the southwestern USA, ranging in age from 100 years before present (BP) to approximately 10 000 years BP. They found a systematic change in $\delta^{13}C$ from recent values of -7 to values of $-19‰$ at 10 000 years (Fig. 4.3). This shift was related to a floral transition that followed change from the colder, wetter climate of the late Pleistocene to the drier conditions of the Holocene. Bombin & Muehlenbachs (1985) searched for evidence of C_4 plants in the diets of late Pleistocene bison, mammoths and other herbivores from Beringia. All herbivores had $\delta^{13}C$ values

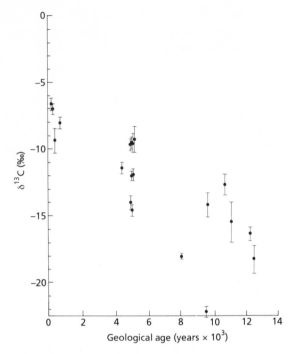

Fig. 4.3 Carbon isotope values for fossil *Bison* sp. collagen from the southern High Plains (USA) plotted against geological age. The rise in $\delta^{13}C$ values from 13 000 years ago to present indicates the increasing importance of C_4 grasslands in the region. (From Stafford *et al.*, 1994.)

indicative of diets based solely on C_3 plants, confirming that a relatively cold climate was present over the land bridge that linked Asia and North America.

Even in ecosystems that contain only or dominantly C_3 plants, carbon isotopes may supply dietary and habitat information. In relatively closed woodlands, such as dense tropical forests, the $\delta^{13}C$ values of leaves collected near the forest floor may be ^{13}C-depleted by as much as 4 or 6‰ relative to typical terrestrial plants and to leaves from the top of the canopy, due either to the incorporation of respired CO_2 or to changes in plant water-use efficiency at low light levels (see Chapter 1). The consumption of ^{13}C-depleted forest floor foliage has been detected in isotopic studies of modern foodwebs (van der Merwe *et al.*, 1988; van der Merwe & Medina, 1991). Climatically-induced changes in African forest cover have been assessed through carbon isotope analysis of herbivores that fed primarily on forest floor plants. In the Kenyan Rift Valley, the $\delta^{13}C$ values of collagen (−25‰) from prehistoric bushbuck and bushpigs were several per mil more negative than those from contemporary individuals at the same location ($\delta^{13}C = -20‰$; Ambrose & DeNiro, 1989). Ambrose & DeNiro (1989) suggest that these lower $\delta^{13}C$ values are indicative of a denser, more-closed canopy forest in the region prior to 5600 BP.

The $\delta^{13}C$ of animal collagen has also been exploited to investigate changes in the composition of atmospheric carbon dioxide. The isotopic composition of this gas has changed by about 1‰ since the 1850s, owing to the addition of ^{13}C-depleted carbon dioxide ($\delta^{13}C = -35‰$) from the burning of fossil fuels (Friedli et al., 1986). Moose teeth collected from Isle Royale National Park, Michigan, were analysed to study the atmospheric carbon cycle and changes in forest ecology (Bada et al., 1990). The carbon isotopic composition of moose teeth declined from the 1940s to the present. Bada et al. (1990) demonstrated that the isotopic fluctuations in moose teeth were only partially due to changes in the $\delta^{13}C$ of atmospheric CO_2. They attributed residual isotopic shifts either to changes in the forest canopy or moose diet following fires that swept across the island.

Nitrogen isotopes in collagen

The relationship between the $\delta^{15}N$ of diet and collagen is similar to the trend observed for whole organisms. Accordingly, nitrogen in collagen is enriched in ^{15}N by about 3‰ (range 2–5‰) for every step up the trophic ladder (Fig. 4.4; DeNiro & Epstein, 1981a; Schoeninger & DeNiro, 1984; Minagawa & Wada, 1984; Ambrose & DeNiro, 1986; Sealy et al., 1987;

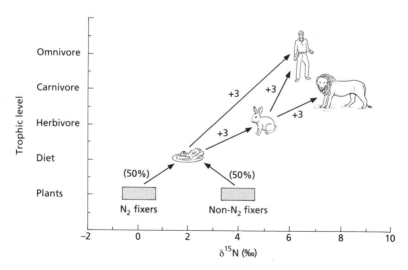

Fig. 4.4 Model of nitrogen isotope variation in plants and the fractionation of nitrogen isotopes between different trophic levels, including herbivores, carnivores and omnivores.

Hare *et al.*, 1991; see Chapter 7). Although several complicating factors influence $\delta^{15}N$ values, isotope analysis has proven successful in assessing: (i) trophic structure; (ii) the contribution of nitrogen-fixing organisms to a palaeoecosystem; and (iii) metabolic or water stress on an organism.

The chief palaeodietary application of nitrogen isotope analysis has been studies of trophic structuring in late Pleistocene and Holocene ecosystems (Keegan & DeNiro, 1988; Katzenberg, 1989; Ambrose & DeNiro, 1989; Bocherens *et al.*, 1991, Tuross *et al.*, 1994). Trophic level increase in $\delta^{15}N$ from primary producer to top carnivore is particularly pronounced in marine ecosystems, which typically have long food chains (Schoeninger & DeNiro, 1984; Keegan & DeNiro, 1988). However, most faunal studies of $\delta^{15}N$ differences between terrestrial herbivores and carnivores have been performed to calibrate the degree of carnivory in humans. For example, Bocherens *et al.* (1991) concluded that Neanderthals were primarily carnivorous, by comparing the isotopic composition of a single specimen of Neanderthal collagen to those of known herbivores and carnivores.

Biological nitrogen fixation, whereby nitrogen is enzymatically converted to NH_3 by microorganisms, results in plants tissues that are ^{15}N-depleted relative to tissues from non-nitrogen-fixing plants (see Chapter 1). In ecosystems where nitrogen fixers are important primary producers, consumers may be ^{15}N-depleted relative to animals that eat materials derived from plants that do not fix nitrogen. An example of this isotopic pattern was documented by Keegan & DeNiro (1988) in modern organisms inhabiting a coral-reef system in the Bahamas. They detected ^{15}N-depletion in the tissues of certain reef invertebrates and fish relative to the typical values for marine fish. This ^{15}N-depletion presumably resulted from the fixation of molecular nitrogen by marine microorganisms found in association with corals and mangroves (Macko *et al.*, 1984). Keegan & DeNiro (1988) also examined fossil bones and verified a similar dependence on nitrogen-fixing organisms. The nitrogen isotope ratio of a prehistoric parrotfish, a carnivorous reef fish (+3.9‰), was 8‰ more negative than expected for a typical carnivorous, marine fish (Schoeninger & DeNiro, 1984).

Evidence is mounting that the physiological status of an animal may also affect the $\delta^{15}N$ value of its collagen. Modern African herbivores from arid regions have higher $\delta^{15}N$ values than closely related herbivores from areas with high rainfall (Heaton *et al.*, 1986; Sealy *et al.*, 1987). Although plants exhibit a similar pattern of ^{15}N-enrichment (Heaton, 1987b), the trend in herbivores can not be completely explained by isotopic changes in food plants. In essence, the trophic level fractionation between plants and herbivores is greater in arid regions. Ambrose & DeNiro (1986) argued that $\delta^{15}N$ increase resulted from isotopic affects induced by urea concentration in water-stressed animals from arid regions. An alternative hypothesis (Sealy *et al.*, 1987) relates $\delta^{15}N$ increase to more severe nutritional stress in arid regions. Under conditions of protein deprivation, animals may recycle nitrogen-rich components, such as urea and proteins, within their bodies. Each time nitrogen is recycled, it would be enriched in ^{15}N due to trophic level effects, potentially causing an increase in ^{15}N in the internal nitrogen pool.

4.2.2 Non-collagenous proteins, amino acids and bulk organic matter

Although collagen is the most easily identifiable molecular fossil, other types of organic matter are much more abundant in a geological context. The chemical structure and isotopic relationship to diet of geochemical organic matter are incompletely known, however. One significant characteristic of non-collagenous proteins in calcified tissues is their higher solubility relative to vertebrate type I collagen. In general, the non-collagen calcified matrix is very heterogeneous, with many protein products closely associated with either calcium carbonate or calcium phosphate. In the bones of living vertebrates, indigenous non-collagenous proteins are varied and number as many as 200 (Delmas *et al.*, 1984). In calcitic bivalves and phosphatic brachiopods, multiple soluble protein products are found in extracts following decalcification of shells (Weiner *et al.*, 1980; Tuross & Fisher, 1989).

Non-collagenous biomolecules that persist in the fossil record are of great interest as both palaeodietary and palaeoecological indicators. In the bones and teeth of fossil vertebrates, an amino acid-containing residue often persists after all traces of collagen are gone (i.e., no hydroxproline and altered glycine:aspartic acid values; Wycoff, 1972; Hare, 1980; Masters, 1987; Ajie *et al.*, 1991; Tuross & Stathoplos, 1993). Calcitic organisms also retain amino acids in peptide linkages for long periods of time (Abelson, 1954; Weiner *et al.*, 1976). Recent studies of ancient ostrich eggshells indicate that this material is unusually resistant to loss of organic matter with time (Brooks *et al.*, 1990; Miller *et al.*, 1991). The isolation, characterization and isotopic utility of non-collagenous molecules from ancient vertebrate and invertebrate calcified tissues is in development. There are several confounding factors in the quest. In bone, each non-collagenous protein is present in very small amounts, necessitating destructive analysis of large samples. In many bones, non-collagenous proteins are 'hidden' beneath a large amount of collagen degradation products (Tuross *et al.*, 1980). Decalcifying and solubilizing agents must be presumed to be incompatible with isotope analysis until proven otherwise. In calcium carbonate structures such as bivalve shells or ratite egg shells, the protein molecules associated with contemporary specimens are incompletely characterized. Therefore no immunological criteria or protein sequence data are available for positive identification of ancient molecules.

It can be difficult to determine whether a fossil-derived organic extract is made up of resistant products of degradation, remnant indigenous non-collagenous proteins, exogenous insoluble protein contamination, or some combination of all three possibilities. Attempts to utilize non-collagen amino acid-containing moieties for isotope analysis have met with ambiguous results. The purification of any non-collagenous protein from bone in quantities sufficient for traditional isotope ratio mass spectrometry remains a largely elusive goal. Further, the relationship of the isotopic value of this non-collagen residue to that of the diet is unknown, although interpretable carbon and nitrogen isotope values (i.e., within the range of biological possibility) sometimes result from its analysis (e.g., Masters, 1987; Ajie *et al.*, 1991). Here, we consider isotope analysis of four types of non-collagenous organic matter: (i) bulk organic matter in crystalline aggregates from bones; (ii) osteocalcin in fossil bones; (iii) invertebrate shell proteins; and (iv) individual amino acids.

Methods

A variety of methods have been utilized in extraction of non-collagenous molecules from fossil bones, teeth and shells. Because of the friable nature of these materials, care is taken in demineralizing the bone or shell sample, such as the use of dilute acids and gentle chelating agents (e.g., EDTA) at low temperatures (4°C; Tuross, 1989). Because these molecules tend to be soluble in mineral acids or solutions of chelating agents, samples are reacted within dialysis tubing rather than directly in the demineralizing agent. A high-molecular-weight fraction is purified by demineralization followed by exhaustive dialysis or desalting on chromatographic columns (Tuross, 1989; Ajie *et al.*, 1991).

DeNiro & Weiner (1988b) utilized sodium hypochlorite to oxidize most of the organic matter in contemporary and fossil bones and generated a preparation that they called 'crystalline aggregates'. They reasoned that the organic material contained within apatite crystals had been protected from contamination and microbial attack. The amino acid compositions of the organic matter associated with the aggregates were variable, however. In modern organisms, the carbon isotopic composition of aggregate organic matter did not bear a strong relationship to that of collagen, perhaps because this material contained a large fraction of non-collagenous proteins. In addition, amino acid enantiomer compositions of 'crystalline aggregate' organic matter gave no indication of the actual age in bones with known ages (Elster *et al.*, 1991).

Individual amino acids have also been utilized for isotope analyses (Tuross *et al.*, 1988; Hare *et al.*, 1991). Large amounts of protein (approximately 100 mg) are hydrolysed to break peptide bonds, and amino acids are then separated by ion-exchange chromatography through elution with increasingly concentrated hydrochloric acid. Each individual amino acid must be monitored carefully for complete collection, as substantial fractionation, especially of nitrogen isotopes, occurs during chromatographic separation (Hare *et al.*, 1991).

Isotopic studies of NCPs, bulk organic matter and amino acids
In an ambitious attempt to extend palaeodietary reconstruction back 75 million years to the study of dinosaur food webs, Ostrom *et al.* (1990) measured the nitrogen and carbon isotopic compositions of high-molecular-weight organic matter retrieved from a crystalline fraction of bone that was insoluble in mineral acids. Their results are ambiguous, however, in that two tests to confirm the indigeneity of the organic material were inconclusive. First, amino acids in living organisms are primarily in the L form; after death, amino acids convert slowly to the D form until a 50:50 or racemic mixture is reached (Hare, 1980). The D:L value of amino acids from dinosaur organic matter was low for a material millions of years old. However, the D:L value is itself a function of numerous diagenetic factors. Kimber & Hare (1992) have documented that D:L values for a single amino acid, such as aspartic acid, can vary tremendously in a single bone, and that high-molecular-weight fractions are significantly less racemic than smaller peptide chains. Second, the hydrolysable fraction of the high-molecular-weight organic matter in the study of Ostrom *et al.* (1990) had significant levels of glycine and serine, which are consistent with recent contamination (Hare, 1980). Despite these ambiguities, $\delta^{15}N$ values exhibit the expected pattern of enrichment between herbivorous and carnivorous dinosaurs. However, $\delta^{13}C$ values exhibited no apparent trends, and were generally too negative to be derived solely from collagenous material.

Osteocalcin is a non-collagenous protein that contains γ-carboxyglutamic acid (Gla) residues (Ajie *et al.*, 1991). Osteocalcin is very tightly bound to the apatite mineral in bone, and thus might be resistant to degradation and removal during diagenesis. Osteocalcin has been detected in ancient bones through immunological assays and amino acid concentration analysis (Ulrich *et al.*, 1987; Ajie *et al.*, 1991). Ajie *et al.* (1991) measured the isotopic compositions of organic material that they considered 'osteocalcin' from a suite of fossil bones. When the collagen from these bones was not seriously degraded, the collagen and osteocalcin fractions gave carbon and nitrogen isotope values that agreed within analytical error. In samples wherein the collagen was seriously degraded, the 'collagen' and 'osteocalcin' fractions had different isotopic compositions. It is not clear from this study, however, whether the 'osteocalcin' fraction was pure osteocalcin. Ajie *et al.* (1991) reported the molecular weight distribution of the 'osteocalcin' fraction as $6-17 \, kDa$, whereas intact osteocalcin in modern bones has a molecular weight of $5.2-5.9 \, kDa$. Furthermore, it is difficult to reconcile the amounts of protein needed for both accelerator mass spectrometer (AMS) radiocarbon dates and stable carbon and nitrogen isotopic compositions with the low fossil protein levels reported by Ajie *et al.* (1991). Osteocalcin concentrations in modern bone are low ($0.28 \, mg \, g^{-1}$), and immunoassays revealed that only tens of nanograms of intact osteocalcin per gram of fossil bone were preserved. The most sophisticated gassource accelerators require at least $3 \, mg$ of protein for radiocarbon determinations, and stable isotope analysis typically uses $1-2 \, mg$ of protein. However, demonstration of both Gla residues and even small amounts of intact osteocalcin in buried bone is an important contribution to the study of non-collagenous molecules in vertebrate calcified tissues.

The amount of protein associated with modern calcite shells is very small (approximately 1–3% by weight), and the proteins that are associated with many shells are quite heterogeneous (Weiner *et al.*, 1977; Cariolou & Morse, 1988; Tuross & Fisher, 1989). Consequently, attempts to use organic matter from invertebrate calcified tissues for isotopic dietary studies are rudimentary when compared to the efforts on mammalian bones and teeth (DeNiro & Epstein, 1978; Cobabe, 1991). Goodfriend (1988) utilized $\delta^{13}C$ values from modern and fossil land snail organic matter to track climatically driven shifts in floral zonations in the Negev Desert during the mid-Holocene. In a study that involved contemporary and fossil bivalve material, the use of symbiotic bacterial substrates could be distinguished from non-symbiotic food sources through carbon isotope analyses (Cobabe, 1991).

Hare *et al.* (1991) measured $\delta^{13}C$ and $\delta^{15}N$ values from individual amino acids derived from collagen in laboratory-reared and wild modern and fossil vertebrates. They discovered consistent patterns of carbon and nitrogen isotope fractionations between different amino acids. These patterns varied in modern mammals between well-fed laboratory animals and a zebra that died of natural causes during a drought in East Africa (Fig. 4.5), and ultimately may prove useful in assessing nutritional or water stress. For example, the nitrogen isotopic differences between eight major amino acids from both a modern and fossil whale were roughly identical (Hare *et al.*, 1991). Deviations from this pattern could indicate an unusual diet, stress or contamination by exogenous proteins. Moreover, the isotopic values of certain amino acids that are found in abundance only in collagen, such as proline and hydroxyproline, may provide isotopic dietary information for samples with degraded or contaminated collagen. The use of individual amino acids in isotopic studes should greatly expand,

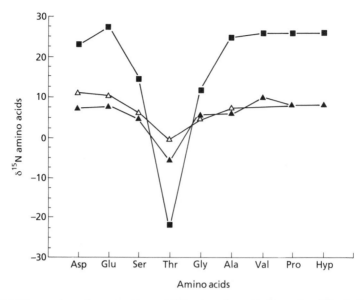

Fig. 4.5 Nitrogen isotopic compositions of different amino acids from zebra (■), pig (▲) and fossil bison (△) bone collagen. Note the difference in pattern between the zebra, which died of starvation during a drought, and the pig, which was raised on a plentiful, controlled diet. (From Hare *et al.*, 1991.)

particularly given the new analytical technique of gas chromatography–combustion–isotope ratio mass spectrometry (GC–C–IRMS), which allows more rapid analysis of individual amino acids than liquid chromatographybased techniques (Engel *et al.*, 1990; Silfer *et al.*, 1991).

4.2.3 Isotopic integrity of fossil organic matter

Unlike measurements of tissue from modern animals, determinations of the isotopic composition of organic matter from fossils must be scrutinized carefully to assure fidelity to the original signal. We have outlined the chemical criteria for assessing the nature of collagen taken from fossil samples (see Section 4.2.1). The primary assumption underlying these criteria is that if organic matter in fossils has the molecular composition of modern collagen, with no contamination evident from preservatives or humic substances, then the isotopic composition of the material will reflect that of the animal before death.

Surficial weathering and diagenesis does not compromise the use of collagen for isotopic dietary studies (Koch *et al.*, 1990). The $\delta^{13}C$ and the $\delta^{15}N$ values of bones collected from carcasses from 1 to 15 years after death were within analytical measurement error. After burial, however, the preservation of collagen in bones is extremely variable as a function of time. Decomposition is especially swift if bone is buried in wet, warm soils (Hare, 1980). Conversely, bones that have been buried in permafrost (e.g., Pleistocene whales, mammoths and rhinoceros) or clay may contain >50% of their original collagen after 75 000 years (Tuross *et al.*, 1988). Tuross *et al.* (1988) demonstrated that at least 90% of the collagen can be degraded without changing the isotopic composition of the remaining material. However, when loss of collagen is >90%, the isotopic composition of the recovered material usually does not reflect lifetime values (Ambrose, 1990).

Often, the most interesting specimens are those in which less than 10% of the original collagen is preserved. A number of techniques have been proposed to overcome the inadequacies of these specimens, especially GC–C–IRMS. Isotopic compositions of individual amino acids may be used to test for faithful preservation or indigeneity of the organic material in two ways. First, the characteristic pattern of isotopic differences between the eight major amino acids in modern mammals may be ubiquitous (Hare *et al.*, 1991). If so, deviations from this pattern in fossil material would indicate isotopic alteration. Second, Serban *et al.* (1988) proposed that comparison of the isotopic compositions of D and L enantiomers of amino acids in fossils can serve as a monitor of indigeneity. As the carbon skeleton of an amino acid is identical after racemization, the $\delta^{13}C$ values of D and L enantiomers of amino acids from fossil protein should be similar if the fossil has not been contaminated by amino acids from exogenous sources. Further testing of this approach is also necessary and will be greatly facilitated by GC–C–IRMS techniques. Uncharacterized organic matter analysed by either or both of these techniques might then be studied to define fossil food webs. However, until it has been demonstrated unambiguously that original isotopic signals can be retrieved from uncharacterized organic matter, this material should be used with caution.

4.3 Isotope analysis of biogenic minerals from the fossil record

Despite progress in isotope analysis of organic residues, it is unlikely that well-preserved organic material from ancient organisms will ever be abundant. In contrast, the mineralized hard parts of organisms are often preserved for hundreds of millions of years, and potentially bear palaeodietary information over this vast time span. The mere preservation of the mineralized parts of bones, teeth and shells, however, is no guarantee of their usefulness for isotopic analysis. The post-mortem processes that convert an animal carcass into a geologically inert fossil may erase or alter biogenic isotope signatures. Each type of mineralized tissue bears a unique palaeodietary signal, and each has a different potential for maintaining isotopic fidelity in fossils.

4.3.1 Hydroxyapatite

The mineral in bones and tooth dentin and enamel is a form of hydroxy-apatite $[Ca_{10}(PO_4)_6(OH,F)_2]$. When compared to geological apatites, biogenic hydroxyapatite is poorly crystalline, composed of tiny crystals with numerous defects and high strain. Tooth enamel apatite has larger crystals, however, and is more highly crystalline than either bone or dentin apatite (LeGeros, 1981). Significant amounts of carbonate (2–5 wt%) occur in biogenic hydroxyapatite, either as a structural substitute for PO_4^{3-} or OH^-, or as HCO_3^- in hydration layers adsorbed to crystal surfaces (Chickerur *et al.*, 1980; LeGeros, 1981). Total carbonate content is highest in bone and lowest in tooth enamel. The carbon and oxygen in carbonate and the oxygen in phosphate from hydroxyapatite have been used for isotope analysis.

As discussed above (see Section 4.2), throughout much of an animal's life, bone is dissolved and redeposited during growth and remodelling, whereas dentin and enamel grow by accretion with little reworking. Thus as with organic components, the apatite mineral from different tissues records dietary information from different portions of an animal's life.

Methods

Successful isotope analysis of biogenic apatite depends on isolation of original structural carbonate and phosphate from potential contaminants. For example, original organic matter may remain and exogenous organic debris may invade after burial (Fig. 4.1). Diagenetic minerals, such as calcite and dolomite, may precipitate in pore spaces (Sillen, 1989). The carbonate in hydration layers is highly labile and may exchange with carbonate in burial fluids over time (Lee-Thorp, 1989). Finally, recrystallization following burial (Sillen, 1989) might allow sediment-derived ions to infiltrate the structural carbonate and phosphate fractions of apatite in a fossil.

Different oxidants, including sodium hypochlorite (NaOCl), hydrogen peroxide (H_2O_2) and hydrazine (NH_2NH_2), have been used to remove organic matter (Termine *et al.*, 1973; Koch *et al.*, 1989; Lee-Thorp & van der Merwe, 1991). Acetic acid (CH_3COOH), hydrochloric acid (HCl) and triammonium citrate ($[NH_4]_3C_6H_5O_7$) have been used to strip diagenetic calcium carbonate and adsorbed carbonate from samples (Hassan *et al.*, 1977; Sillen, 1986; Lee-Thorp & van der Merwe, 1991). Different pretreatments may profoundly influence the mineralogy and isotopic composition of the remaining apatite. For example, prolonged treatment with even relatively dilute (1.0 N) acetic acid causes biogenic apatite to recrystallize to the mineral brushite ($CaHPO_4 \cdot 2H_2O$) with a coincident loss of carbonate and isotopic alteration (Lee-Thorp & van der

Merwe, 1991). We recommend either treatment with NaOHCl followed by a brief dilute acetic acid wash (0.1–1.0 N; Lee-Thorp & van der Merwe, 1991) or with hydrogen peroxide (Koch *et al.*, 1989).

To release CO_3^{2-} as CO_2 gas for isotope analysis, we react apatite with 100% H_3PO_4 in sealed, evacuated vessels at 50°C for 5 h (McCrea, 1950; Koch *et al.*, 1989). CO_2 is isolated cryogenically prior to mass spectrometry. During the reaction of apatite with acid, only two of three oxygen atoms from carbonate are converted to CO_2 ($2H^+ + CO_3^{2-} \rightarrow H_2O + CO_2$). The fractionation of oxygen isotopes associated with this process has not been directly measured. When estimating the oxygen isotopic composition of the original apatite mineral, the fractionation associated with dissolution of calcite in phosphoric acid is used (Kolodny & Kaplan, 1970; McArthur *et al.*, 1980). All carbon atoms from carbonate are converted to CO_2, thus $\delta^{13}C$ values can be calculated without correction.

Apatite phosphate has attracted attention in recent years, because it is thought to be more resistant to alteration than other oxygen-bearing biominerals (Shemesh *et al.*, 1983). Analysis of phosphate oxygen involves (i) isolation of PO_4^{3-} in a mineral phase and (ii) release of oxygen from PO_4^{3-} and conversion into CO_2. Two methods are available for isolation of phosphate. Samples can be dissolved in nitric acid, and then phosphate is purified through a series of reprecipitation steps and collected as bismuth phosphate (Tudge, 1960; Kolodny *et al.*, 1983). This method is time consuming and produces a material, $BiPO_4$, that is hygroscopic and easily contaminated by atmospheric water vapour. In the alternative method (Wright & Hoering, 1989; Crowson *et al.*, 1991), samples are dissolved in HF, then added to an ion-exchange column to isolate phosphate. Following elution of phosphate from the column, silver phosphate, a non-hydrating mineral, is precipitated. To release oxygen from either bismuth or silver phosphate, samples are reacted under vacuum with excess BrF_5. The oxygen generated by this procedure is converted to CO_2 by hot graphite, collected, and analysed on a mass spectrometer (Clayton & Mayeda, 1963).

Carbon isotopes in hydroxyapatite

It is generally accepted that carbonate in apatite is derived from blood bicarbonate. An isotopic enrichment of c. 10‰ is expected in the system $CO_2/HCO_3^-/CO_3^{2-}$ at thermodynamic equilibrium at mammalian body temperatures (Emrich *et al.*, 1970; Lee-Thorp, 1989). In field and laboratory studies of mammalian herbivores, the $\delta^{13}C$ value of diet (the source of blood bicarbonate) and apatite differ consistently by 10–12‰ (DeNiro & Epstein, 1978; Sullivan & Krueger, 1981; Lee-Thorp & van der Merwe, 1987). A browsing herbivore, with a C_3 diet (i.e., trees and shrubs), produces apatite with a $\delta^{13}C$ value of $-14‰$, whereas a grazing herbivore, with a C_4 diet (i.e., grasses and sedges), has apatite with a value of $0‰$ (Fig. 4.2). Factors influencing the $\delta^{13}C$ of carnivore and omnivore apatite are more complex, because heterogeneous dietary resources (i.e., plant carbohydrates and lipids vs. animal proteins and lipids) may have different carbon isotopic compositions. These differences, however, must be either small or compensatory, because carnivores, omnivores and herbivores in a region usually have similar apatite $\delta^{13}C$ values (Lee-Thorp et al., 1989a). Finally, because diet controls the carbon isotopic composition of both apatite and collagen, there is a characteristic difference in $\delta^{13}C$ value between these two carbon sources within an animal. This difference ($\Delta^{13}C_{apatite-collagen}$) is $7 \pm 1.5‰$ in herbivores, $4 \pm 1‰$ in carnivores, and is intermediate in omnivores (DeNiro & Epstein, 1978; Sullivan & Krueger, 1981; Lee-Thorp et al., 1989a). Krueger & Sullivan (1984) suggested that these differences are due to different dietary sources for blood bicarbonate and collagen carbon. A recent study of rats raised on controlled diets indicates that apatite carbonate is derived from near uniform mixing of atoms from all sources, whereas collagen is strongly controlled by the isotopic composition of dietary protein (Ambrose & Norr, 1993).

Because of early failures in radiocarbon dating of bone apatite due to diagenetic addition of carbon (Tamers & Pearson, 1965), stable carbon isotope analysis of bone apatite has been suspect as well. To test the isotopic fidelity of apatite, Sullivan & Krueger (1981) compared collagen and apatite carbonate $\delta^{13}C$ values in modern and fossil herbivores, and argued that, in the absence of diagenetic alteration, $\Delta^{13}C_{apatite-collagen}$ should be constant. They discovered a regular fractionation of $8‰$ and concluded that both apatite and collagen bore original isotopic compositions. Ericson et al. (1981) measured the $\delta^{13}C$ value of enamel from 2 million-year-old herbivores, and reconstructed diets that were in agreement with those inferred by analogy to closely related modern herbivores. In contrast, Land et al. (1980) compared fossil and modern deer bones and discovered that $\Delta^{13}C_{apatite-collagen}$ values for fossils were unlike those measured for recent deer. Similarly, Schoeninger & DeNiro (1982) could not obtain consistent values for $\Delta^{13}C_{apatite-collagen}$ from human skeletons.

There are non-diagenetic considerations that may explain these contradictory results. In the early 1980s, little was known about potential variations in $\Delta^{13}C_{apatite-collagen}$ between omnivores, herbivores and carnivores. Many fossils that were originally thought to be altered were subsequently found to be within the range of modern $\Delta^{13}C_{apatite-collagen}$ values (Krueger & Sullivan, 1984; Lee-Thorp et al., 1989a). Sample pretreatment to remove diagenetic carbonate differed among the studies. Treatments with very concentrated acids, such as those used by Schoeninger & DeNiro (1982) and Nelson *et al.* (1986) are now thought to produce extensive recrystallization, which can alter $\delta^{13}C$ values (Lee-Thorp & van der Merwe, 1991).

The isotopic fidelity of apatite carbonate has subsequently been tested using pretreatments that avoid recrystallization (e.g., Lee-Thorp & van der Merwe, 1991). The strongest test is to demonstrate isotopic discrimination among animals with different feeding habits from the same stratigraphic unit. Using tooth enamel, isotopic differences between browsers (C_3 feeders) and grazers (C_4 feeders) have been detected in mammals ranging in age from 1000 to 3 million years old, despite minor invasion of structural CO_3^{2-} by sediment-derived carbonate in the oldest samples (Lee-Thorp & van der Merwe, 1987; Lee-Thorp *et al.*, 1989b). As discussed below, bone apatite is an unreliable isotopic source material for even relatively young specimens (5000 to 10 000 years old).

Testing the isotopic fidelity of very ancient tooth enamel is more difficult. Before the widespread appearance of abundant C_4 plants in the Miocene, there are no easily predictable mechanisms for generating large isotopic differences between animals living in a single time period. Isotopic studies have been considered successful when predicted temporal variations between animals of different age have been discovered. For example, in Miocene mammals, spanning from 16 to 7 million years, from Pakistan, $\delta^{13}C$ values change with age, indicating a dietary switch from C_3 to newly arrived C_4 plants (Quade *et al.*, 1992). The increase in C_4 grasslands has been tied to the onset of monsoonal climates in southern Asia (Quade *et al.*, 1989). While this isotopic shift coincides with changes in soil geochemistry and in tooth morphology that indicate increased grass cover in the region, the sequence contains no layers where C_3 and C_4 feeders co-existed. In another example, enamel $\delta^{13}C$ values from therapsid mammal-like reptiles shifted by 7‰ in the late Permian (c. 250 million years old), tracking changes in the $\delta^{13}C$ of marine carbonates (Thackeray *et al.*, 1990). Terrestrial animals and marine carbonates exhibit coincident isotopic fluctuations because they are linked via atmospheric CO_2 and land plants. In both examples, the authors examined other carbonatebearing minerals, in order to strengthen their arguments that the trends observed between different stratigraphic intervals were not merely due to early diagenetic alteration while enamel apatite was near the surface in soil horizons.

Oxygen isotopes in apatite phosphate and carbonate
Apatite phosphate and carbonate are thought to precipitate in oxygen isotopic equilibrium with body water, but this assumption remains untested. Blood carbonates and phosphates are presumed to be in oxygen isotopic equilibrium with body water, owing to rapid, exchange reactions

that are catalysed by enzymes such as carbonic anhydrase and ATPases (Dahms & Boyer, 1973; Faller & Elgavish, 1984; Nagy, 1989). There are no experimental studies of oxygen isotopic equilibrium between fluids and apatite in inorganic systems. The accepted equilibrium relationship for phosphate oxygen was itself determined through analysis of fish and invertebrate organisms (Longinelli & Nuti, 1973; Kolodny et al., 1983). If there are kinetic fractionations associated with unidirectional, enzyme-driven precipitation of biogenic apatite, the isotopic relationship measured in animals will not represent equilibrium. If it is assumed, however, that the relationship measured in aquatic animals does represent isotopic equilibrium, then mammalian body water and apatite PO_4^{3-} are indeed in equilibrium at typical mammalian body temperatures (Luz & Kolodny, 1985). Even less information is available about the equilibrium assumption for oxygen in apatite carbonate.

Assuming equilibrium, the $\delta^{18}O$ values of apatite PO_4^{3-} and CO_3^{2-} are determined by (i) the temperature at which the tissue forms and (ii) factors influencing the composition of body water. The factors affecting land vertebrates are presented in Fig. 4.6. Oxygen enters the body as ingested water (drinking water and water in plant and animal tissues), inspired O_2 gas and dietary solids. Within the body, metabolism of food produces H_2O and CO_2 as by-products. Oxygen is lost from the body as liquid water in urine, sweat and faeces, and as water vapour and CO_2 in respiratory gases. When an isotopic mass balance is calculated for an animal at steady state, a predictable linear relationship exists between $\delta^{18}O$ of ingested water and that of body water (Luz et al., 1984). The slope of the relationship is a ratio (Fig. 4.6) that relates the ingested water flux to the flux of metabolically processed oxygen. In animals that drink large volumes of water and have low metabolic rates, such as the elephant and hippopotamus, the slope approaches one. Consequently, changes in the isotopic composition of body water are closely correlated to changes in environmental water. This model relationship has been verified experimentally (Luz & Kolodny, 1985), and these experiments also demonstrated that environmental stresses, which alter metabolic rate and other physiological parameters, produced only slight effects on the composition of body water. Thus, within any particular mammalian species, which grows apatite at a constant body temperature, variations in the $\delta^{18}O$ of apatite PO_4^{3-} and CO_3^{2-} dominantly reflect changes in $\delta^{18}O$ of ingested water. In non-homeothermic vertebrates (i.e., fish, amphibians, reptiles) both temperature and ingested water can have an impact.

At least three factors influence the oxygen isotopic composition of ingested water and, consequently, body water and apatite in terrestrial animals. Globally, the $\delta^{18}O$ value of precipitation (rainfall and snow) is strongly correlated to mean annual temperature, with low values in cold

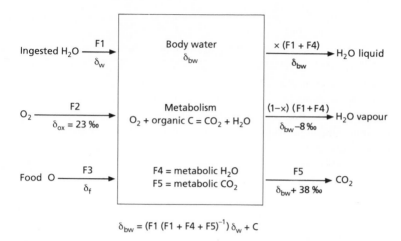

$$\delta_{bw} = (F1 (F1 + F4 + F5)^{-1}) \delta_w + C$$

Fig. 4.6 Model of oxygen fluxes entering and leaving the body of an animal. The term for the magnitude of each flux is represented by letters above each arrow (F1, F2, etc.). F4 and F5 refer to fluxes of water and CO_2 produced by metabolism within the animal, respectively. The isotopic composition of each oxygen pool entering or leaving the body is represented by a δ value below each arrow. Atmospheric oxygen has a uniform value of 23.5‰ (standard mean of ocean water). Values for isotopic composition of respired CO_2 (δ_{bw} + 38) and water vapour (δ_{bw} − 8) are derived from Luz *et al.* (1984). The equation relating δ_w to δ_{bw} is obtained by determining an isotopic mass balance for an animal at steady state.

climates and more positive values in warm climates (Fig. 4.7a; Dansgaard, 1964). Locally, the $\delta^{18}O$ value of precipitation may vary with the seasons, with low values in cold months and more positive values in warm months (Fig. 4.7b; Gat & Gonfiantini, 1981). Surface and ground waters supply the drinking and plant-included waters that animals ingest. Consequently, the $\delta^{18}O$ of mammalian apatite tracks the composition of local waters, which may ultimately reflect climatic temperature (Land *et al.*, 1980; Longinelli, 1984).

Second, the water in plant leaves is enriched in ^{18}O relative to ground water through evaporative transpiration. This enrichment is most intense in arid climates (Dongmann *et al.*, 1974). The impact of ingested leaf water is evident when the $\delta^{18}O$ values are compared among mammals that inhabit regions with similar drinking water, but different humidities (Ayliffe & Chivas, 1990; Luz *et al.*, 1990).

Finally, within a habitat, plant water may differ among species due to morphological and physiological differences. Aquatic plants contain water similar in $\delta^{18}O$ value to their aquatic medium, whereas water in terrestrial C_3 and C_4 plants is more enriched in ^{18}O than local ground water (Sternberg *et al.*, 1986). Furthermore, there may be isotopic differences in plant water among terrestrial plants related to photosynthetic pathway or plant

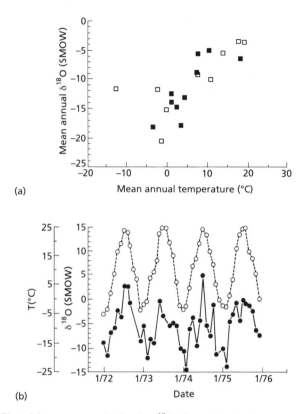

Fig. 4.7 (a) Plot of the mean annual value for $\delta^{18}O$ of precipitation (rain and snow) vs. mean annual temperature for 19 IAEA stations in North America. Low oxygen isotope values correspond to cold stations, whereas higher values are obtained at warmer stations. □ = Coastal stations; ■ = continental stations. (b) Plot of monthly values for $\delta^{18}O$ precipitation and temperature in Chicago vs. month over a 4-year period. Note the strong relationship between monthly temperature and oxygen isotope value. ○ = temperature; ● = $\delta^{18}O$. (From Koch, 1989.)

morphology (see Chapter 1; Sternberg, 1989; Ziegler, 1989; Flanagan *et al.*, 1991). Differences in dietary plant water may contribute to the oxygen isotope segregation among C_3, C_4 and aquatic feeding mammals from Amboseli National Park, Kenya (Koch *et al.*, 1990). If consistent differences in oxygen isotope values are discovered between modern C_3 and C_4 feeders, oxygen isotopes may provide an independent measure of carbon isotope fidelity in ancient tooth apatite.

Phosphate oxygen isotope analysis of fossil vertebrates has been limited to marine palaeothermometry with fish (Kolodny *et al.*, 1983; Kastner *et al.*, 1990). There are scattered analyses of fossil land vertebrates (Longinelli, 1973; Kolodny *et al.*, 1983), but no detailed investigations.

Land *et al.* (1980) analysed recent and fossil deer and suggested that the $\delta^{18}O$ of apatite CO_3^{2-} could serve as a diagenetic monitor for radiocarbon analyses. Koch *et al.* (1989) examined $\delta^{18}O$ variations across mastodont and mammoth tusks. Tusks grow throughout an animal's life, and they contain conspicuous annual growth bands. They discovered cyclic fluctuations in $\delta^{18}O$ that were attributed to seasonal changes in the $\delta^{18}O$ value of the water that proboscideans drank (Koch, 1989; Fig. 4.8). They reconstructed the time of year when the animals died from either natural causes or human hunting in order to investigate seasonal mortality patterns.

Preservation of hydroxyapatite

Although hydroxyapatite has been used successfully in a number of isotopic studies, problems from diagenetic alteration remain. Land *et al.* (1980) used a mild acid pretreatment, but still were unable to obtain consistent values for $\Delta^{13}C_{apatite-collagen}$ for deer bones. Lee-Thorp & van der Merwe (1987) documented that structural carbonate in fossil bones did not retain original carbon isotope values on time scales as short as 1 million years. In our own comparisons of $\delta^{13}C$ values between collagen and apatite in enamel, dentin and bone from human and proboscidean skeletons (Fig. 4.9; Koch *et al.*, 1990), bone and dentin were unreliable sources of carbonate $\delta^{13}C$ values even in very young specimens. Finally, Kastner *et al.* (1990) discovered that phosphate oxygen from Miocene fish bones and teeth had exchanged isotopically with pore-water oxygen following burial. Clearly, bone apatite is not a favourable choice for isotope

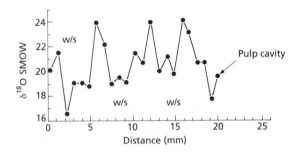

Fig. 4.8 Oxygen isotope variation in apatite carbonate across a mammoth tusk. Tusks grow by accretion and contain conspicuous lamellae that have been intrepreted as annual growth bands. Oxygen isotope values vary in phase with growth bands, with low values from regions thought to represent winter growth and higher values from regions thought to represent spring and summer growth. The last sample (≈ 20 mm) was collected from directly off the pulp cavity and thus formed immediately before the animal's death in the late winter or early spring. w/s indicates transition from winter to spring growth based on growth line analysis. (From Koch, 1989.)

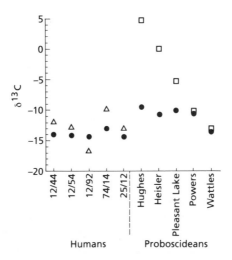

Fig. 4.9 Comparisons of apatite carbonate carbon isotope values between enamel (●), dentin (□) and bone (△) for different specimens of humans and proboscideans. Enamel yields very consistent isotopic values between individuals. Bone and dentin, in contrast, are highly variable, suggesting that they are subject to isotopic alteration. Human specimens are ≈5000 years old, proboscideans are ≈11 000 years old.

analysis, whereas tooth enamel is apparently much more resistant to diagenetic change.

Factors controlling the isotopic alteration of biogenic hydroxyapatite are poorly constrained. Exogenous carbonate that infiltrates pore spaces and apatite hydration layers should be removable by pretreatment with acidic rinses. Although hydration layer carbonate may be replaced by carbon-bearing ions from the rinsing solution (e.g., acetate or citrate), these ions do not yield CO_2 gas when the mineral is dissolved in phosphoric acid.

The effects of recrystallization during diagenesis are more vexing. Following death, the small, poorly organized apatite crystals in bone and dentin rapidly begin to fuse into larger, more-organized crystals (Eanes & Posner, 1970; Schoeninger, 1982; Tuross *et al.*, 1989b). During recrystallization, carbonate and phosphate from pore solutions can be incorporated into structural positions in the apatite lattice; these exogenous ions may have isotopic values unrelated to biogenic values. Enamel apatite is much more highly crystalline than bone and dentin, and thus may not suffer as extensively from recrystallization. Higher crystallinity most likely explains the lower susceptibility to diagenetic alteration of enamel apatite.

There have been two attempts to circumvent the problems of recrystallization. Sillen (1986, 1989) has argued that biogenic apatite is more soluble than recrystallized apatite, but less soluble than diagenetic calcite. It should be possible, therefore, to selectively leach biogenic apatite from samples containing mixtures of all three phases. This technique has been employed in trace element studies, with mixed success (Sillen, 1986; Francalacci, 1989; Tuross *et al.*, 1989a; Sealy *et al.*, 1991; Koch *et al.*, 1992). There have been no tests using selective leaching for stable isotope analysis.

The second approach has been to identify and avoid recrystallized apatite. The extent of recrystallization in bone and dentin has been monitored with X-ray diffraction and infrared spectrometry (Hassan *et al.*, 1977; Schoeninger, 1982; Lee-Thorp, 1989; Sillen, 1989; Bartsiokas & Middleton, 1992). Using infrared spectrometry, Shemesh (1990) has demonstrated that with increasing crystallinity the amount of structural CO_3^{2-} decreases, the amount of fluorine increases, and the diagenetic exchange of oxygen between pore water and apatite increases. By monitoring crystallinity it may be possible to screen bones for alteration before isotope analysis. Unfortunately, changes in apatite crystallinity have only been studied in fish 'debris' (mainly bone and dentin) and mammal bones. There are presently no studies of potential diagenetic changes in the crystallinity of enamel apatite. Crystallinity indices must be developed for enamel apatite, which is the only vertebrate hardpart likely to hold its isotopic composition for great lengths of time. In addition, recent advances in resolution-enhanced Fourier transform infrared spectrometry suggest that it may be possible to determine the crystallographic environment of carbonate in tooth enamel (Rey *et al.*, 1991). Thus shifts in the position of carbonate in the enamel apatite lattice may eventually serve as a monitor of diagenetic alteration.

4.3.2 Calcium carbonate

Many organisms secrete shells of calcium carbonate, as either aragonite or calcite. An extensive literature describes the relationship of oxygen and carbon isotopes in marine and freshwater invertebrates to temperature, salinity and productivity (e.g., Keith *et al.*, 1964; Mix, 1987; Stott & Kennett, 1989). We will consider only studies of land snail carbonate, because diet may have an influence on its composition. Although several groups of vertebrates secrete calcium carbonate (otoliths of some teleost fishes, urinary tract calcifications in mammals), only bird eggshells have received much attention in isotopic studies of diet and habitat. Finally, methods for isotope analysis of calcium carbonate and for detection of diagenetic alteration are well established and will not be reviewed (McCrea, 1950; Brand & Veizer, 1981; Carpenter & Lohmann, 1989).

Land snail shells

The shells of land snails are composed of aragonite, which is secreted from extrapallial fluid under the control of the mantle. Extrapallial bicarbonate is the source of oxygen and carbon to the growing mineral. Bicarbonate oxygen is supplied by non-equilibrium exchange with body-water oxygen. Equilibration with atmospheric water vapour and ingestion of this vapour as dew are important controls on the $\delta^{18}O$ value of snail body water (Magaritz et al., 1981; Goodfriend et al., 1989). Thus, as is the case for continental vertebrates, oxygen in land snail carbonate varies in relation to temperature and other climatic and geographical factors that influence the $\delta^{18}O$ value of local precipitation (Magaritz et al., 1981; Lécolle, 1985; Goodfriend & Magaritz, 1987; Goodfriend et al., 1989).

Carbon in extrapallial bicarbonate can be derived from metabolism of plant carbon, from exchange with atmospheric CO_2, and from ingestion of detrital carbonates (Rubin et al., 1963). Goodfriend & Hood (1983) determined that from 0 to 33% of shell carbonate carbon is derived from limestone, whereas 25–40% may come from plant carbon and 30–62% may be atmospheric. Thus while dietary influences on the carbon isotopic composition of shell carbonate are possible, they are complicated and masked by these non-dietary factors (Goodfriend et al., 1989).

Because of these complications, there are no dietary interpretations of fossil land snails using carbon isotopes in shell carbonate. Indeed, carbon data are rarely reported in studies of fossil snail carbonate. Generally, oxygen isotopes have been examined to estimate climatologic parameters such as temperature and humidity (Yapp, 1979). Oxygen isotopes have also been used to explore differences through time in the sources and trajectories of storms shedding precipitation on a region (Goodfriend, 1991).

Eggshells

The eggs of birds, crocodiles and dinosaurs are covered by shells made predominantly of calcite crystals that are secreted around fibrous sheets of organic matter. In bird eggs, calcite occurs in three layers that are covered externally by cuticle and anchored internally to a shell membrane (Silyn-Roberts & Sharp, 1986, 1989). Carbon in shell calcite is supplied by metabolic CO_2, which is released by cells in the uterus and then converted to HCO_3^- by carbonic anhydrase in the oviduct (Simkiss, 1961; Taylor, 1970). Birds incorporate carbon chiefly from metabolism of recently ingested or stored foods, rather than from ingested carbonate minerals (Folinsbee et al., 1970; von Schirnding et al., 1982). The fractionation of carbon isotopes between diet and eggshell calcite is >16‰ in ostriches, whereas the fractionation between diet and eggshell protein is >2‰ (von Schirnding et al., 1982). Modern bird eggshells exhibit carbon isotope differences that result from ingestion of plants with different photosynthetic pathways by either the birds or their prey. In addition, birds feeding on marine versus terrestrial foods can be distinguished (von Schirnding et al., 1982; Schaffner & Swart, 1991).

Oxygen isotopes in eggshell carbonate are controlled by the isotopic composition of body water, because of the rapid exchange between oxygen atoms in body water and HCO_3^-, which is catalysed by carbonic anhydrase. Eggshell carbonate is in oxygen isotopic equilibrium with blood serum at 20.6°C (Folinsbee et al., 1970; Erben et al., 1979). Consequently, as with mammal bones, oxygen in bird eggshells varies in response to factors that affect the composition of body water, the most important being the composition of ingested water. Folinsbee et al. (1970) demonstrated that blood serum and eggshell carbonate equilibrate with changes in drinking water in approximately 14 days. Schaffner & Swart (1991) could distinguish marine and terrestrial feeding in shore birds, because oxygen isotopes values in freshwater systems are usually more negative than values in nearby marine settings.

Isotopic studies of fossil eggshells from birds and their near relatives, dinosaurs and crocodiles, have revealed both climatic and dietary information. Carbon in fossil ostrich eggshells recorded the proportion of C_3 to C_4 plants in the diet (von Schirnding et al., 1982; Freundlich et al., 1989). The carbon and oxygen isotopic composition of fossil aquatic bird, reptile and dinosaur eggshells have also been measured (Folinsbee et al., 1970; Erben et al., 1979; Sarkar et al., 1991). As expected, the $\delta^{18}O$ values of the dinosaur eggs varied in relation to inferred temperatures at fossil sites. Protoceratops shells from the early Upper Cretaceous of Mongolia had lower values than Hypselosaurus eggs from the latest Cretaceous of France. Erben et al. (1979) examined oxygen and carbon isotope trends in late Cretaceous dinosaur eggshells from France and Spain in order to reconstruct palaeoclimatic conditions near the time of dinosaur extinctions. They concluded that climates became cooler and/or more humid toward the end of the Cretaceous.

Carbon isotope values vary rather widely among different Cretaceous dinosaur eggshells: c. −6‰ for Mongolian Protoceratops; −13.5‰ for European Hypselosaurus; and −10‰ for Indian sauropods (Folinsbee et al., 1970; Erben et al., 1979; Sarkar et al., 1991). If the fractionation measured in modern ostriches is assumed to apply to dinosaurs, these animals ate food with $\delta^{13}C$ values of −22 to − 29‰. Such large differences in dietary composition could result from: (i) climatic differences between sites, which might affect the carbon isotopic composition of C_3 plant tissues (O'Leary, 1981; Tieszen & Boutton, 1989); (ii) the presence and consumption of C_4 or CAM plants; or (iii) changes in the $\delta^{13}C$ value of atmospheric CO_2 during the Cretaceous. It is impossible, at present, to rule out any of these causes.

Diagenetic alteration of such ancient material is another important consideration. In most of the studies, X-ray diffraction patterns and scanning electron microscope photographs were examined, and trace element and isotopic compositions of eggshells were compared to those of surrounding sediments. As a result, the potential for strong diagenetic overprinting was considered low (Erben *et al.*, 1979; Sarkar *et al.*, 1991). While further tests of diagenetic modification must be developed before carbonate from such ancient material can be used routinely, eggshell carbonate is a promising isotopic source material that has not been greatly exploited.

4.4 Conclusions

In the past, carbon isotopes received the most attention as a potential source of dietary information, chiefly to track the input of C_3 versus C_4 plants in faunal and hominid diets. Nitrogen isotopes provided powerful insights into trophic structuring and, perhaps, marine vs. terrestrial inputs into the diet. Oxygen isotopes remained the least exploited tool, in part because oxygen isotope variations in animals are strongly influenced by drinking water and, ultimately, climate. In the future, through analysis of different carbon-, nitrogen- or oxygen-bearing components within the same fossil animal, we may gain more insight into the physiological status of the animal. For example, we may be able to determine whether an animal was under nutritional or water stress at the time of death, which could figure importantly in different scenarios for evolution and extinction.

The carbon, nitrogen and oxygen isotopic compositions of the plants and water that animals ingest are influenced by features of the climate and habitat. At present, nearly all attempts to monitor palaeoclimates using the isotopic composition of continental organisms yield very qualitative results (e.g., the climate was wetter or drier, hotter or colder). The ultimate goal of such studies, however, should be to supply quantitative estimates of climatic parameters, in order to evaluate evolutionary and ecological scenarios, as well as the results of computer climatic models. If isotopic studies of continental palaeoclimates are ever to yield quantitative results, the relationship between the isotopic values in plants and animals and different climatic factors must be established through experimental and modern ecological studies.

Palaeodietary reconstruction may be attempted using the isotopic compositions of different biochemical and mineralogical constituents isolated from bones, teeth and shells. The carbon and nitrogen isotopic compositions of collagen remain the most promising for dietary reconstruction, because a variety of techniques are available to purify collagen and to confirm its chemical integrity. Owing to diagenesis, however, the collagen molecule rarely survives past 100 000 years. Recent reports of high-molecular-weight material with amino acid abundances similar to collagen in dinosaur bones are intriguing and deserve further investigation

(Bocherens *et al.*, 1988). Uncharacterized organic matter in fossil tissues may some day prove useful for palaeodietary analysis, yet all attempts to date that use this material have produced ambiguous results. Past (c. 150 000 years) $\delta^{13}C$ values for this material often converge to a single value from widely separated trophic pairs (e.g., Ostrom *et al.*, 1990; Cobabe, 1991). Well-characterized biological molecules other than collagen, such as noncollagenous proteins and individual amino acids, may ultimately extend the reach of isotopic palaeodietary analysis back many millions of years.

Tests have shown that carbonate and phosphate in fossilized tooth enamel yield isotopic dietary and climatic tracers for specimens up to 3 million years old, and tooth enamel shows promise as an isotopic source material for even very ancient (>100 million year old) specimens. Apatite in bone and tooth dentin, in contrast, is an unreliable dietary indicator past 1000 to 5000 years. The key to reliable use of mineralized tissues lies in development of analytical monitors of diagenesis that are independent of the isotopic signal, including data from elemental composition and crystallinity. Ecological associations and geological context also provide vital clues as to the integrity of isotopic signals in ancient biominerals.

Acknowledgements
We thank A.K. Behrensmeyer, B.J. Johnson, B. Marino, J.A. Silfer, D.J. Velinsky and the editors for helpful reviews of the manuscript.

CHAPTER 5

Carbon dioxide, carbon monoxide and methane in the atmosphere: abundance and isotopic composition

M. WAHLEN

5.1 Introduction

Most of the Earth's carbon is stored in sedimentary rocks and ocean sediments. Other pools are soil organic matter, fossil fuels and the biomass. Only a very small fraction of carbon is present in atmospheric gases. Atmospheric carbon compounds include the radiatively important species CO_2 and CH_4. CO_2 exchanges with the ocean and the biosphere, and is affected by anthropogenic emissions. Atmospheric CH_4 is mostly of biogenic origin, with some abiogenic contributions; both sources are strongly influenced by human activities. These compounds are thought to play an important role in climate forcing, and their recent anthropogenic alterations are thought to contribute to global warming. Ice core measurements show that the atmospheric concentrations of these gases have changed substantially during the last glacial cycle, and during industrial times. Carbon monoxide and non-methane hydrocarbons (NMHC) are not greenhouse gases, but, along with CH_4, they essentially control the oxidative state of the atmosphere by their interactions with hydroxyl (OH) radical.

This chapter will focus on the atmospheric trace gases CO_2, CO and CH_4, and on the use of the stable isotopes of carbon (ratio $^{13}C/^{12}C = 0.011$), oxygen ($^{18}O/^{16}O = 0.002$), hydrogen (D/H $= 0.00015$) and the radionuclide ^{14}C (modern $^{14}C/^{12}C$ about 10^{-12}) in elucidating the workings of the global cycles of these gases, and their ocean–atmosphere and biosphere–atmosphere interactions.

The isotopic compositions of the carbon compounds, variously produced and destroyed in the global carbon cycle, are affected by kinetic isotope effects in chemical and enzymatic reactions (e.g., atmospheric CO_2 to plant carbon), as well as equilibrium isotope effects (e.g., CO_2 exchange between ocean and atmosphere). Generally, due to energy considerations, the lighter isotopic species will react slightly faster than the heavier ones (although there are exceptions) resulting in a change in isotopic composition going from reactant to product compounds. These small differences can be measured very precisely by isotope ratio mass spectrometry. Thus stable isotopes can be used as tracers in the various biogeochemical cycles to study global source–sink relationships, as well as mechanistic processes. In many cases the isotopic composition of a gas from a specific source is quite distinct and characteristic. For a particular trace compound, comparison of the isotopic composition of the sources to that of the atmospheric inventory, in principle, allows one to derive global source budgets. The range of the isotopic composition of a gas from a specific source is often found to be quite small (e.g., $\delta^{13}CH_4$ of 12 samples collected over different spots from a tundra site over 3 years averaged $-60 \pm 2‰$), which makes the isotopic budget approach attractive. Methane flux measurements from a given source can range over orders of magnitude, and thus make global extrapolations difficult.

The cosmogenic radionuclide ^{14}C produced in the upper atmosphere is eventually incorporated into all living carbon. Its measurement has wide applications in ^{14}C dating. The large increase in atmospheric $^{14}CO_2$ by nuclear weapons testing in the late 1950s to early 1960s provided an atmospheric input variation which could be traced through the terrestrial and oceanic carbon reservoirs; it produced a wealth of information on the interaction between different carbon reservoirs. With respect to atmospheric CO_2 and CH_4, ^{14}C measurements (in conjunction with $\delta^{13}C$ measurements) allow one to quantify the split between biogenic sources and fossil carbon sources.

5.2 Carbon dioxide

5.2.1 Atmospheric mixing ratios
The most convenient unit for the abundance of an atmospheric constituent (i) is the mixing ratio (moles (i)/moles (air); molecules (i)/molecules (air) in ppmv (10^{-6}), ppbv (10^{-9}), etc.), since it is independent of temperature and pressure (altitude).

Recent time records
The most complete account on recent atmospheric mixing ratio measurements and their interpretation in terms of the global carbon cycle is given in a series of papers (Heimann & Keeling, 1989; Heimann et al., 1989; Keeling et al., 1989a,b) collected in the Geophysical Monograph 55 (Peterson, 1989). Additional data has been collected by Trivett & Worthy (1989), and world-wide by the NOAA Climate Monitoring and Diagnostics Laboratory (Tans et al., 1990).

The longest record for CO_2 is from Mauna Loa, Hawaii, starting in 1957 (Fig. 5.1). It is complemented by shorter records at other latitudes. All stations in the studies by Keeling et al. (1989a) are remote from fossil fuel combustion and major plant activity. These records show a pronounced seasonality and a monotonous increase in the mixing ratio. The mean mixing ratio increase for the longest record (Mauna Loa) is

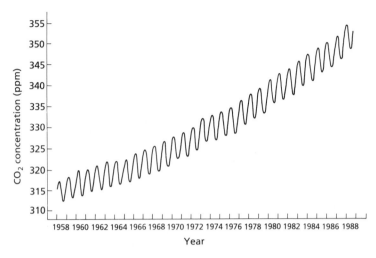

Fig. 5.1 Atmospheric CO$_2$ mixing ratio at Mauna Loa Observatory, Hawaii (fit to weekly average concentrations). (From Keeling *et al.*, 1989b.)

12% over 31 years (from 315 ppmv in 1958 to 352 ppmv in 1988). The atmospheric life time of CO$_2$ against exchange with the oceans and the biosphere is relatively long compared to interhemispheric atmospheric mixing times. Carbon dioxide is therefore well mixed in the global atmosphere. The increase observed at Mauna Loa thus is representative of the global atmosphere. This increase is mostly due to fossil fuel combustion, but contributions from the changing biosphere (deforestation and biomass burning) are also important. For the last two decades the growth rate of atmospheric CO$_2$ has been slowing down or holding steady, but the reasons for this are not clear.

The annual mixing ratio variations (up to 15 ppmv in high northern latitudes) are attributed to seasonal changes in carbon uptake, or photosynthesis (causing late summer CO$_2$ minima) and respiration of the land biota (causing spring CO$_2$ maxima). The amplitude of the seasonal cycle decreases with decreasing latitude in the northern hemisphere. This decrease continues with increasing latitude in the southern hemisphere (1 ppmv at 90°S) and is attributed to the decreasing distribution of land biomass and population from north to south. There are secular variations in the records which are thought to be partially caused by El Niño events.

Detailed and precise measurements from all latitudes show that CO$_2$ is not completely mixed in the global atmosphere. These data reveal latitudinal fine structure in the annual mean mixing ratios, which also change in time. The pole-to-pole difference in the annual mean mixing ratio is about 3 ppmv, higher in the north than in the south. This is attributed in part to the latitudinal distribution of fossil fuel consumption, but the fine structure of the latitudinal profiles is also influenced by atmospheric transport and latitudinally different exchange of CO_2 between atmosphere and oceans. The analysis of the interhemispheric gradients over time suggests that the southern hemisphere CO_2 concentration might have been higher in pre-industrial times, for which the oceans would have been responsible. Tans *et al.* (1990) argue, based on the present inter-hemispheric gradient, that the biosphere is a major sink. This sink could be increasing due to CO_2 fertilization that stimulates plant growth.

Long-term records
Long-term records of the atmospheric CO_2 mixing ratio over centuries, and over the last 160 000 years covering the last glacial/interglacial cycle, have been obtained from ice cores recovered from the Greenland and Antarctic ice sheets. Small bubbles which contain air from the times the bubbles were formed are occluded in this ice. From analyses of this air one can reconstruct the history of the atmospheric mixing ratios of CO_2 and other trace gases over long periods of time (Delmas *et al.*, 1980; Neftel *et al.*, 1982; Neftel *et al.*, 1985; Pearman *et al.*, 1986; Barnola *et al.*, 1987; Neftel *et al.*, 1988). Data obtained from an Antarctic core for the last two centuries (Fig. 5.2) reveal a gradually steepening increase in the atmospheric CO_2 mixing ratios starting about AD 1740, and they smoothly connect to the records of direct atmospheric measurements around 1958 (Neftel *et al.*, 1985; Siegenthaler & Oeschger, 1987).

The pre-industrial atmosphere contained about 280 ppmv CO_2, which rose to 352 ppmv in 1988 due to anthropogenic sources. The amount of CO_2 added by fossil fuel combustion since the onset of the Industrial Revolution (around 1850) is believed to be quite accurately known (Rotty & Marland, 1984). By combining ice core data for the CO_2 increase and fossil fuel derived emissions with model estimates of oceanic CO_2 uptake, it is possible to derive the residual time-dependent source or sink of CO_2 from the terrestrial biosphere (Siegenthaler & Oeschger, 1987). These model calculations suggest a substantial increase of the biospheric con-tribution to the atmospheric CO_2 prior to the fossil fuel combustion input. This is attributed to deforestation and expanding agriculture. The growth of this biogenic input has slowed down during this century, perhaps due to a fertilization effect on biota by increased atmospheric CO_2. A summary of the most likely split between biogenic and fossil CO_2 contributions for different times is given in Table 5.1.

We have recently measured CO_2 in samples from the new deep ice core GISP 2 (Greenland Ice Sheet Project 2) which is being drilled in Central Greenland. The initial results (Wahlen *et al.*, 1991) show no increase of CO_2 mixing ratios prior to AD 1800. The average mixing ratio

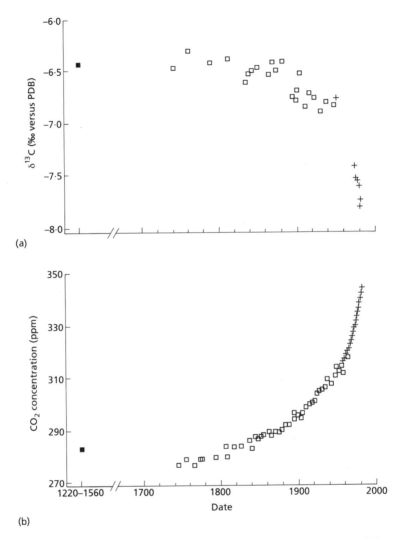

Fig. 5.2 Atmospheric CO$_2$ mixing ratio in the past 200 years from measurements of air trapped in the Siple Station (□) ice core (b) and δ^{13}C in CO$_2$ (a) in the same core. ■ = South Pole Station; + = results from direct atmospheric samples at Mauna Loa, Hawaii. (From Siegenthaler & Oeschger, 1987.)

between 1530 and 1800 is constant at 280 ± 5 ppmv. The subsequent rise in the 19th century is somewhat steeper in this core, but the data meet the directly measured atmospheric mixing ratios well. The nature of the process which incorporates air into the bubbles of ice leads to an age difference between occluded air and surrounding ice. The air is younger by several hundred to several thousand years depending on accumulation

Table 5.1 Model results for CO_2. (From Siegenthaler & Oeschger, 1987)

(a) Production (in Gt C or Gt C yr^{-1}) obtained by deconvolution of CO_2 concentration history of Fig. 5.2 by means of box-diffusion (BD) and outcrop-diffusion (OD) model

	BD		OD	
	1900	1980	1900	1980
Cumulative production (Gt C):				
fossil (observed)	12	160	12	160
non-fossil	62	89	81	153
total	74	249	93	313
Average production rate, 1959–83 (Gt C yr^{-1}):				
fossil (observed)	4.0		4.0	
non-fossil	0.0		0.9	

(b) Cumulative non-fossil production (in Gt C) obtained by deconvolution of shifted CO_2 concentration histories

	BD		OD	
	1900	1980	1900	1980
Shifted by −6 ppm/−10 yr	77	121	100	195
Shifted by +6 ppm/+10 yr	50	58	66	113

(c) Airborne fraction = ratio of atmospheric increase to total production in same period

	BD	OD
Deconvolution:		
1770–1980	0.510	0.405
1959–83	0.581	0.479
Only fossil CO_2 input:		
1770–1980	0.612	0.519
1959–83	0.625	0.533
Observed airborne fraction of fossil input:		
1959–83	0.588	

rate and temperature. Dating of the occluded air in this core has been performed experimentally by analyses of $^{14}CO_2$ in bubble air, and by locating the depth at which the peak in the atmospheric bomb produced $^{14}CO_2$ (see Section 5.2.3) activity appears (A. Wilson, personal communication 1990) which corresponds to the year 1963.

The impressive data obtained from deep ice cores extend the record back into the last ice age. The Vostok ice core in particular goes back to the penultimate ice age (Barnola et al., 1987). The Vostok record (Fig. 5.3) shows that the pre-industrial CO_2 mixing ratio of 280 ppmv persisted throughout the last warm period during the Holocene, and was very

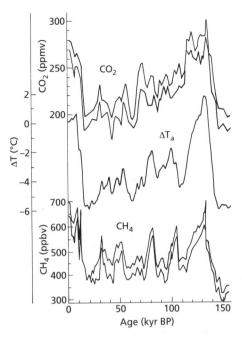

Fig. 5.3 Record of CO_2 and CH_4 mixing ratios from the Vostok ice core, along with the reconstruction of the temperature change over Antarctica derived from deuterium/hydrogen measurements in the ice. (From Lorius *et al.*, 1990.)

similar to ratios during the previous warm period 130 000 years ago. At the onset of the last glacial about 110 000 years ago the CO_2 mixing ratio drops abruptly, reaches levels of 180–200 ppmv by the beginning of the last glacial maximum (LGM) some 60 000 years ago and about 18 000 years ago it rapidly rises to Holocene levels. The levels observed for the penultimate GM are the same as for the last LGM. This observed change for glacial to interglacial atmospheric CO_2 amounts to about 90 ppmv when the average temperature change was about 6°C in Antarctica, and possibly more in the northern hemisphere (Lorius *et al.*, 1990). Comparing this glacial to interglacial difference to the recent >70 ppmv anthropogenic input suggests the possibility of global warming in the future.

Great efforts have been and are being made in explaining the mechanisms that can produce the observed glacial to interglacial CO_2 concentration change. At present, none of the numerous models can fully explain the entire concentration difference. It is thought that changes in the oceanic uptake of CO_2 are involved (Broecker & Peng, 1989). During warm periods, the ocean circulation is thought to be driven by salinity in a conveyer belt mode. Warm equatorial surface water moves to the north

Atlantic, delivering heat to the atmosphere, and then sinks to become deep water returning to the Pacific. Changes in this circulation system during cold times might change the ocean alkalinity, increase the high latitude oceans productivity, and thus increase the oceanic CO_2 uptake, resulting in lower atmospheric CO_2 concentrations.

Of great interest in view of potential warming is the magnitude of climate forcing by CO_2 and CH_4. Climatic changes are thought to be driven by changes in the Earth's orbital parameters (Milankovitch, 1941). This orbital forcing, however, is relatively weak, and the changing trace gas content is thought to amplify climate forcing. The contribution by greenhouse gases is estimated to be 40–65% (Lorius *et al.*, 1990).

5.2.2 $\delta^{13}C$ in atmospheric CO_2

Analysis of $\delta^{13}CO_2$ provides additional information on the carbon cycle. Plant carbon is depleted in ^{13}C with respect to atmospheric $\delta^{13}CO_2$, with C_3 plants more depleted than C_4 plants (see Chapter 1). The average value expected for CO_2 from terrestrial plant activity is about −25‰ (Degens, 1969). Atmospheric $\delta^{13}CO_2$ today is about −8‰ , and $\delta^{13}CO_2$ of oceanic origin is closer to the atmospheric value; ocean carbonate is about 0‰. Therefore $\delta^{13}CO_2$ can be used to distinguish between CO_2 contributions of oceanic and biogenic origin. It cannot be used to distinguish between biogenic and fossil fuel combustion sources of CO_2, as the latter CO_2 is equally light. This distinction can be made with ^{14}C (see Section 5.2.3).

The first extensive measurements of atmospheric $\delta^{13}CO_2$ in remote locations were made in 1955–56 by Keeling (1958, 1961). Later time series measurements made at locations from various latitudes are summarized by Keeling *et al.* (1989a). These data show pronounced seasonal variations as well as long-term trends, with $\delta^{13}CO_2$ becoming lighter toward the present. The seasonality in atmospheric $\delta^{13}CO_2$ with values heavier in summer, lighter in winter is similar to that of the CO_2 mixing ratio, with similar amplitude and phase. One can thus extract the isotopic composition of CO_2 which is seasonally removed from or added to the atmosphere from the analysis of $\delta^{13}CO_2$ versus $1/CO_2$. The result is approximately −25‰ and thus points to a biogenic origin for the seasonality. The seasonal amplitude of $\delta^{13}CO_2$ is about 1‰ in high northern latitudes and, similarly to CO_2, decreases toward the equator and further into the southern hemisphere.

$\delta^{13}CO_2$ at Mauna Loa changed from −6.69‰ in 1956 to −7.24‰ in 1978, with a $\delta^{13}CO_2/CO_2$ change intermediate between that expected for biogenic (including fossil) and oceanic CO_2. There are distinct latitudinal differences from the records of various stations, which also vary with time. The interhemispheric difference is as large as 0.2‰ with the northern hemisphere lighter than the southern hemisphere, and with a substantial gradient north of 30°N. This again reflects the larger biogenic and human activity in mid-to-high northern latitudes.

The longer term trend of $\delta^{13}CO_2$ has been reconstructed from the Antarctic Siple Station ice core (Friedli *et al.*, 1986) back to AD 1740, and from a few data from a core at the South Pole for the period AD 1200 to 1550 (Friedli *et al.*, 1984). The average $\delta^{13}CO_2$ before 1800 is −6.4‰. After that it changes gradually towards lighter values, and the change accelerates after 1950 (Fig. 5.2). The overall decrease between 1740 and 1980 is −1.14‰. Model calculations by Siegenthaler & Oeschger (1987) based on the Siple and Mauna Loa records yield the results shown in Table 5.1. Thus from pre-industrial times to today a cumulative 90–150 Gt C of biogenic carbon have been released, possibly due to deforestation and enhanced soil respiration from increased agriculture, as compared to a cumulative input of 160 Gt C from fossil carbon. The biogenic input for 1953 to 1983 was $0-0.9 \, Gt \, C \, yr^{-1}$, compared to $4.0 \, Gt \, C \, yr^{-1}$ from fossil carbon. Another estimate for the biogenic input by Bolin (1986) is somewhat larger ($1.6 \, Gt \, C \, yr^{-1}$) but it does not include possible recent carbon storage by plants from the hypothesized CO_2 fertilization effect.

The Siple core $\delta^{13}CO_2$ data from ice air, however, have not been corrected for the effect of gravitational fractionation which was discovered later by Craig *et al.* (1988a). The air bubbles only close off at a certain depth (e.g., 100 m) at the transition from firn (increasingly denser aggregate of snow) to solid ice. Until then the air in the firn is in contact with the atmosphere through air channels. Bubble air is consequently younger than the surrounding ice. Until close-off of the air bubbles, constituent gases and their isotopic species gravitationally fractionate depending on mass difference, with the heavier species enriched with respect to the atmosphere. This correction for the Siple core $\delta^{13}CO_2$ amounts to about −0.4‰. When corrected for gravitational fractionation these data do not smoothly connect to the atmospheric record in the 1950s. However, from work on the GISP 2 ice core, we found that the extraction process for small amounts of air and CO_2 depletes $\delta^{13}CO_2$. The absolute magnitude of this depletion is similar to the gravitational enrichment, and thus these two processes might effectively cancel each other. The Siple data as reported therefore might be representative for the atmospheric $^{13}CO_2$ evolution, with somewhat larger uncertainty.

Long-term atmospheric $^{13}CO_2$ records have also been reconstructed from measurements of $\delta^{13}C$ in cellulose of tree rings (Stuiver *et al.*, 1984; Freyer, 1986), assuming that there is a relationship between $\delta^{13}C$ of atmospheric CO_2 and that of tree cellulose. However, individual records can be rather different because various tree species fractionate the carbon isotopes differently, and this fractionation is influenced by physiological and environmental factors. Stuiver *et al.* (1984) obtained a change of about $-0.9‰$ for atmospheric $\delta^{13}CO_2$ between 1800 and 1975 when normalizing the data to tree-ring area, compatible with the ice-core data. Freyer (1986) compiled an average trend from all available tree-ring data and obtained a change of $-1.6‰$ between pre-industrial times and 1980, somewhat larger than the data from ice cores.

An interesting approach to reconstructing atmospheric $\delta^{13}CO_2$ has recently been proposed by Marino & McElroy (1991), Marino *et al.* (1992) and B.D. Marino (personal communication 1991). They extract atmospheric $^{13}CO_2$ from cellulose ^{13}C of stored *Zea mays* corn kernels. The results for corn kernels grown between 1948 and 1986 in Ames, Iowa (42°N) bridge the gap between the isotopic ice record (ending about 1950) and the direct atmospheric observations from 1978 to 1986 (Siegenthaler & Oeschger, 1987; Keeling *et al.*, 1989a), and there is excellent agreement for the respective time periods.

5.2.3 $\delta^{18}O$ in atmospheric CO_2

A large asymmetric meridional gradient in $\delta^{18}O$ of atmospheric CO_2 has been reported by Francey & Tans (1987) from samples of six remote stations (90°S to 71°N). The southern hemisphere values are rather constant up to 40°S, and then become progressively lighter going north to 71°N, with differences of almost 2‰ between these latitudes in 1984 and 1985. These gradients are extremely large. Enormous amounts of CO_2 must be exchanged per year in order to maintain this isotopic gradient against the vigorous inter- and intrahemispheric atmospheric mixing. If one assumes that CO_2 is equilibrating with a source that is 10‰ lighter in ^{18}O in the northern half than in the southern half of the northern hemisphere, this exchange would have to amount to $200 \, Gt \, C \, yr^{-1}$, an enormous amount when compared to the net primary production of $40-70 \, Gt \, C \, yr^{-1}$ or the exchange with the ocean of about $90 \, Gt \, C \, yr^{-1}$. Neither exchange with the ocean, given the limiting factors and the isotopic fractionations involved, nor cloud water exchange can explain the isotopic gradient observed. Furthermore, the interhemispheric difference in $\delta^{18}O$ of CO_2 of fossil origin could only maintain about 15% of the observed gradient. Atmospheric oxygen does not seem to exchange at all, and it is not in isotopic equilibrium with either oceans or CO_2. $\delta^{18}O$ of atmospheric O_2 appears to be constant ($+23.5‰$ relative to standard mean ocean water; Dole, 1935; Kroopnick & Craig, 1972), and is determined by isotopic fractionation associated with production and consumption of O_2 by photosynthesis and aerobic respiration.

The most obvious explanation for the latitudinal gradient of $\delta^{18}O$ of CO_2, according to Francey & Tans (1987), is isotopic exchange with plant leaf water, and possibly with soil water. Approximately one-third of the CO_2 entering the stomata of plants is fixed as gross primary production, with the rest diffusing back out. During the short time that CO_2 is inside the leaf one would expect isotopic equilibration with leaf $H_2^{18}O$, because the catalytic enzyme carbonic anhydrase is associated with one of the fastest enzymatic reactions. Thus atmospheric $\delta^{18}O$ in CO_2 could reflect the gradual depletion of ^{18}O in precipitation when going north and inland which would influence leaf and soil water (see also Chapter 1). The necessary exchange of carbon estimated via gross primary production of equilibrating CO_2 is between 200 and 400 Gt C yr^{-1}.

5.2.4 ^{14}C in CO_2

Although a radionuclide and not a stable isotope of carbon, the use of ^{14}C as tracer should be briefly mentioned. Produced in the upper atmosphere by cosmic ray-derived neutrons reacting with nitrogen ($^{14}N(n,p)^{14}C$), it is oxidized to $^{14}CO_2$ and mixes into the troposphere, and consequently into all carbon reservoirs. It provides for the ^{14}C dating technique (Libby, 1955). Living matter stops incorporating contemporary ^{14}C upon death, and ^{14}C decays with the half-life of 5730 years. By measuring the remnant ^{14}C activity in organic carbon, the age can then be inferred. Since cosmic ray intensities are somewhat variable (e.g., modulation by the 11-year solar cycle, etc.) the atmospheric ^{14}C production rate is not constant. From extensive ^{14}C measurements in dendro-chronologically-dated tree rings, an absolute calibration for ^{14}C dating has been established for almost the last 10 000 years.

Between about 1955 and 1963, numerous nuclear weapon tests were conducted in the northern hemisphere atmosphere, and large amounts of ^{14}C were injected into the lower stratosphere. From there $^{14}CO_2$, exchanged into the troposphere and into the southern hemisphere. The bomb produced $^{14}CO_2$ almost doubled the northern hemisphere tropospheric $^{14}CO_2$ inventory by 1963, at which time nuclear atmospheric testing stopped. Long-term monitoring of the tropospheric $^{14}CO_2$ activity in both hemispheres over more than 30 years (Nydal & Lovseth, 1983; Levin et al., 1985; Manning et al., 1990) has revealed the characteristics of stratosphere–troposphere and interhemispheric air exchange. Today tropospheric $^{14}CO_2$ activity is only slightly higher than in pre-bomb times. This time-dependent atmospheric $^{14}CO_2$ input provided a unique tracer which could be followed into the biospheric and oceanic carbon reservoirs. A wealth of information was obtained on the mechanism of the carbon cycle, especially for the exchange of atmospheric CO_2 with the biosphere (plants, soils), and with surface and ultimately deep oceans (Broecker & Peng, 1982).

^{14}C in CO_2 can distinguish between sources of recent biogenic and fossil origin. Fossil CO_2 contains no ^{14}C, while biogenic CO_2 contains ^{14}C at modern atmospheric levels. Tree-ring records of ^{14}C reveal the so-called Suess effect (Suess, 1970), a depletion of the wood ^{14}C content with increasing fossil fuel consumption.

Long-term records of atmospheric $^{14}CO_2$ at various latitudes (Levin et al., 1991) show pronounced seasonality in the northern hemisphere, with maxima in late summer and minima in late spring. Possible explanations for this seasonality are seasonality of the fossil fuel combustion, the strong fractionation during photosynthesis in summer causing enriched atmospheric $^{14}CO_2$ values, and/or the seasonally varying air exchange between stratosphere and troposphere. Similar to patterns seen in $^{13}CO_2$, the seasonality in $^{14}CO_2$ vanishes in the southern hemisphere. Latitudinal variations are observed, with a maximum $^{14}CO_2$ around 30°N (from biogenic CO_2), decreasing towards higher northern latitudes due to CO_2 from fossil fuel burning. The lowest values of atmospheric $^{14}CO_2$ are found in the high-latitude southern hemisphere. This is due to oceanic exchange with upwelling circumpolar waters which are depleted in $^{14}CO_2$.

5.3 Carbon monoxide

Atmospheric CO seems to be increasing in concentration by $1-2\%\ yr^{-1}$ (Khalil & Rasmussen, 1988), mainly due to increasing combustion of fossil fuel and biomass. Part of this increase may also be due to increasing oxidation of methane or decreasing concentration of atmospheric OH, by which CO and CH_4 are oxidized in the troposphere (Thompson & Cicerone, 1986). There is considerable uncertainty with respect to the CO increase. The atmospheric mixing ratios for CO scatter widely due to the short lifetime of about $1-3$ months. It is thought that natural and anthropogenic sources currently contribute about equally to the atmospheric burden (Seiler, 1974; NASA/WMO, 1985). Table 5.2 shows a current global budget for CO.

The atmospheric mixing ratio of CO varies considerably. There is a strong interhemispheric gradient with levels of about 35 ppbv in the southern hemisphere, and between 100 ppbv (remote) and 200 ppbv (polluted) in the northern hemisphere. Major anthropogenic sources are from combustion of fossil fuel and biomass. Natural sources are predominantly from the oxidation of CH_4 and NMHC. The oceanic CO source remains highly uncertain (Conrad et al., 1982).

The major sink for atmospheric CO is the oxidation reaction with OH radical, a reaction that has been extensively studied (e.g., Hynes et al., 1986). A smaller sink for CO is biological consumption by soils (Seiler & Conrad, 1987). δ^{13}-Carbon in atmospheric CO ranges between -22 and $-31‰$, reflecting the ^{13}C composition of the major CO sources and the fractionation during OH oxidation and soil consumption (Stevens et al.,

Table 5.2 Carbon monoxide (1984 concentrations 30–200 ppb). (From Logan *et al.* (1981), updated by Logan *et al.* (private communication 1984))

Atmospheric burden (10⁶ tons as C)	200
Sinks + accumulation (10^6 tons as $C yr^{-1}$)	
Reaction with OH	820 ± 300
Soil uptake	110
Accumulation (5.5% yr^{-1})	10
Total	940 ± 330
Sources (10^6 tons as $C yr^{-1}$)	
Fossil fuel combustion	190
Oxidation of anthropogenic hydrocarbons	40
Wood used as fuel	20
Oceans	20
Oxidation of CH_4	260
Forest wild fires (temperate zone)	10
Agricultural burning (temperate zone)	10
Oxidation of natural hydrocarbons (temperate zone)	100
Burning of savanna and agricultural land (tropics)	100
Forest clearing (tropics)	160
Oxidation of natural hydrocarbons (tropics)	150
Total	1060
Tropical Contribution (10^6 tons as $C yr^{-1}$)	
Burning	100
Forest clearing	160
Oxidation of hydrocarbons	150
Total	410

1972, 1980). $\delta^{13}CO$ from fossil fuel combustion is $-27‰$, similar to the average of fossil fuels (Tans, 1981). The isotopically lightest CO is produced from the oxidation of methane ($\delta^{13}CH_4$ about $-47‰$) with OH radical. One of the unresolved problems concerning the isotopic budgets of both CO and CH_4 is uncertainty over the relative contribution of C_3 and C_4 plants to the biomass burning source (Marino *et al.*, 1990). Uncertainties in the estimates of the isotopic composition of CO and CH_4 from biomass burning probably exceed 5‰.

[14]CO has been used to deduce spatial and temporal variations of atmospheric OH. The amount of atmospheric [14]CO is determined by the production by cosmic rays in the upper troposphere and lower stratosphere, by terrestrial sources of contemporary ^{14}C, and by sinks such as oxidation of CH_4 (and NMHC) in the atmosphere and on soils. Cosmic ray production varies with altitude and latitude, but shows no seasonality. Volz *et al.* (1981) analysed both their own [14]CO data and those of

MacKay *et al.* (1963) which exhibit a strong seasonality of atmospheric ^{14}CO concentration for mid-northern hemisphere latitudes. Minima occur in late summer and maxima in winter, with a peak-to-peak amplitude of about 50%. These atmospheric values are high ($\Delta^{14}CO$ of 1100–1300‰) as compared to present atmospheric $^{14}CO_2$ (200‰). The strong ^{14}CO concentration seasonality is attributed to the seasonally varying oxidation by OH.

Recently, Brenninkmeijer *et al.* (1990) measured ^{14}CO in the southern hemisphere by accelerator mass spectrometry and established that the southern hemisphere follows the same annual pattern as in the northern hemisphere, but that the southern hemisphere concentration values (^{14}CO molecules cm^{-3} air) are only about half of that in the northern hemisphere. This result is puzzling. Possible explanations might be that southern hemisphere OH concentrations are considerably higher, or that the exchange with upper atmosphere air is strongly reduced.

5.4 Methane

5.4.1 CH₄ mixing ratios

Methane is a strong infrared absorber, and affects atmospheric temperature directly and indirectly (Lacis *et al.*, 1981; Ramanathan, 1988; Hansen *et al.*, 1988). It influences tropospheric and stratospheric ozone levels (Johnston, 1984), and is a major source of stratospheric water (Ehhalt, 1979; Pollock *et al.*, 1980). Methane is produced by bacteria under anaerobic conditions in wet environments such as wetlands, bogs, tundra, rice fields, in the stomachs of ruminants and possibly by termites. Most biogenic CH_4 is produced by two major pathways, via acetate fermentation and via CO_2 reduction (from $CO_2 + H_2$) (Whiticar *et al.*, 1986; Wolin & Miller, 1987; Cicerone & Oremland, 1988). Other sources of CH_4 are from leakage of natural gas from drilling and distribution, from coal mining and from biomass burning, where CH_4 is produced by incomplete combustion. The annual production rate and the magnitude of the different sources are still being debated (Cicerone & Oremland, 1988; Wahlen *et al.*, 1989; Fung *et al.*, 1991; Quay *et al.*, 1991). The main sink for CH_4 is oxidation in the troposphere which is initiated by OH radical. In a series of reactions it is ultimately oxidized to CO_2 and H_2O. The reaction constant for $CH_4 + OH$ has been measured (Vaghjiani & Ravishankara, 1991). The atmospheric lifetime of CH_4 is 9–12 years. A small additional sink is from bacterial oxidation on soils under aerobic conditions. Some CH_4 is exported to the stratosphere where it is oxidized by OH at lower altitudes, and by O^1D and chlorine at higher altitudes.

The history of atmospheric CH_4 mixing ratios has been reconstructed from measurements of air occluded in ice cores, and large natural and anthropogenic variations have been observed. The atmospheric mixing ratio was about 350 ppbv during the last glacial and at the end of the previous glacial, and rose to about 650 ppbv in the Holocene (Chappellaz *et al.*, 1990). During the last 200 years the mixing ratio has risen rapidly to about 1700 ppbv (Rasmussen & Khalil, 1981a,b; Craig & Chou, 1982; Stauffer *et al.*, 1985). Direct measurements in the global atmosphere since 1978 demonstrated that atmospheric CH$_4$ increases about 1% per year (Blake & Rowland, 1988). Using data from a larger network of global sampling stations, Steele *et al.* (1992) recently determined that the rate of increase in the atmospheric mixing ratio has been slowing down considerably since about 1983. The reason for this is unknown. The average CH$_4$ mixing ratio in the northern hemisphere today is about 90 ppbv higher than in the southern hemisphere (Steele *et al.*, 1987, 1992; Blake & Rowland, 1988) as CH$_4$ emissions occur predominantly in the northern hemisphere. There is a small seasonal variation in the atmospheric mixing ratio of methane in both hemispheres with an amplitude of about 2%, with a minimum in local summer and a maximum in local winter (Steele *et al.*, 1987, 1992; Quay *et al.*, 1991). This is attributed to the seasonal variation in the destruction reaction with OH.

The large increase in CH$_4$ mixing ratio between pre-industrial times and today is thought to be due to anthropogenic increases in methane production by ruminants, rice production and increases in biomass burning. This increase (650 to c. 1700 ppbv) when compared with the natural glacial–interglacial variations (350–650 ppbv) illustrates the importance of understanding the global methane budget in view of global warming. If current global anthropogenic CH$_4$ emissions are scaled down according to population growth, one can approximately explain the observed preanthropogenic Holocene atmospheric mixing ratios. However, if one then reduces the remaining natural CH$_4$ sources (wetlands) according to the glacial ice coverage, the remaining wetland sources as they exist today seem not strong enough to maintain the CH$_4$ mixing ratio of the glacial atmosphere. This suggests that during glacial times some ice free areas must have been considerably wetter than they are today.

Stevens & Rust (1982) first proposed the isotopic approach to a CH$_4$ budget using [13]C. We carried this approach further to include ^{14}C and deuterium in methane (Wahlen *et al.*, 1988, 1989, 1990a,b). If the isotopic compositions of methane from various sources are distinctly different, then the strengths of individual sources can be determined from the comparison of the isotopic composition of methane from individual sources to that of atmospheric methane. With more isotopic species investigated, the budget should become more constrained. The isotopic composition of product CH$_4$ is influenced by many steps, i.e., conversion of atmospheric CO$_2$ to organic matter by photosythesis, decomposition of organic matter and generation of CH_4 by bacteria from the decomposition products, and it includes aspects of the hydrologic cycle by incorporation of hydrogen from water. All of these transformations will introduce isotopic fractionations. While most of the details are not known, it appears that the isotopic composition of CH_4 from a particular source is mainly determined by the composition of the substrate material and the specific methane generating pathway. Isotopic CH_4 budgets, however, require knowledge of the fractionation associated with the atmospheric sink reactions.

5.4.2 ^{14}C in methane

^{14}C in CH_4 is particularly useful in assessing the contributions from fossil CH_4 containing no ^{14}C. Fossil methane is believed to be from releases of natural gas during exploration and distribution, or from seepage from reservoirs on land and near shore, as well as from coal mining. ^{14}C in atmospheric methane has been measured by accelerator mass spectrometry (Wahlen *et al.*, 1989), and was found to be about 120 pMC (percent Modern Carbon), with a small interhemispheric gradient, reflecting different hemispheric source strengths. The derivation of the fossil contribution is somewhat complicated by an anthropogenic source of $^{14}CH_4$ from nuclear pressurized water reactors (Kunz, 1985). Since the atmospheric $^{14}CH_4$ inventory is small, the accumulated and remaining contribution from this source in today's atmosphere is about 15%. This source is increasing and will lead to a noticeable increase of atmospheric $^{14}CH_4$ in the future. Boiling water reactors, on the other hand, discharge mainly $^{14}CO_2$ (Kunz, 1985). Given the much larger atmospheric $^{14}CO_2$ inventory, this source is insignificant on a global scale.

$^{14}CH_4$ from biogenic sources has also been investigated (Wahlen *et al.*, 1989). CH_4 from cows and sheep contains ^{14}C at the level of contemporary $^{14}CO_2$, because these animals are fed modern grass, corn or hay. Biogenic $^{14}CH_4$ emitted from wetlands, peat bogs, rice fields and tundra, however, is variously depleted in ^{14}C with respect to contemporary $^{14}CO_2$ by as much as about 10%. Besides converting relatively recent organic carbon (containing ^{14}C at contemporary levels), methane-generating bacteria must also partially utilize older, more refractory organic carbon (containing ^{14}C at lower levels) stored in these systems. This is especially true for CH_4 from peat bogs and from the tundra. The latter source could be subject to a potentially strong positive feedback mechanism upon global warming. Large amounts of carbon are stored in permafrost, which, upon lowering of the permafrost level, would become available for CH_4 generation.

In order to deduce the contribution of fossil CH_4 to the atmospheric inventory, both current atmospheric $^{14}CH_4$ and the response of atmospheric ^{14}CH$_4$ to the varying levels of atmospheric bomb produced ^{14}CO$_2$ must be considered (Wahlen *et al.*, 1989). A model was used to simulate the ^{14}C transfer (CO$_2$ to organic carbon to CH$_4$) for biogenic CH$_4$ sources, together with data for the remaining sources, in order to best fit the time trend in atmospheric ^{14}CH$_4$ reconstructed from data on stored dated samples for the period 1966–87. The model also had to satisfy the CH$_4$ mass balance and the ^{13}CH$_4$. The best model fit and the data are displayed in Fig. 5.4. Results of this study suggest that 21% of annual methane production is from fossil CH$_4$.

5.4.3 *$\delta^{13}C$ and δD in CH$_4$*

δ^{13}CH$_4$ in clean air is close to −47‰ (Stevens & Rust, 1982; Tyler, 1986; Stevens & Engelkemeir, 1988; Quay *et al.*, 1988; Wahlen *et al.*, 1989). This value is depleted with respect to organic carbon, and is the result of the CH$_4$ source mix which contains some highly depleted biogenic components. This depletion is caused by isotope fractionation in carbon degradation and in bacterial CH$_4$ production. The δ^{13}CH$_4$ of atmospheric methane is also influenced (enriched) by the fractionation due to the reaction of CH$_4$ with OH radical. This enrichment factor has been measured in the laboratory, and values range between 3 and 10‰ (Rust & Stevens, 1980; Davidson *et al.*, 1987; Cantrell *et al.*, 1990). This quantity needs to be firmly established. Measurements of δ^{13}CH$_4$ from the lower stratosphere, where the OH oxidation causes the concentration

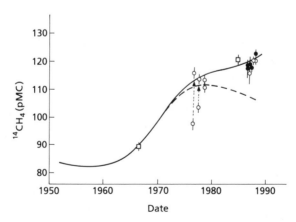

Fig. 5.4 Atmospheric ^{14}CH$_4$ in percent Modern Carbon (pMC) for 81 clean air samples collected over oceans (1986–88; ● northern, ○ southern hemisphere) and for five stored and dated samples (1966–84; □ methane from laboratory tank air, ○ methane from urban tank air sources). Solid line is the best fit of a one-box model prediction, including ^{14}CH$_4$ contribution from pressurized water reactors, broken line is without the reactor contribution. The model also satisfies the CH$_4$ mass balance and the ^{13}CH$_4$ balance. The resulting source strength distribution is given in Table 5.3. (From Wahlen *et al.*, 1989.)

to decrease with altitude and the $\delta^{13}CH_4$ to become progressively more enriched, indicate an enrichment factor which is somewhat higher (Wahlen et al., 1990a).

Biogenic CH_4 is variably depleted in ^{13}C with respect to the carbon source. Rust (1981) found that CH_4 from ruminants differed in $\delta^{13}CH_4$ according to the difference in the feed (C_3 versus C_4 feed plants), indicating that the isotopic composition of ruminant CH_4 is dependent on the isotopic composition of the substrate carbon and the methanogenic production pathway, which in this case is mainly by CO_2 reduction (Wolin & Miller, 1987). Biogenic CH_4 from wet, anaerobic environments (wetlands, rice production and tundra) is also variously depleted in ^{13}C (Stevens & Engelkemeir, 1988; Quay et al., 1988; Wahlen et al., 1989). The isotopic signatures for sources seem fairly narrowly defined (Fig. 5.5). CH_4 from biomass burning appears to be unfractionated with respect to organic carbon as expected for a high-temperature combustion process.

The isotopic fractionation of aerobic CH_4 consumption by dry soils, which is a global sink for methane of about 1–10% (Born et al., 1990), was investigated by King et al. (1989). They found enrichment factors of about 20‰ for this process. We have observed bacterial CH_4 consumption in the soil cover over a municipal landfill and on dry forest soils with enrichment factors of 29 and 15‰, respectively (M. Wahlen, unpublished data).

Time-series measurements of atmospheric $\delta^{13}CH_4$, especially at high latitudes, reveal a seasonality, anticorrelated to the seasonal variation in the CH_4 concentration (Quay et al., 1991). Values are more enriched during summer, when CH_4 mixing ratios are lower. This effect again is explained by stronger OH oxidation during summer, and the related isotope fractionation effect in the oxidation. Longer term variations of $\delta^{13}CH_4$ have been evaluated by Craig et al. (1988b) from air trapped in polar ice cores. They found that the carbon isotopic composition of atmospheric CH_4 in pre-industrial times was about 2.5‰ lighter than today, and attributed this change mainly to increased CH_4 from biomass burning, which is isotopically heavy.

The global average δD of atmospheric methane is $-83‰$. There is an interhemispheric difference of 10‰, with the southern hemisphere being heavier than the northern hemisphere (Wahlen et al., 1990a,b; M. Wahlen, unpublished data). This is thought to represent different δD source mixtures for the two hemispheres, caused perhaps by more biomass burning CH_4 (heavy in δD) in the southern hemisphere, and lighter biogenic CH_4 from the northern hemisphere, where both land distribution and population are more abundant.

δD of CH_4 of biogenic origin is highly depleted due to large fractionation effects (Fig. 5.5). To best illustrate the power of the isotopic

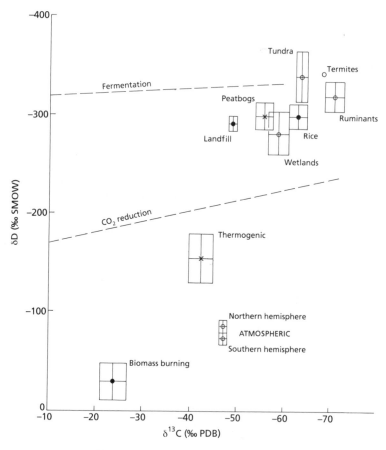

Fig. 5.5 δD in CH_4 vs. $\delta^{13}CH_4$ for various methane sources. (M. Wahlen, unpublished data.)

approach, consider Fig. 5.5, which shows the stable isotope plot, δD in CH_4 versus $\delta^{13}CH_4$. The boxes contain sample groups from the indicated sources, with the 1σ uncertainty (Wahlen *et al.*, 1989, 1990a,b; M. Wahlen, unpublished data). Also indicated are the values for atmospheric CH_4. The isotopic composition of CH_4 from the different sources is characteristically distinct. Samples from different ecosystems in water-logged environments can be isotopically distinguished (rice fields, peat bogs, wetlands, tundra). The broken lines in Fig. 5.5 indicate the isotopic boundaries for methane production from acetate and from CO_2 reduction, as derived from isotope systematics in natural gases of biogenic origin (Schoell, 1980; Whiticar *et al.*, 1986). The mix of methanogenic pathways (acetate fermentation and CO_2 reduction) in different ecosystems from waterlogged environments seems fairly similar, with about

70% of CH_4 derived from acetate fermentation and the remainder from CO_2 reduction. The isotopic signature of ruminant methane and the one data point for methane from termites are puzzling. Enteric fermentation is thought to produce methane predominantly through CO_2 reduction (Wolin & Miller, 1987), yet the data resemble those for acetate fermentation. This may be caused by some dynamic effect in food processing associated with a large fractionation.

Examination of the data for deuterated methane emitted from inundated ecosystems (Fig. 5.6) reveals a statistically significant correlation between δD in CH_4 and δD in local precipitation (the latter is obtained from Jouzel *et al.*, 1987). The data are from the Florida Everglades (Burke *et al.*, 1988), Louisiana rice fields, Appalachian peat bogs, wetlands from western New York, wetlands from northern Ontario, and tundra in northern Manitoba (M. Wahlen *et al.*, unpublished data). As

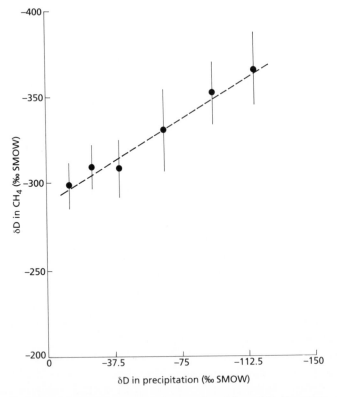

Fig. 5.6 δD in CH_4 vs. δD in precipitation. Data are from left to right: Everglades, Florida (Burke *et al.*, 1988); Louisiana rice fields; peat bogs, West Virginia; wetlands in western New York; wetlands in northern Ontario; tundra in Manitoba (M. Wahlen *et al.*, unpublished data).

Table 5.3 Global methane budget

Source	Percentage (approx.)
Wetlands and tundra	24
Rice production	24
Ruminants	21
Fossil methane	21
Biomass burning	10

CH$_4$ production $= 585 \times 10^{12}\,\mathrm{g\,yr}^{-1}$.
Atmospheric lifetime $= 8.8$ years.

the precipitation δD becomes lighter, so does the δD of product methane. The number of hydrogen atoms, derived from meteoric water and incorporated into CH$_4$ produced by bacteria, is different for acetate fermentation and CO$_2$ reduction pathways. Thus, for the isotopic signature of precipitation to carry through into that of the CH$_4$ produced, it is necessary that the pathway mix (acetate fermentation versus CO$_2$ reduction) is fairly constant for these diverse ecosystems. The deuterium correlation allows for a fairly reliable global extrapolation. One can apply the observed correlation for δD to the detailed world-wide database for methane emissions from wetlands and rice production (Aselmann & Crutzen, 1989) and the global precipitation distribution for δD given by Jouzel *et al.* (1987). From this, and considering the isotopic hydrogen composition, source strengths and global distribution for the other major sources, one can derive the average δD for the global methane sources in both hemispheres. The result suggests that CH$_4$ emissions in the southern hemisphere are significantly heavier in δD than in the northern hemisphere. Table 5.3 summarizes the global methane budget obtained from isotopic considerations described above.

5.5 Conclusions

Measurements of the mixing ratios in the global atmosphere reveal the working of the biogeochemical cycles of radiatively important trace gases. Reconstruction of the history of the abundance of atmospheric trace gases over various time scales in the past has helped in the understanding of the climatic implications of the recent anthropogenic increases of these gases. Recent advances in the measurement of the isotopic composition of the these gases, both in their sources and in the atmosphere, have aided in the elucidation of the global cycles of these gases and their ocean–atmosphere and biosphere–atmosphere interactions. Further analyses will allow scientists to make more precise estimates of global sources and sinks, and thus a more exact picture of their global cycles.

CHAPTER 6

The use of stable isotopes for the study of gaseous nitrogen species in marine environments

T. YOSHINARI & I. KOIKE

6.1 Introduction

Gaseous nitrogen species have important roles in the oceanic nitrogen cycle through atmospheric N_2 fixation and denitrification ($NO_3^- \rightarrow NO_2^- \rightarrow NO \rightarrow N_2O \rightarrow N_2$), where N_2, NO and N_2O are eventually returned to the pool of atmospheric N_2. Chemical characteristics of nitrogenous gases and physical processes that affect the distribution of gaseous nitrogen in the marine environment have been reviewed (Scranton, 1983).

The application of natural abundance ^{15}N as a tracer in the marine environment has been reviewed by Owens (1987) and Wada & Hattori (1991). The application of tracer techniques using ^{15}N-labelled substrate in various aspects of nitrogen metabolism in the marine environment (Harrison, 1983) and in marine denitrification (Hattori, 1983; Seitzinger, 1988; Koike & Sørensen, 1988; Koike, 1990; Wada & Hattori, 1991) has also been reviewed. However, the cycling and isotopic composition of gaseous nitrogen species in the marine environment have not been as extensively reviewed.

In this chapter we review (i) the distribution of natural abundance gaseous ^{15}N in the marine environment and (ii) the tracer techniques that have been applied for identifying mechanisms of gaseous nitrogen metabolism, and for quantifying the rate of gaseous nitrogen formation or consumption in the marine environment. Our discussion will focus mainly on N_2 and N_2O. Other gaseous species such as NO and NH_3 are discussed marginally, since information on the isotopic signatures of these species is rather limited. Biogeochemical aspects of nitrogen and oxygen in N_2O are discussed to a much greater extent than those of N_2. While there are some other methods for studying nitrogen transformations in these systems, such as ^{13}N tracers and the acetylene reduction technique, we included these topics only for the comparison with ^{15}N methodology.

6.2 Gaseous nitrogen in the sea

6.2.1 Dinitrogen

Dinitrogen enters the marine nitrogen cycle through the activity of nitrogen-fixing micro-organisms in oligotrophic waters and returns to the atmosphere through denitrification. The suboxic environments in the eastern tropical Pacific and in the Arabian Sea (Codispoti & Christensen, 1985; Codispoti, 1989) and continental shelf (Hattori, 1983; Christensen *et al.*, 1987a,b) are considered as the main sites for the production of N_2 by denitrification. The importance of N_2 production through the nitrification–denitrification couple by the pathway ($NH_4^+ \rightarrow N_2O \rightarrow N_2$), has been suggested (Codispoti & Christensen, 1985). Generally, the process of N_2 formation by denitrification in the sea is very slow, primarily due to carbon limitation (Hattori, 1983). In contrast, the rate of N_2 production in marine sediments is controlled by supply of nitrate from either the overlying water or nitrification at the sediment–water interface.

Current estimates of N_2 produced by denitrification (c. $120 \, Tg \, yr^{-1}$; Codispoti & Christensen, 1985; Christensen *et al.*, 1987a,b) are far greater than estimates of N_2 fixation (c. $20 \, Tg \, yr^{-1}$; Capone & Carpenter, 1982). Each is, respectively, less than 0.0004% and 0.0001% of the total dissolved N_2 in the sea ($3.4 \times 10^7 \, Tg \, yr^{-1}$).

6.2.2 Nitric oxide

Nitric oxide (NO) is formed by photochemical degradation of NO_2^- in surface seawater (Zafiriou *et al.*, 1980; Zafiriou & McFarland, 1981) and by microbial formation in suboxic waters. Since it is chemically reactive and easily oxidized to NO_2 and NO_2^-, and the amount that can be detected at any time is minute (c. $1 \, nmol \, l^{-1}$), the isotopic determination of NO in the environment is not currently feasible. The observation that NO existing exclusively in oxygen limiting waters ($10–100 \, \mu mol \, l^{-1}$) is 13% of the average integrated nitrification flux (Ward & Zafiriou, 1988) suggests that NO plays a significant role as an intermediate in the nitrification process.

The role of NO in denitrification, which was a controversial issue for a long time, has been recently resolved. NO is an obligate intermediate (Zafiriou *et al.*, 1989; Goretski & Hollocher, 1990), and NO reductase is a discrete enzyme that reduces NO to N_2O (Carr & Ferguson, 1990a,b; Zumft & Kroneck, 1990). While the conditions for NO production by nitrification have also been controversial (Lipschultz *et al.*, 1981; Anderson & Levine, 1986), Remde & Conrad (1990) have clarified that (i) similar to N_2O, the main source of NO is from the reduction of NO_2^- and (ii) unlike N_2O, the formation of NO by nitrification is unaffected by the presence of O_2.

As shown above, the reductive pathways of NO_2^- to NO and N_2O are commonly shared by denitrifiers and nitrifying denitrifiers under oxygen limiting conditions, and the conditions for N_2O production by these two different processes are quite similar (Conrad, 1990; Remde & Conrad,

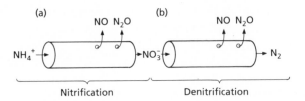

Fig. 6.1 A conceptual model of the two levels of regulation of nitrogen trace gas production via nitrification and denitrification: (a) flux of nitrogen through the process 'pipes' and (b) holes in the pipes through which trace nitrogen-gases 'leak'. (From Firestone & Davidson, 1989.)

1990). Pathways of gaseous nitrogen formation by these processes are shown schematically in Fig. 6.1.

6.2.3 Nitrous oxide

N_2O is an interesting gaseous species in that it is: (i) a gas that can be produced by both nitrification and denitrification; (ii) a species returning to the atmosphere in the global nitrogen cycle ($20 \pm 2.4 \, \text{Tg yr}^{-1}$) with an annual increase of $0.25–0.31\% \, \text{yr}^{-1}$ (Prinn et al., 1990); (iii) a greenhouse gas (Lacis et al., 1981; Dickinson & Cicerone, 1986) about 200 times more powerful than CO_2 on a molecular basis (Rodhe, 1990); (iv) a natural NO source in the stratosphere (Crutzen, 1970; McElroy & McConnel, 1971; Johnston, 1972; Cicerone, 1989); and (v) a stable gas against photochemical and other chemical reactions in the troposphere.

A linear correlation between apparent production of N_2O ($\Delta N_2O = [N_2O]_{\text{measured}} - [N_2O]_{\text{saturation}}$) and apparent oxygen utilization in the open ocean (Yoshinari, 1976; Elkins et al., 1978; Cohen & Gordon, 1979; Cline et al., 1987; Butler et al., 1989; Oudot et al., 1990), and an inverse correlation between N_2O production by marine nitrifying bacteria and the dissolved oxygen concentration (Goreau et al., 1980) have supported the hypothesis that most of N_2O produced in the open ocean originates from nitrification. However, some of the N_2O once produced by nitrification and/or denitrification seems to be re-utilized by denitrifers in regions that contain both suboxic and low-oxygen waters (Elkins, 1978; Cohen & Gordon, 1978; Pierotti & Rasmussen, 1980; Kaplan, 1984). The ocean as a whole is a source for atmospheric N_2O and its source strength has been estimated to be c. $4 \, \text{Tg yr}^{-1}$ (Elkins et al., 1978; Cohen & Gordon, 1979; Cline et al., 1987; Butler et al., 1989; Oudot et al., 1990). Productive regions underlain by low O_2 and suboxic waters may be a source of oceanic N_2O that was not previously considered in the budget (Codispoti & Christensen, 1985; Law & Owens, 1990). Considering the contribution from upwelling regions of the eastern Pacific and northern Indian Oceans, a revised estimate of N_2O production could become as much as $6\,Tg\,yr^{-1}$ (Naqvi & Noronha, 1991; Codispoti *et al.*, 1992). The annual flux of N_2O to the atmosphere by denitrification is about 5% of that of N_2O (6 Tg vs. 120 Tg). The annual flux of N_2O is 1% of the total amount of dissolved N_2O in the sea (5 Tg/500 Tg).

There are no reliable estimates for net production of N_2O in coastal environments. The rates of N_2O production at the sediment–water interface and sediment column are largely dependent on the concentrations of dissolved oxygen and available organic carbon (Sørensen, 1978b; Kasper, 1982; Nishio *et al.*, 1983; Jørgensen *et al.*, 1984; Seitzinger *et al.*, 1984; Seitzinger, 1988). While significant amounts of N_2O were suspected to be produced by nitrification at the suboxic sediment–water interface, Jørgensen *et al.* (1984) demonstrated in laboratory studies that overall production of N_2O in coastal sediments was mostly by denitrification. Microelectrode-measured N_2O profiles in the top few millimetres of sediments (Christensen *et al.*, 1988; Revsbech *et al.*, 1988) indicate that the N_2O fluxes from sediments to the water column are of minor significance when looking at nitrogen dynamics in coastal marine ecosystems. This is because N_2O is most likely reduced further to N_2 in sediments (Revsbech *et al.*, 1988). The removal of up to 85% of regenerated ammonium-nitrogen through nitrification–denitrification coupling in continental shelf sediments (Christensen *et al.*, 1987a,b) could well be mostly in the form of N_2, since N_2O is most likely reduced further to N_2 (Revsbech *et al.*, 1988).

6.3 Factors that influence gaseous nitrogen metabolism

The energy required for the nitrogen cycle in the sea is provided by primary producers which fix inorganic carbon. Subsequent transformations of organic carbon and nitrogen are carried out by other biological processes (i.e., secondary producers, autotrophic and heterotrophic bacteria). Therefore, the extent of metabolism in surface and subsurface waters is also closely correlated with primary productivity (Karl *et al.*, 1984; Karl & Knauer, 1984; Lipschultz *et al.*, 1990).

6.3.1 Particulate organic nitrogen (PON)

The composition of PON may vary but the main component is hydrolysable proteinaceous materials, which amount to as much as 40–65% of the flux of the PON (Lee & Cronin, 1982). When organic nitrogen as PON, typically composed of living cells, detritus and faecal pellets, is mineralized in the euphotic zone, NH_4^+ is likely to be re-utilized by phytoplankton (Checkley & Miller, 1989). Also, a part of the mineralized NH_4^+ out of PON in the surface water may be further oxidized by nitrification. During this process N_2O is likely to be produced. Because of the O_2 constraint, however, N_2O yield in shallow water should be very low. While it is to be determined from future studies, the production of N_2O by nitrification in shallow water may turn out to be significantly large because the turnover rate of organic nitrogen to NH_4^+ is expected to be much higher than in deeper water (Checkley & Miller, 1989).

A part of PON sinks below the euphotic zone. An increase in organic carbon and nitrogen at mid-depths (700–900 m) has been reported to correlate with *in situ* production by chemoautotrophic micro-organisms in deep waters that are associated with sinking particles (Karl *et al.*, 1984; Karl & Knauer, 1984). For NH_4^+-oxidizing chemoautotrophs at these depths, the supply of NH_4^+ will arise only from the hydrolysis of PON. If the reported process is prevalent in the entire ocean, then the localized particulate organic matter (POM) at the oxygen minimum layer, which is primarily transported from the surface water, seems to have a much more significant role for *in situ* O_2 utilization and N_2O production than previously assumed.

6.3.2 POM and dissolved oxygen

The relationship between the size of POM and the oxygen concentration in the surrounding water above $4 \mu mol\, l^{-1}$ may be critically important for gaseous metabolism through nitrification, denitrification and nitrification–denitrification coupling, because the $[O_2]$ gradient within the particle is likely to be regulated by the size of POM. The relationship between the diameter of POM and the oxygen consumption rate (Kaplan & Wofsy, 1985) suggests that denitrification can take place in oxygenated water as long as the diameter of POM can compensate to establish an anaerobic condition in the centre of the particle (Fig. 6.2). This relationship will vary under different oxygen concentrations. Production or reduction of N_2O by different microbial processes could take place in O_2-limiting sites within the matrix of POM, which is suspected to be a source of ammonia for nitrifiers or as an electron donor for denitrifiers. However, it is clear that N_2 and N_2O are unlikely to form by denitrification where predominantly fine POM particles are present in highly oxygenated water (see the curves at the upper left corner in Fig. 6.2). This agrees with the conclusion that the sites and rates of many nitrogen transformations depend primarily on the exact nature of the distribution of dissolved oxygen (Codispoti & Christensen, 1985).

6.4 Isotopic fractionation in nitrogen metabolism

6.4.1 Isotopic fractionation during nitrification and denitrification

The use of stable isotopes of gaseous nitrogen, and oxygen in the case of N_2O and NO, as tracers is based on the principle that there exists an

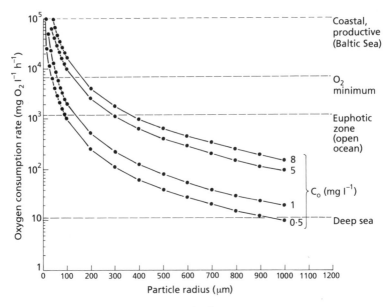

Fig. 6.2 Oxygen consumption rate $(mg\, O_2\, l^{-1}\, h^{-1})$ needed to cause anoxia at the centre of particles, plotted as a function of particle radius (μm) and ambient oxygen tension $(mg\, O_2\, l^{-1})$. (From Kaplan & Wofsy, 1985.)

isotopic order (kinetic isotopic fractionation) in biochemical processes in which the substrate is enriched and the product is depleted in heavy isotopes when the substrate pool is infinite, since organisms preferentially utilize lighter nitrogen or oxygen. In the process of biological fixation of atmospheric nitrogen, an estimated average value of the discrimination (α) of ^{15}N over ^{14}N is 1.002 (Mariotti *et al.*, 1982; Wada & Hattori, 1991). In the processes of nitrification ($NH_4^+ \rightarrow NO_3^-$), and denitrification ($NO_3^- \rightarrow N_2O \rightarrow N_2$), ammonium-N and nitrate-N, respectively, are enriched with heavy nitrogen, ^{15}N, and the α values for both processes are in the range of 1.01–1.03 (Wada & Hattori, 1991). Using available data of isotopic fractionation factors of these processes and ^{15}N abundance in nitrogen of naturally occurring substances in different natural environments, Wada *et al.* (1975) assessed the magnitude of denitrification in the marine environment to be approximately one-third of global denitrification.

6.4.2 Factors affecting the isotopic composition of N₂O in the sea

The ^{15}N of N_2O produced by *Nitrosomonas europaea* under different concentrations of dissolved oxygen and ammonium was depleted ($\varepsilon =$ -60), which was attributed to the product of kinetic nitrogen isotopic fractionation factors of the reactions of ($NH_4^+ \rightarrow NO_2^-$) and ($NO_2^- \rightarrow$

N_2O) (Yoshida, 1988). The value ε for ($NH_4^+ \rightarrow NO_2^-$) calculated by Yoshida at $Po_2 = 0.19$ and $38\,mmol\,l^{-1}\,NH_4^+$ was in agreement with that by Mariotti *et al.* (1981). The magnitude of fractionation can be different under laboratory and natural conditions. The values for $\delta^{18}O$ of N_2O, mostly derived from nitrification in a nitrification plant of a sewage treatment facility where the range of [NH_4^+] as the substrate for nitrification was $0.4–0.9\,mmol\,l^{-1}$, was about 22‰ as compared with 45.5‰ in N_2O in ambient air (Yoshinari & Wahlen, 1985). However, $\delta^{15}N$ or $\delta^{18}O$ values of N_2O derived from nitrification in the open ocean may not be as low as that shown above; the degree of isotopic fractionation in the ocean could well be very small because of substrate limitation. For example, a small degree of isotopic fractionation ($\varepsilon = -9.1$) was observed by *Skeletonema costatum* in the estuary during the uptake of NH_4^+, whereas ε in laboratory culture was -20 (Cifuentes *et al.*, 1989). These differences were attributed to a positive correlation between the isotopic fractionation and substrate concentration.

Biochemical processes

N_2O derived either by nitrification or denitrification proceeds by steps in which different kinetic fractionation is involved. The isotopic signature of N_2O in oxic waters is likely to retain its signature since N_2O is neither metabolized by nitrifiers nor reduced by denitrifiers because of high [O_2]. On the other hand, isotopic enrichment of N and O in N_2O will take place in suboxic waters with a moderate supply of organic carbon, where a part of N_2O, irrespective of its origin, is further consumed by denitrifiers. Thus, the isotopic signatures of N_2O from nitrification, denitrification or the combination of nitrification and denitrification, should be characteristically different. If such is the case, it should be possible to understand formation and/or consumption of N_2O occurring in a given water mass, or the origin of N_2O and/or the extent of process involved for *in situ* consumption out of the N_2O pool, from the isotopic signature of N_2O.

The isotopic signature of N_2O produced by denitrification could be correlated to that of oxidized nitrogen in either dissolved NO_3^- ($\delta^{15}N$ +6 to +8‰; Miyake & Wada, 1967; Cline & Kaplan, 1975; Wada & Hattori, 1976; Liu & Kaplan, 1989) or NO_2^- produced from nitrification using NH_4^+ that came from POM. In the Peru upwelling area, selective removal of ^{15}N-depleted NO_3^- by denitrifiers increased the $\delta^{15}N$ of the remaining nitrate by 11–16‰ (Liu & Kaplan, 1989). In turn, the $\delta^{15}N$ of N_2O produced is generally far less than that of surrounding nitrate-nitrogen.

POM

$\delta^{15}N$ values of POM vary seasonally, reflecting the magnitude of NO_3^- flux into the euphotic zone and subsequent uptake by phytoplankton

(Altabet & Deuser, 1985). While the fast sinking PON-N is depleted in ^{15}N ($\delta^{15}N$ = 2.9–4.4‰; Wada & Hattori, 1976; Saino & Hattori, 1980, 1985), $\delta^{15}N$ of PON in deep water, which is largely composed of fine, slow sinking particles, is enriched by approximately 6‰ relative to PON in shallow waters. These high values of $\delta^{15}N$ are mainly due to isotopic fractionation during remineralization and subsequent export of ^{15}N-rich particles (Altabet & McCarthy, 1985; Saino & Hattori, 1987; Checkley & Miller, 1989).

Temperature
Generally, the rate of metabolism in deep waters is expected to be slow due to low temperature and a limiting supply of electron donors. Since the isotopic fractionation of any biochemical reactions should be increased under such conditions, it is possible that the $\delta^{15}N$ of N_2O would be enriched if there is a slow process of denitrification which reduces N_2O to N_2 in the matrix of POM in deep waters.

6.5 Methods

6.5.1 Extraction and purification of N_2O in seawater
There are several different methods for the extraction of N_2O from seawater samples. Yoshida *et al.* (1984) used a 150 l stainless-steel tank in which approximately 100 l of seawater was filled. A column packed with a molecular sieve 5A was used to adsorb N_2O that was extracted from the water using a pump and a sintered spherical glass bubbler. In another procedure, an apparatus developed for the extraction of a large volume of seawater (c. 2000 l) was used for the collection of dissolved N_2O in the sea (Kim & Craig, 1990). By inserting a bubbler with a fritted glass fitting (5 cm in diameter) into a 20-l GoFlo sampler and adding double end shut-off valves on both ends, dissolved N_2O of seawater collected in the sampler was extracted without transferring the water to another container (Yoshinari *et al.*, 1990). The N_2O collected on molecular sieve 5A was thermally desorbed and purified by gas chromatography (Yoshida *et al.*, 1984; Kim & Craig, 1990; Yoshinari *et al.*, 1990).

Recently, a new method for the extraction of dissolved methane in natural waters has been reported (Schmitt *et al.*, 1991). The method is based on the principle of extracting the dissolved gas in a vacuum system while the water sample is exposed to ultrasonic energy. Although it used about 1 l of seawater, the apparatus could be developed to process a much larger volume of water for the extraction of dissolved N_2O in seawater.

6.5.2 Determination of the isotopic composition of N₂O

There are several different methods for determining the isotopic composition of N_2O (Table 6.1). The $\delta^{15}N$ of about $1–2\,\mu mol$ N_2O in seawater was determined for the first time using isotope ratio mass spectrometry (Yoshida & Matsuo, 1983; Yoshida et al., 1984). These authors could not, however, determine $\delta^{18}O$ simultaneously since N_2O was converted to N_2 by removing O_2 with Cu and CuO. $\delta^{18}O$ of N_2O from different environments was thus subsequently determined by high-resolution infrared spectroscopy with a tunable diode as a detector (Wahlen & Yoshinari, 1985). However, these authors could not determine $\delta^{18}O$ of N_2O in seawater since the method requires about $20\,\mu mol$ $(0.5\,ml)$ of N_2O for analysis, which corresponds to a $500–3000\,l$ seawater sample.

Recently, a new method was developed which can determine the isotopic composition of both nitrogen and oxygen of N_2O by mass spectrometry after converting the sample to N_2 and CO_2 over a heated graphite rod (Yoshinari, 1990; Kim & Craig, 1990). Using their method, Kim & Craig (1990) were the first to report oceanic data sets of N_2O-N and N_2O-O isotopic composition in the oceanic water column. They converted $12–25\,\mu mol$ of N_2O to N_2 and CO_2 in the presence of a graphitic carbon rod wrapped with platinum wire. The reaction was carried out at 700°C. After 7 minutes reaction, CO_2 and the remaining N_2O were frozen in liquid nitrogen, and N_2 was transferred into a sample tube. The N_2O and CO_2 were then warmed up and reacted again. To

Table 6.1 Methods for the determination of ^{15}N, ^{17}O and ^{18}O in N_2O

Method	N₂O sample size (μl)	Isotope species	Reference
High resolution infrared spectroscopy with tuneable diode lasers	500	^{18}O	Wahlen & Yoshinari (1985)
Mass spectrometry			
1 $N_2O + Cu \rightarrow N_2 + CuO$	20–30	^{15}N	Yoshida & Matsuo (1983)
2 $N_2O + C \rightarrow N_2 + CO_2$	500	$^{15}N, {}^{18}O$	Kim & Craig (1990)
	20–30	$^{15}N, {}^{18}O$	Yoshinari et al. (1990)
3 $N_2O + Ni \rightarrow N_2 + NiO$	200	$^{15}N, {}^{17}O, {}^{18}O$	Thiemens & Trogler (1991)
$NiO + BrF_5 \rightarrow O_2 + NiF_2$			
4 Direct injection of N_2O to mass spectrometer	20–30	$^{15}N, {}^{18}O$	Tanaka et al. (1992)

ensure a complete conversion, three such cycles were repeated. CO existing in the N_2 sample as contaminant was removed by using I_2O_5, whereby CO_2 and I_2 are formed (Schutze, 1944; Nagashima & Suzuki, 1984).

Using the same principle that was used by Kim & Craig (1990), Yoshinari (1990) developed a method that requires approximately 50 µl (c. 2 µmol) of N_2O for the analysis of $\delta^{15}N$ and $\delta^{18}O$. After transferring the sample cryogenically into quartz tubing in which a graphitic carbon rod wrapped with gold wire was previously placed, N_2O was converted to N_2 and CO_2 by heating at 615°C for 90 min. By lowering the reaction temperature to 450°C and prolonging the reaction period to 24 hours (T. Yoshinari, S. Ueda & B. Fry, unpublished results), it was found that the reaction forming N_2 and CO_2 was quantitative, with the level of CO and NO so low that these species did not affect the isotopic composition of N_2 and CO_2. The standard error of 1σ (1σ SE) for N_2 and CO_2 was 0.04 and 0.03, respectively, with 2 µmol of standard N_2O. Results from seawater from Monterey Bay, California using this method (Yoshinari *et al.*, 1990; Codispoti *et al.*, 1992) were comparable to those previously reported by Kim & Craig (1990). The advantage of this method is that it requires 20–100 l of seawater, depending on the concentration of N_2O, as compared with 2000 l of seawater that was used for a single analysis of N_2O by Kim & Craig (1990).

The isotopic composition of N_2O can also be determined by first converting the sample to NiO and N_2 by heating overnight at 700°C to 750°C in a nickel tube. NiO is subsequently reacted overnight with BrF_5 at 500°C (Thiemens & Trogler, 1991). This method also requires about 10 µmol N_2O (M.H. Thiemens, personal communication 1993).

It would be desirable to develop a method that can determine the isotope ratios of both nitrogen and oxygen with <0.5 µmol N_2O. Considering that the contamination of atmospheric N_2 becomes more critical as the sample size becomes smaller, determining the mass ratios of 45:44 ($^{15}N^{14}N^{16}O : ^{14}N^{14}N^{16}O$) rather than 29:28 ($^{15}N^{14}N : ^{14}N^{14}N$) for nitrogen and 46:44 ($^{14}N^{14}N^{18}O : ^{14}N^{14}N^{16}O$) for oxygen seems more advantageous (N. Tanaka, personal communication 1993). For this reason, a new method based on direct injection of N_2O into the isotope ratio mass spectrometer has been developed (Tanaka *et al.*, unpublished manuscript). This method can also determine nitrogen and oxygen isotope ratios of N_2O with 1σ SE of 0.03 and 0.04, respectively. The only disadvantage of this method is that it requires a mass spectrometer dedicated solely to the analysis of N_2O, since the residue of trace N_2O in the system could be a serious source of error for the subsequent determination of the ratios of $^{13}CO_2 : ^{12}CO_2$ in other samples.

6.6 Stable isotopic composition of N_2 and N_2O in the marine environment

6.6.1 Dinitrogen

Nitrogen in the sea is slightly enriched in ^{15}N relative to the atmosphere (Benson & Parker, 1961; Miyake & Wada, 1967). Measurement of the isotopic composition of dissolved N_2 cannot be used as a tracer for quantitative estimation of annual denitrification and nitrogen fixation, because existing isotope ratio mass spectrometers cannot accurately determine such minute changes in $^{15}N:^{14}N$. However, such an approach may become possible in the future if the precision of mass spectrometers is greatly improved. Only a slight enrichment in ^{15}N of dissolved N_2 was found in the anoxic waters of the Cariaco Trench in the Caribbean Sea and the Dramsfjord in Norway (Richard & Benson, 1961). While $\delta^{15}N$ as low as $-22‰$ was measured for N_2 produced in the eastern tropical North Pacific, where the water mass is deficient in O_2 (as inferred from $\alpha_{(NO_3^- \rightarrow N_2)} = 1.03-1.04$; Cline & Kaplan, 1975), actual measurements of $\delta^{15}N$ in these waters have not been made.

6.6.2 Nitrous oxide

Oxic shallow waters

Yoshida et al. (1984) reported the first $\delta^{15}N$ values of N_2O in various parts of the Pacific Ocean. $\delta^{15}N$ of N_2O in oxic shallow waters (average $\delta^{15}N = 5.2 \pm 1.3‰$) was significantly depleted compared to values of atmospheric N_2O (average $\delta^{15}N = 8.0‰$). This result was later confirmed by Kim & Craig (1990), who measured isotopic compositions ($\delta^{15}N$ and $\delta^{18}O$) of N_2O in the Pacific Ocean (Fig. 6.3). Kim & Craig (1990) found that both isotopes of N_2O in waters from the surface to c. 600 m were lighter than the values of atmospheric N_2O. However, this was not observed for $\delta^{15}N$ of N_2O at either Stations C and E (Fig. 6.4; Yoshida et al., 1989).

The lower values of $\delta^{15}N$ and $\delta^{18}O$ of N_2O in shallow waters may result from the production of N_2O through nitrification, utilizing ^{15}N-depleted NH_4^+ derived from PON (Miyake & Wada, 1967; Checkley & Entzeroth, 1985), since the $\delta^{15}N$ of PON in shallow waters is on average 6‰ lower than that at depth (Wada & Hattori, 1976; Saino & Hattori, 1980, 1985, 1987; Checkley & Miller, 1989). The main cause for the depletion of ^{15}N of PON in the euphotic zone is due to the activity of pelagic heterotrophs, especially particle grazing copepods (Checkley & Entzeroth, 1985; Checkley & Miller, 1989; Altabet, 1988, 1989). POM in oligotrophic waters, where N_2 is fixed by cyanobacteria, is also depleted in ^{15}N (Wada, 1980; Minagawa & Wada, 1986).

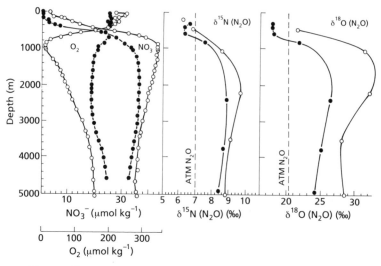

Fig. 6.3 $\delta^{15}N$ and $\delta^{18}O$ profiles in N_2O in two Pacific Oceans Stations (Stations 204 and 314). The NO_3^- and dissolved O_2 profiles from the same locations are shown in the left panel. It should be noted that values of $\delta^{18}O$ in this figure are expressed relative to that of atmospheric O_2 ($\delta^{18}O_{ATM}$). Values of $\delta^{18}O$ relative to standard mean of ocean water (SMOW) can be obtained from the equation $\delta^{18}O_{SMOW} = (\delta^{18}O_{ATM} + 23.0) \times 1.0235$. $\bigcirc$ = station 204 (North Pacific); $\bullet$ = station 314 (South Pacific). (From Kim & Craig, 1990.)

Oxic subsurface and deep waters

Similar depletions would be seen for N_2O in deep waters if it were formed through nitrification that depends on the release of NH_4^+ from PON by deamination. Kim & Craig (1990), however, found that N_2O was enriched in ^{15}N and ^{18}O in deep and bottom waters. They also noticed that both heavy-isotope enrichments increase and $[O_2]$ decreases from the South to the North Pacific. These enrichments may be due to microbial reduction of N_2O produced by nitrifiers, or they may reflect a prior enrichment of ^{15}N and ^{18}O in an intermediate compound like NH_2OH, from which the light isotope is preferentially channeled into NO_2^- and NO_3^- during nitrification (Kim & Craig, 1990).

Suboxic and O_2-minimum layer waters

Distinctively high values of $\delta^{15}N$ in N_2O (average $\delta^{15}N = 12.6 \pm 1.1\%$) were found in the waters below $20\,\mu mol\,l^{-1}$ O_2 in the central North Pacific, suggesting that the enrichment by denitrification is taking place in low-O_2 water (Yoshida et al., 1984).

Based on results of the distribution of $\delta^{15}N$ of N_2O and NO_3^- in the water column in the Pacific Ocean, and of the estimate for the rates of

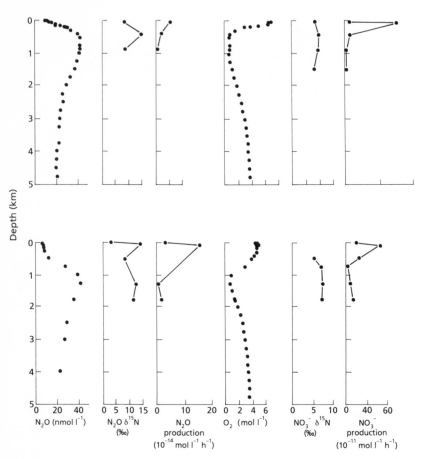

Fig. 6.4 The vertical profiles at Station C (top row) and Station E (bottom row) (from left to right) N_2O, $\delta^{15}N$ of N_2O, the potential production rate of N_2O by nitrification, dissolved O_2, $\delta^{15}N$ of NO_3^- and potential production rate of NO_3^- by nitrification. (From Yoshida *et al.*, 1989.)

N_2O production in the ocean from incubation experiments using tracer techniques (Fig. 6.4), Yoshida *et al.* (1984, 1989) concluded that N_2O production in the ocean is derived from denitrification rather than from nitrification. The main basis for their argument is that the N_2O yield ($\Delta N_2O/NO_3^-$) is very low ($0.04-0.19 \times 10^{-3}$ at Station C and $0.08-0.27 \times 10^{-3}$ at Station E), compared with the range estimated in the ocean ($0.1-0.4 \times 10^{-3}$; Yoshinari, 1976; Cohen & Gordon, 1978; Elkins *et al.*, 1978). However, as Kim & Craig (1990) have pointed out, their rates of nitrification ($V_{(NH_4^+ \rightarrow NO_3^-)}$) determined from incubations as $4.8 \, \text{mmol m}^{-2} \text{day}^{-1}$, could have been much higher than *in situ* rates. The substrate level of $1 \, \mu\text{mol l}^{-1}$ $^{15}NH_4^+$ used in their incubation experiment was rather high when compared with a half-saturation constant (K_s) for

$[NH_4^+]$ in a natural population of ammonia oxidizers $[K_s < 0.07\,\mu mol\,l^{-1}$ (Olson, 1981); c. $0.15\,\mu mol\,l^{-1}$ (Hashimoto *et al.*, 1983)]. By adding $0.2\,\mu mol\,l^{-1}$ $^{15}NH_4^+$ as a tracer, the rate of nitrification in the oxygen minimum layer of the eastern tropical North Pacific was estimated to be $1.5\,mmol\,m^{-2}\,day^{-1}$ (Ward & Zafiriou, 1988). Assuming that a maximum nitrification rate is $1.4–2 \times 10^{-12}\,mol\,NH_4^+\,cell^{-1}\,day^{-1}$, Karl *et al.* (1984) derived a potential rate of $V_{(NH_4^+ \to NO_3^-)}$ as $0.16–0.23\,mmol\,m^{-2}\,day^{-1}$. Based on the above results and on the fact that the rate of nitrification is correlated to substrate concentration, especially between 0 and $0.1\,\mu mol\,l^{-1}\,NH_4^+$ (Hashimoto *et al.*, 1983), it is possible that the potential rate of $V_{(NH_4^+ \to NO_3^-)}$ derived by Yoshida *et al.* (1989) deviates considerably from the *in situ* rate. Also, Yoshida *et al.* (1989) made no distinction between the N_2O yield calculated from direct measurements of the rates of nitrate and N_2O production $[V_{(NH_4^+ \to NO_3^-)}$ and $V_{(NH_4^+ \to N_2O)}]$, and the reported values based on $\Delta N_2O/NO_3^-$ (which is derived from apparent N_2O production and the concentration of NO_3^- observed in the water). Based on their premise that high or low ΔN_2O proportion (defined as the percentage of nitrate produced) corresponds to a younger or older water mass, Oudot *et al.* (1990) derived an average N_2O production rate of $0.027\,\mu l\,l^{-1}\,yr^{-1}$ in the convergence zone of the tropical Atlantic Ocean. The maximum rates of N_2O production (0.011 and $0.037\,\mu l\,l^{-1}\,yr^{-1}$ at Stations C and E, respectively; Fig. 6.4) calculated by Yoshida *et al.* (1989) are in the same range as rates computed by Oudot *et al.* (1990), suggesting the rate determined by using the ^{15}N-tracer technique is close to the *in situ* rate. Based on the above points, the hypothesis by Yoshida *et al.* (1989) that N_2O in the ocean is formed mainly by denitrification does not seem to be fully warranted.

In contrast, Kim & Craig (1990) contend that the source of heavy N_2O in the deep waters of the Pacific Ocean (Fig. 6.3) was nitrification rather than denitrification. They suggested that NH_2OH, an intermediate during the oxidation of ammonia by nitrifiers, serves both as a substrate to form N_2O (presumably through NH_2OH oxidase) and as precursor to NO_2^- and N_2O that is formed from NO_2^- by the denitrifying enzyme. They speculated on the possibility that the reaction $NH_2OH \to NO_2^-$ is faster than the reaction $NO_2^- \to N_2O$, thus leaving the heavy fraction of NH_2OH behind. If N_2O is produced from this NH_2OH, then it will likely be isotopically enriched. However, the production of N_2O (Falcone *et al.*, 1963), and N_2O and NO (Anderson, 1964) from the oxidation of NH_2OH, presumably via HNO, has so far only been confirmed from *in vitro* experiments with the extracts of *Nitrosomonas europaea*. The observation that NH_2OH present in water is positively correlated with the rate of ammonia oxidation (Butler *et al.*, 1987) suggests that NH_2OH is closely associated with nitrification. However, the production of N_2O by the mechanism proposed by Kim & Craig (1990) does not agree with the findings that N_2O produced by nitrifiers arises mostly from the reduction of nitrite rather than from the decomposition of an unstable intermediate during the oxidation of NH_2OH to NO_2^- (Poth & Focht, 1985; Remde & Conrad, 1990).

In view of these recent reports on N_2O isotopes, Naqvi (1991) proposed a possible mechanism for NO as a key intermediate in nitrification–denitrification coupling by which heavy N_2O may be derived. Because an increased production of N_2O in low-O_2 waters is difficult to explain by the mechanism offered by Kim & Craig (1990), Naqvi suggested that NO has a significant role as a precursor of N_2O ($NH_4^+ \rightarrow NH_2OH \rightarrow NO \rightarrow N_2O$), assuming that NO is a major product from the oxidation of NH_2OH during nitrification at low O_2. This assumption is based, in part, on the findings of Ward and Zafiriou (1988) that the integrated flux of NO in the water column was c. 13% of the nitrification flux, and that the maxima in NO concentration and NO turnover rate are located at about the same depths as those of N_2O. Naqvi also pointed out that NO produced during nitrification (Yoshida, 1988) has been enriched by as much as 20‰ relative to NH_4^+. The proposed scheme emphasizes a reduction step of NO $\rightarrow$ N_2O rather than that of $NO_2^- \rightarrow N_2O$ since N_2O produced by the latter reaction is likely to be depleted in ^{15}N. In either case, if NH_2OH or NO is assumed to be a precursor of N_2O, a substantial portion of these intermediates must undergo further oxidation, leaving behind heavy nitrogen that will be used for the production of N_2O.

6.6.3 Future research directions on the isotopic composition of N_2O
Owing to its great potential for providing important information that cannot be obtained by conventional means, the natural variation of the isotopic composition of N_2O in the marine environment can be used (i) to analyse long-term processes, (ii) to determine major sources and sinks and (iii) to obtain information on the relative significance of nitrification and denitrification. A few specific areas of study that may be of interest for further investigation are outlined below.

Determining the rate of N_2O production
Both measurements of natural abundance signatures and applications of tracer techniques (Yoshida *et al.*, 1989) will be useful in obtaining quantitative information on the rate of N_2O production and the rate of flux of gaseous nitrogen. The possibility that the low N_2O yield observed by Yoshida *et al.* (1989) is due to the production of N_2O through the nitrification–denitrification couple can also be tested. If a large proportion of NO_2^- produced within the O_2-limiting matrix is reduced to N_2O without further oxidation to NO_3^-, then the isotopic composition of N_2O should be correlated to the relative rates at which primary $NH_4^+ \rightarrow NO_2^-$ and secondary $NO_2^- \rightarrow NO_3^-$ nitrification occurs within particulates. In

this case, the $\delta^{15}N$ of N_2O will be equal to or less than that of PON or nitrite, and $\delta^{15}N$ values will not be larger than that of PON. However, if the simultaneous reduction of N_2O to N_2 takes place within the particulate matrix such that depletion of N_2O does not occur before its escape to the surrounding water, the remaining N_2O will be isotopically enriched.

Mechanisms of N₂O formation

A close examination of the pathways of N_2O production from NH_2OH or NO may be interesting for the following reasons. As oxygen exchange between NO_2^- and H_2O is very rapid in ammonium oxidizing bacterium (Andersson *et al.*, 1982), $\delta^{18}O$ of N_2O derived from NO_2^- reduction is likely to be influenced by a $\delta^{18}O$ similar to the surrounding water. In contrast, since NH_2OH-O is derived from dissolved oxygen (Hollocher *et al.*, 1981), the $\delta^{18}O$ of NO that is the product of NH_2OH oxidation is likely to be similar to that of dissolved oxygen. As the $\delta^{18}O$ of dissolved O_2 at 25% saturation or less in seawater is c. 35‰ (Kroopnick & Craig, 1976), the $\delta^{18}O$ of N_2O of NO origin in deep waters that are undersaturated with O_2 would be significantly larger than that from the reduction of NO_2^-. Thus, the values of $\delta^{18}O$ in N_2O may carry important information regarding the source of N_2O (i.e., from NH_2OH, NO or NO_2^-). Simultaneous determination of $\delta^{18}O$ of O_2 and H_2O from *in situ* incubation experiments with or without the addition of labelled $H_2^{18}O$ should be done to obtain critical information.

Nitrification–denitrification coupling

For the production of N_2O during nitrification–denitrification coupling, the following reaction sequence may be worthwhile examining in productive regions:

$$NO_3^- \rightarrow PON \rightarrow NH_4^+ \rightarrow NO_2^- \rightarrow N_2O \rightarrow N_2.$$

At the top of the thermocline in the Guinea Dome area, a zone where nitrate in the euphotic layer is elevated and thus primary productivity enhanced, Oudot *et al.* (1990) found that $\Delta\dot{N}_2O$ was not correlated with the production of NO_3^-. Furthermore, they observed a small amount of N_2 production in the same water that contained 160–$200\,\mu mol\,l^{-1}$ of dissolved oxygen. While denitrification is generally considered to occur at $[O_2] \leqslant 4\,\mu mol\,l^{-1}$ (Hattori, 1983), their observation can be explained by nitrification–denitrification coupling occurring within POM, according to the sequence shown above, rather than the result by phytoplankton assimilatory nitrate reduction. The anomalies observed by Oudot *et al.* (1990) may not be uncommon in regions of upwelling where significant amounts of large diameter POM is formed through high productivity, since the inner core of a large-diameter particle can be easily depleted in oxygen (Kaplan & Wofsy, 1985).

Heavy N_2O in deep water
The origin of heavy N_2O in deep waters may be mostly from *in situ* production. However, since water with a high concentration of N_2O appears to be transported from the edges of the low-oxygen 'active zone' (Oudot *et al.*, 1990; Naqvi & Noronha, 1991), the isotopically heavy N_2O in deep water could well originate from the mixture of N_2O produced in the active zone with low $[O_2]$ and N_2O produced in the surrounding environment with slightly higher $[O_2]$. Thus it appears worthwhile to examine the possibility that some of heavy N_2O in deep water is laterally transported from such a source.

6.7 Studies of gaseous nitrogen metabolism by using ^{15}N-labelled substrates

6.7.1 Rate measurements of biological nitrogen fixation
In the marine environment, various bacteria of different trophic categories, such as phototrophs, heterotrophs and chemolithotrophs, have been found to possess the nitrogenase system for nitrogen fixation. The distribution of those micro-organisms extends from shallow water to deep sea sediments (Howarth *et al.*, 1988a). Since the combined nitrogen input to the upper ocean through biological nitrogen fixation is a form of 'new nitrogen', defined as the amount of nitrogen available for export from the system at steady state (Dugdale & Goering, 1967), reliable measurements of nitrogen fixation in various marine environments are the key components in understanding biogeochemical cycling. Previous estimates of nitrogen fixation in oceanic waters based on the marine cyanobacteria *Trichodesmium* indicated that the contribution by nitrogen fixation to the total nitrogen supply was quite minor compared with the total nitrogen demands of primary producers in the whole ocean (Carpenter, 1983). However, several lines of recent evidence suggest that nitrogen fixation may be a significant source of nitrogen in nitrogen-depleted tropical and subtropical oceanic waters (Legendre & Gosselin, 1989). The previous estimates also neglected the importance of benthic nitrogen fixation in shallow tropical marine environments, such as coral reefs and seagrass beds (Capone, 1983, 1988; Corredor & Capone, 1985; Moriarty & O'Donohue, 1993). For example, bacterial nitrogen fixation in tropical seagrass beds was estimated to support a significant portion of seagrass nitrogen demands (Capone & Budin, 1982; Moriarty & O'Donohue, 1993).

For the quantitative measurement of biological nitrogen fixation in

Table 6.2 Summary of methods for the quantitative determination of N_2 fixation. (From Capone, 1988)

Method	Comment	Relative sensitivity	References
Increase in total nitrogen (Kjeldahl)	Imprecise	1	Burns & Hardy (1975)
Manometry	Imprecise		Burns & Hardy (1975)
$^{15}N_2$ reduction	Precise, tedious	10^3	Burris (1974)
C_2H_2 reduction	Simple, precise	10^6	Burris (1974); Burns & Hardy (1975)
$^{13}N_2$ reduction	Direct, exotic	$10^{10}-10^{12}$	Cooper et al. (1985)
^{15}N fertilizer isotope dilution	Untested, marine environment	?	Boddey et al. (1985)
Membrane-leak mass spectrometry		?	Jensen & Cox (1983)
Natural abundance ^{15}N		?	Amarger et al. (1979)

water or sediment, several methods have been proposed (see Capone, 1988; Table 6.2). Among these methods, incorporation of ^{15}N-tagged dinitrogen into cellular organic nitrogen is a straightforward and precise method (Burris & Wilson, 1957). Laboratory applications of tracer ^{15}N to study physiological and biochemical aspects of nitrogen fixation were begun quite early (Burris et al., 1943). Another sensitive method involves the measurement of the rate of acetylene reduction by nitrogen fixing organisms (Stewart et al., 1967). Use of the acetylene reduction technique for the field assessment of nitrogen fixation quickly became very popular because of the simple analytical procedure and easy access to gas chromatographs with a flame ionization detector (FID), as compared with the mass spectrometer required for ^{15}N analysis. This method is based on the low specificity of the biological nitrogenase system, which can reduce various chemical analogues including acetylene (Stewart et al., 1967). Although a theoretical ratio of acetylene reduced per nitrogen fixed ($3:1\,\text{mol}^{-1}$) can be obtained from the microbial nitrogen fixation reaction, calibration of this ratio against a more direct technique, i.e., the ^{15}N tracer method, is commonly required, but not always conducted, in natural systems.

Tracer ^{15}N methodology for the assessment of nitrogen fixation in the marine environment was used during a rather short period of the 1960s to survey the capacity and ecological significance of nitrogen fixation by

Trichodesmium and other autotrophic nitrogen-fixing communities in tropical and subtropical pelagic waters, such as the Sargasso Sea and Arabian Sea (Dugdale *et al.*, 1961, 1964; Goering *et al.*, 1966). This technique was also used to assess nitrogen fixation by biological communities associated with seagrass beds and other shallow water environments (Stewart, 1965, 1967; Goering & Parker, 1972; Patriquin & Knowles, 1972). Shortly after the acetylene reduction methodology was developed, the ^{15}N tracer procedure was mostly used for the calibration of this new technique (Patriquin and Knowles, 1972; Mague *et al.*, 1974, 1977; Carpenter *et al.*, 1978; Seitzinger & Garber, 1987; Larkum *et al.*, 1988). Although the ^{15}N tracer technique for nitrogen fixation is a tedious and expensive method, the advantage of such a technique includes the tracing of the internal transfer of root-fixed ^{15}N nitrogen into the whole-plant system of seagrass (Capone, 1988). Recent comprehensive reviews of nitrogen fixation in marine environments, including ^{15}N techniques, are Howarth *et al.* (1988a,b) and Capone (1988).

The experimental protocol of tracer ^{15}N methodology for nitrogen fixation in the water column is quite simple and straightforward:

1 a seawater sample containing the desired organisms is placed in a glass flask with a gas-exchange unit;

2 the air in the flask is replaced with a mixture of ^{15}N dinitrogen gas (c. 50–99% ^{15}N) and other gases depending on the experimental purpose;

3 the sample is incubated under specified conditions;

4 the organic nitrogen in the sample is converted to dinitrogen and δ^{15}N of the dinitrogen is measured with a mass spectrometer or a ^{15}N emission spectrometer (Fielder & Proksch, 1975).

For the measurement of nitrogen fixation by planktonic organisms such as *Trichodesmium*, collection of the samples on a glass-fibre filter and conversion to nitrogen gas by the Dumas method can minimize the dilution of ^{15}N enrichment in the sample (Dugdale *et al.*, 1964). However, separation of epiphytes from seagrass is not easy, resulting in significant dilution of enriched ^{15}N in epiphytes (Patriquin & Knowles, 1972). The assessment of nitrogen fixation by microbes in very organic-rich sediments using the ^{15}N addition method is difficult, as the ^{15}N signal is diluted by the nitrogen-rich organic matter. Recently Seitzinger & Garber (1987) separated exchangeable ammonium and labile organic nitrogen from other sedimentary nitrogen after incubating coastal anaerobic sediments with ^{15}N dinitrogen to improve the detection of ^{15}N enrichment.

The long incubation time necessary to measure nitrogen fixation by the ^{15}N addition method is another problem for its *in situ* assessment. With the exception of *Trichodesmium* experiments in tropical waters (length of incubation, 3–24 hours; Dugdale *et al.*, 1964; Goering *et al.*, 1966), more than 1 day of incubation (seagrass and sediments) has been necessary to obtain significant ^{15}N enrichment (Patriquin & Knowles, 1972; Seitzinger & Garber, 1987). The rates obtained from prolonged incubation under rather artificial conditions are regarded at best as a potential activity.

The results of ^{15}N addition and acetylene reduction intercalibration trials for nitrogen fixation measurements are useful for gaining insights into the mechanisms of nitrogen fixation in marine environments. Two different sources of deviation from theoretical values have been recognized, i.e., that originating from the physiological condition of individual microbes having the nitrogenase system, and that from the effect of acetylene on various microbial communities in the assay system. The former includes nitrogen starvation of nitrogen-fixing organisms during the assay and the difference in hydrogen metabolism under nitrogen-fixing and acetylene-reducing conditions (Schubert & Evans, 1976; David & Fay, 1977). The variation originating from the physiological condition is rather small compared with that originating from the effect of acetylene. Using the ^{15}N addition method as a control, the reported conversion factor (acetylene : nitrogen gas) for planktonic blue green algae (*Oscillatoria*) was 6.3 (Carpenter & Price, 1977), and 1.9 (Mague *et al.*, 1977), while values for benthic algae ranged from 3 to 7 (Carpenter *et al.*, 1978; Potts *et al.*, 1978). On the other hand, the conversion factor is quite variable for more heterogeneous marine samples. The conversion factor ranged from 3.3 to 56 when nitrogen fixation of *Richelia interacellularis* occurring within the cells of a diatom (*Rhizosolenia*) was measured in the north Pacific Ocean (Mague *et al.*, 1974). A similar large variation of the factor (12–94) was reported for nitrogen fixation by anaerobic microbial communities in subtidal sediments (Seitzinger & Garber, 1987). The large deviation from the theoretical ratio in sedimentary nitrogen fixation might be attributed to the combination of several factors including the effect of prolonged incubation (48 hours to 26 days) and endogenous production of ethylene by other microbes. Using sediments consisting of the rhizosphere of seagrasses, Patriquin & Knowles (1972) also obtained a large variation but a much smaller conversion ratio (1–16), suggesting complicated factors controlling the ratio in the various sediments.

6.7.2 Rate measurements of denitrification

Application of enriched ^{15}N nitrate/nitrite to enclosed sediment subsamples from many marine systems and tracing the production of gaseous ^{15}N nitrogen is a sensitive and direct method to evaluate the occurrence of denitrification in the marine environment. However, *in situ* rate measurements of denitrification require a homogeneous distribution of the ^{15}N substrate, such as nitrate, around the denitrifying bacterial communities, which must be achieved without disturbing their micro-environments.

For denitrifying bacteria in the water column, disturbance of micro-environmental conditions through ^{15}N addition can be minimized (Koike *et al.*, 1972; Shaffer & Ronner, 1984; Ronner & Sorensson, 1985).

A typical denitrification measurement in the water column can be conducted as follows:

1 a water sample is collected without exposure to air and transferred to a vacuum-tight bottle;

2 the bottle is sealed with a rubber stopper;

3 a small volume of ^{15}N-enriched substrate is added through the rubber stopper using a hypodermic syringe;

4 incubation is begun under simulated *in situ* conditions after gentle shaking of the bottle to achieve a homogeneous distribution of ^{15}N-enriched substrate (Koike & Hattori, 1978b).

After the incubation, the denitrifying activities are terminated by addition of $HgCl_2$; dissolved ^{15}N dinitrogen in the bottle is extracted by connecting the bottle to a vacuum gas sampler. After purification of the extracted dissolved gases in a vacuum line by passing them through a CuO and Cu furnace and liquid nitrogen trap in sequence, dinitrogen is introduced into a mass spectrometer (Koike *et al.*, 1972). Mass peaks of 28, 29 and 30 are used for calculation of the denitrification rate following the method of Hauck *et al.* (1958).

The headspace technique is another common method for collecting ^{15}N nitrogen gas after incubation of water or sediment samples (Goeyens *et al.*, 1987), Because of the large amount of dissolved dinitrogen in water (c. $400\,\mu mol\,kg^{-1}$ in 35 ppt seawater at 20°C), dilution of denitrified ^{15}N dinitrogen by dissolved dinitrogen is critical in the detection of ^{15}N enrichment in dinitrogen by mass spectrometry. If we assume that a high-precision mass spectrometer can detect 0.0002 atom% difference in ^{15}N, then c. 4 nmol of N_2 derived from denitrification per litre of seawater can be detected. On the other hand, a small sample of seawater is required since dinitrogen is plentiful, i.e., c. 10 ml is enough to measure the ^{15}N content of dinitrogen using an isotope ratio mass spectrometer (Nishio *et al.*, 1982). Addition of ^{15}N nitrate/nitrite inevitably increases the concentration of substrate in the sample seawater or pore water of sediments, which can be corrected by measuring the correlation between substrate concentration and denitrification rates (Oren & Blackburn, 1979).

Except for slurry type incubations, homogeneous distribution of the ^{15}N-enriched substrate around the denitrification sites in sediments is difficult to achieve. In addition, the micro-environments in the original sediment could be destroyed during preparation of the homogenized slurry. Thus the measured denitrifying activity may be regarded at best as a potential denitrification rate (Koike & Hattori, 1978a). To avoid disturbance of three-dimensional microstructure of sediments, ^{15}N-tagged nitrate was continuously passed over the sediment and measurements of ^{15}N nitrogen gas produced at the boundary layer of the sediment–water interface was used as the *in situ* rate of denitrification (Nishio *et al.*, 1982, 1983). The assumptions here are that the top layer of the sediment is the dominant site of denitrification, and that nitrate/nitrite in the overlying water is the major substrate for denitrification in the sediments. Those assumptions can be applied to organic-rich sediments with high-nutrient overlying water (Nishio *et al.*, 1982, 1983). Injection of ^{15}N-enriched substrates into the undisturbed sediment core, which was originally developed by Sørensen (1978a) for the acetylene block technique for the measurement of denitrification, is another method to evaluate *in situ* sedimentary denitrification, but has been used only for terrestrial soil systems (Christensen *et al.*, 1990).

Currently, acetylene block methods are more widely used than the ^{15}N tracer method for the evaluation of denitrification in marine environments (Seitzinger, 1988). The method is based on the observation that presence of acetylene inhibits nitrous oxide reduction to dinitrogen by the denitrifying enzyme (Balderston *et al.*, 1976; Yoshinari & Knowles, 1976). Measurement of N_2O by a gas chromatograph equipped with an electron-capture detector (ECD) is quite sensitive because of the low ambient concentration of nitrous oxide. Details of the methodological aspects of the acetylene block technique are given by Knowles (1990).

Another ^{15}N tracer application for denitrification studies is the determination of various non-gaseous ^{15}N components after incubation with ^{15}N-tagged substrates such as ammonium. Evaluation of denitrification from the mass-balance approach has been successful in terrestrial soils, since the gaseous nitrogen loss from certain soil systems, especially after the addition of inorganic nitrogen fertilizer, is significant to the total nitrogen budget (Myrold & Tiedje, 1986; Myrold, 1990). Because of the accumulation of analytical error in each nitrogen pool measurement and the calculation of denitrification by the difference of total nitrogen pools with time, denitrification measurements by this approach are far less accurate than that by direct measurement of ^{15}N dinitrogen production. Rates of denitrification in most marine environments are too low for this mass-balance approach (Seitzinger, 1988; Koike & Sørensen, 1988), and thus it has been used exclusively for sedimentary denitrification studies (Jenkins & Kemp, 1984; Sumi & Koike, 1990). However, one obvious advantage of this method is that an enclosed system is not necessary. Another advantage is that the metabolic fate of added ^{15}N substrate can be determined when this approach is combined with a suitable model. Recently this approach was attempted by Sumi & Koike (1990) using a laboratory incubation of aerobic coastal surface sediments after the addition of ^{15}N-NH_4^+. Simultaneous evaluation of five different processes of nitrogen cycling in surface sediments, i.e., denitrification, nitrate reduction to ammonium, nitrification, ammonium assimilation and organic nitrogen mineralization to ammonium, can be assessed by the combination of the ^{15}N-NO_3^- and ^{15}N-NH_4^+ isotope dilution model. For example, dilution of ^{15}N in nitrate is induced by the production of nitrate having a natural ^{15}N abundance through the nitrification process, while the total nitrate pool is reduced by nitrate reduction to ammonium and/or denitrification (Koike & Hattori, 1978b). The total denitrification rate can be evaluated from the mass balance of non-gaseous ^{15}N during the incubation and the rate of nitrate reduction to ammonium, as the difference between the total nitrate/nitrite reduction and denitrification (Sumi & Koike, 1990).

^{15}N-tagged substrates are now commonly used in the laboratory for studies of the physiology and biochemistry of nitrous oxide metabolism, especially in the identification of the pathway of microbial nitrous oxide production (Ritchie & Nicholas, 1972, 1974; Garber & Hollocher, 1982). In field studies, however, accurate measurement of ^{15}N nitrous oxide is not easy because of the very low concentration of nitrous oxide in seawater.

There has been only one report using the ^{15}N tracer method to measure nitrous oxide production in the open ocean (Yoshida *et al.*, 1989). Using 50 l samples, 50–200 nmol nitrous oxide was collected by vacuum extraction after seawater incubation with ^{15}N-NH_4^+. Addition of a nitrous oxide carrier (1 μmol) to the extracted nitrous oxide was necessary to obtain enough nitrogen for measurement by isotope ratio mass spectrometry, yet Yoshida *et al.* (1989) measured 5×10^{-14} mol l^1 h^{-1} of nitrous oxide production in the upper layer of western North Pacific.

6.8 Conclusions

The measurement of natural abundance stable nitrogen isotope ratios in N_2O is proving to be a promising tool to gain insight into aspects of nitrogen cycling in marine ecosystems. Data on both nitrogen and oxygen isotopic signatures in marine N_2O will provide more information than single isotope ratios for studies of the relative significance of nitrification and denitrification on the oceanic source strength for atmospheric N_2O.

Tracer techniques using ^{15}N-labelled substrates have been used to study the mechanism as well as the kinetics of gaseous nitrogen formation and consumption. These techniques are especially useful for studies at sites where the metabolism of nitrogen is very active (e.g., coastal water and sediment systems), and for analysis of systems in which multiple biological reactions are simultaneously taking place. However, because ^{15}N tracer methodology involves a rather complicated analytical procedure, this method should be used mainly for key point calibration when an indirect but easy method, such as the acetylene block technique for denitrification or acetylene reduction technique for nitrogen fixation, is adopted for routine measurement.

Acknowledgement

T.Y. was supported in part by the NSF grant OCE 8911425 for writing this chapter.

CHAPTER 7

Stable isotope ratios as tracers in marine aquatic food webs

R.H. MICHENER & D.M. SCHELL

7.1 Introduction

Scientists concerned with organic matter flow and food web structures in freshwater and marine ecosystems are increasingly realizing the potential of stable isotope ratios as natural tracers. Isotope ratios present a powerful tool to look at the processes, connections and energy flow within aquatic systems. This chapter examines and summarizes some of the research that has been done on marine food webs. We discuss how natural abundance isotope ratios vary within various marine ecosystems and how stable isotopes have been used in feeding experiments using organisms in the laboratory and the field. We will then summarize studies using stable isotope ratios in marine pelagic food webs, in estuarine and nearshore systems, and finally in salt marsh systems.

7.2 Traditional approaches to food web research

Standard approaches to food web analysis include gut contents analysis, direct observation both in the field and laboratory, and radiotracer techniques (Smith *et al.*, 1979; Beviss-Challinor & Field, 1982; Rounick & Winterbourn, 1986; Hopkins, 1987; Sondergaard *et al.*, 1988; Warren, 1989; Kioboe *et al.*, 1990). These methods have been shown to adequately resolve food web structure, yet each has its drawbacks. Analysis of gut contents involves collecting and dissecting a broad range of organisms to determine food web structure, and requires few tools and equipment. However, some organisms digest their prey quickly, making identification difficult, and gut contents may reflect material which is not assimilated. Also, some prey tend to lose their morphological characteristics more quickly than others (Feller *et al.*, 1979). It is also a somewhat tedious technique and it requires the researcher to have a good taxonomic knowledge of organisms found in the system.

Feeding relationships can also be studied in the laboratory, but obtaining an adequate sampling of prey items may be difficult. The laboratory provides an artificial system and may introduce an artifact into the results (Ockelmann & Vahl, 1970; Feller *et al.*, 1979). Other laboratory methods include food or prey exclusion or inclusion, and monitoring predator/prey abundances with time (Gerlach *et al.*, 1976; Arntz, 1977).

The use of radiolabelling in food web analysis involves adding a radiotracer (for example, ^{14}C or ^{3}H) to a food source or prey species and following the label through the food chain (Marples, 1966; Smith *et al.*, 1979; Beviss-Challinor & Field, 1982; Pearcy & Stuiver, 1983; Hessen *et al.*, 1990). Smith *et al.* (1979) used this technique to partially characterize a coral reef and subtropical estuary involving demersal zooplankton. But again, radiolabelling has its drawbacks: recovery of a statistically significant number of labelled species can be difficult, the need for a license to use radioactive isotopes, and the need to use high dosages of isotope to overcome dilution, especially in the case of tritium.

Another technique for aquatic food web studies is the use of immunological methods (Boreham & Ohiagu, 1978; Feller *et al.*, 1979, 1985). It involves developing antisera from whole organism extracts; double immunodiffusion precipitin tests of antiserum specificity are then done. It has been shown that the antisera are usually taxon-specific and can trace trophic relationships. This has an appeal for organisms whose gut contents cannot be identified. Feller *et al.* (1985) used this technique to look at deep sea food web structure, where changes in water pressure frequently deform organisms and make gut contents difficult to identify. However it is limited to the specificities and number of antisera developed and is purely qualitative. For systems with a large number of species, it would be prohibitively expensive and time-consuming to check all possible antisera. For a review of the technique see Boreham & Ohiagu (1978).

Stable isotope analysis has more recently been used as an alternative, and in some cases, better tool for food web analysis. The collection technique is simple and straightforward, and the analysis is becoming relatively inexpensive as automation becomes more prevalent. With the development of semi-automated systems, a broad survey of organisms in a food web system can be performed with a reasonable amount of time and funds. Further details on theory, methods and collection strategies can be found in the Introduction. To interpret the data, we need to next look at how carbon, nitrogen, hydrogen and sulfur stable isotopes circulate within biological systems.

7.3 Stable isotopes and diet

7.3.1 Carbon

Carbon isotopic compositions of animals reflect those of the diet within about 1‰ (Haines, 1976a; DeNiro & Epstein, 1978; Fry *et al.*, 1978a; Haines & Montague, 1979; Teeri & Schoeller, 1979; Rau, 1980; Rau & Anderson, 1981; Fry & Arnold, 1982; Tieszen *et al.*, 1983; Checkley & Entzeroth, 1985; Peterson & Fry, 1987; Fig. 7.1). Overall, there appears

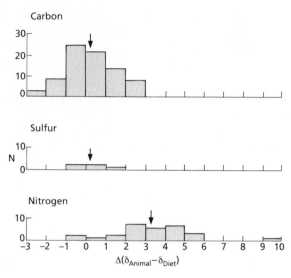

Fig. 7.1 Relationship between organisms and diet for carbon, sulfur and nitrogen isotopes. (From Peterson & Fry, 1987.)

to be a slight (0.5–1‰) enrichment in the animal relative to its diet. There are several possible processes which might contribute to this enrichment: (i) preferential loss of $^{12}CO_2$ during respiration; (ii) preferential uptake of ^{13}C compounds during digestion; or (iii) metabolic fractionation during synthesis of different tissue types (DeNiro & Epstein, 1978; Rau et al., 1983; Tieszen et al., 1983; Fry et al., 1984). There is evidence for each case, but no general consensus. DeNiro & Epstein (1978) studied the grasshopper *Melanoplus* on a wheat diet and found that respired CO_2 was depleted, and whole-body composition and faeces were enriched, relative to the diet. Mass balance showed an approximate 1‰ enrichment. Stephenson et al. (1986) studied lobsters and oysters in a laboratory setting and found a significant enrichment of the animals versus their diet. Their explanation was a selective assimilation of compounds from the diet. They also found an inverse relationship between the calorific value of the diet versus the (δ animal–δ diet) of the lobster tissue. Lobsters fed a high-calorie, high-fat diet will produce tissues with a higher lipid content and lighter $^{13}C/^{12}C$ ratios. This agrees with studies showing that lipids are isotopically lighter than other biochemical fractions (Parker, 1964; DeNiro & Epstein, 1977).

This conservative transfer of carbon isotopic compositions (<1‰) to the animal from the diet can be useful in tracing food webs in systems where there are food sources with large differences in $\delta^{13}C$ values, such as C_3 vs. C_4 plants or marine versus terrestrial systems (Haines, 1976b; Fry *et al.*, 1977, 1978a; DeNiro & Epstein, 1978; Rau, 1981; Schoeninger & DeNiro, 1984; see Chapter 4). However, the researcher must also be aware of isotopic variations in different tissues within an organism, as well as the different rates of tissue turnover when an organism is selectively feeding. This can have implications for estuarine systems where there is more than one food source.

In a feeding study of gerbils, Tieszen *et al.* (1983) switched the diet from a C_4 corn to a C_3 wheat and analysed the major tissues for ^{13}C. They found that the ^{13}C enrichment for the individual tissues fell from hair > brain > muscle > liver > fat. The isotopic composition of tissues changed over time to reflect the new diet, with the more metabolically active tissues turning over more quickly. Other laboratory studies (Teeri & Schoeller, 1979; Rau & Anderson, 1981; Fry & Arnold, 1982; Macko *et al.*, 1982; R.H. Michener, unpublished data) have shown that organisms fed an isotopically distinct diet will approach the dietary value as the organism grows and tissue turns over.

Juvenile animals that migrate offshore from estuaries will tend to change their isotopic composition as they incorporate the new diet. In a field study, Fry (1983) measured the isotopic compositions of juvenile shrimp feeding in south Texas grass flats. Initially juveniles had isotopic values of $\delta^{13}C = -11$ to $-14‰$, and $6-8‰$ for nitrogen and sulfur. As the shrimp migrated offshore, the isotopic values converged toward offshore values of c. $-16‰$, $+11.5‰$ and $+16‰$ for carbon, nitrogen and sulfur.

Depending on tissue turnover, $\delta^{13}C$ values will be biased towards feeding patterns of the recent past. In ecosystem studies, unless the sampling protocol is thorough it is difficult to determine if the isotopic compositions of mobile animals reflect local feeding or food from other sources with different isotopic compositions (Fry & Arnold, 1982). Given these problems and different isotopic compositions of tissues, it would seem prudent to either sample the entire organism to get an integrated isotopic value, or to sample several tissue types covering a range of tissue turnover times. Sampling several individuals of a smaller species would also help eliminate variations within individuals, since animals of the same species fed on the same diet can vary up to $2‰$ (DeNiro & Epstein, 1978).

Despite the seeming difficulties, there are several excellent studies of food web systems using $\delta^{13}C$ measurements (Haines, 1976a; Black & Bender, 1976; Fry *et al.*, 1978a; Fry & Parker, 1979; Haines & Montague, 1979; Rau *et al.*, 1983). Each study was characterized by primary food sources that were isotopically distinct, and again, direct relationships were seen between the organisms and their diets. Rau *et al.* (1983) found roughly a 1.1‰ enrichment per trophic level; here one tissue, muscle, was analysed and organisms were chosen by their clearly identifiable trophic status.

7.3.2 Nitrogen

In contrast to carbon, less has been written about the relationship of $\delta^{15}N$ in the organism and diet (Gaebler et al., 1966; Steele & Daniel, 1978; Rau, 1981b; DeNiro & Epstein, 1981a; Macko et al., 1982; Checkley & Entzeroth, 1985; Peterson & Fry, 1987). As with carbon, DeNiro & Epstein (1981a) found that the $\delta^{15}N$ in the organism reflects the $\delta^{15}N$ of the diet, but in most cases the whole animal is enriched in ^{15}N relative to the diet. In laboratory and field studies of two marine amphipods, Macko et al. (1982) showed a -0.3‰ and $+2.3$‰ fractionation, regardless of the food source. DeNiro & Epstein (1981a) saw similar fractionations for two insect species.

When enrichment occurs, there has been found to be a preferential excretion of ^{15}N-depleted nitrogen, usually in the form of urea and ammonia (Minagawa & Wada, 1984). Differences in ^{15}N retention varies according to species, diet and nutritional stress in birds (Hobson, 1991; see Chapter 4). Isotopic analysis of cattle and their diet found that urine was depleted in ^{15}N relative to diet as well as to blood, faeces and milk (Steele & Daniel, 1978).

When the laboratory mouse *Mus musculus* was fed isotopically distinct diets, various tissues measured were enriched relative to the diet, with $\delta^{15}N$ increasing from kidney to hair to liver to brain (DeNiro & Epstein, 1981a). A similar pattern of enrichment from kidney to hair to liver was seen in a diet-switching experiment with *Mus* (R.H. Michener, unpublished data). Part of these differences in tissue $\delta^{15}N$ may be due to isotopic fractionation during amino acid transamination (Gaebler et al., 1966; Macko et al., 1986). As with carbon, analysis of several tissue types or whole organism $\delta^{15}N$ should be performed when comparing animal and diet $\delta^{15}N$.

Field studies show an average 3.2‰ enrichment in animal $\delta^{15}N$ versus diet (Fig. 7.1), which is reflected as a trophic level effect in food web studies. Minagawa & Wada (1984) found a ^{15}N enrichment of $+3.4 \pm 1.1$‰ per trophic level, independent of habitat. A survey of bone collagen by Schoeninger & DeNiro (1984) for 66 species of vertebrates resulted in an average 3‰ enrichment per trophic level. Hobson (1991) showed that dietary information can be acquired on both short- and long-term intervals in birds through the use of different tissue types, and suggested that nutritional stress was important and must be considered along with diet. In another study of arctic marine food webs, Hobson & Welch (1992) noted a trophic enrichment in $\delta^{15}N$. Rau (1981b), in a study of hydrothermal vent animals, also observed an increase in $\delta^{15}N$ as a function of the presumed trophic level.

7.3.3 Hydrogen and sulfur

There appears to be little or no enrichment in ^{34}S per trophic level or in animal versus diet (Fig. 7.1), although little or no work has been done using laboratory feeding studies (Mekhtiyeva et al., 1976; Peterson & Howarth, 1983). However, the isotopic difference between seawater sulfate and sulfides (c. 21‰ vs. c. −10‰) makes sulfur useful in distinguishing benthic versus pelagic producers and marsh plants versus phytoplankton in estuarine studies (Fry et al., 1982; Peterson & Howarth, 1987). Benthic systems and marsh plants tend to be richer in sulfur derived from sulfides, and thus reflect a lighter $\delta^{34}S$ signal.

Hydrogen also appears to show no enrichment of whole animal δD versus diet (Estep & Dabrowski, 1980; Macko et al., 1983). Although exchangeable hydrogen dilutes the signal between organic tissue and water (DeNiro & Epstein, 1981b), it appears that the δD of the diet is reflected in the animal. This may make hydrogen useful in tracing organisms feeding in freshwater versus marine systems and tracing terrestrial versus marine food webs (Fry & Sherr, 1984).

7.4 Phytoplankton and particulate organic carbon (POC)

In studying any food web, one must first look at the carbon isotopic composition among primary producers, since the isotopic label derived here will be transferred to successive trophic levels. Few measurements of phytoplankton have been done; typically researchers measure POC, assuming that the bulk of POC is, or is derived from, phytoplankton. This section covers carbon isotopes in phytoplankton and POC, and the biological and physical factors that will cause fractionation. More extensive reviews can be found in Fry & Sherr (1984) and in Chapters 9 and 10.

In natural oceanic populations of phytoplankton, there is a large range of isotopic values within and between geographical regions (Degens et al., 1968b; Deuser et al., 1968; Rau et al., 1983; Gearing et al., 1984; Rau et al., 1990). There are several factors that may contribute to this variability, which will be discussed in further detail below:

1 the isotopic composition of the dissolved inorganic carbon (DIC) pool may vary with temperature and location;

2 isotopic discrimination may be related to the morphology of the particular species of phytoplankton;

3 there may be isotopic discrimination by the carboxylating enzyme involved in CO_2 fixation;

4 growth rates of phytoplankton can affect their carbon isotopic composition.

Previous speculation regarding effects of temperature on phytoplankton isotopic composition (Sackett *et al.*, 1965; Eadie & Jeffrey,
1973) has been refined as the role of carbon dioxide pools in marine water
becomes evident. Although the DIC pool in the open ocean is fairly
uniform at 0‰ (Sackett & Moore, 1966; Sherr, 1982; Kroopnick, 1985),
estuarine environments with freshwater input will vary, since DIC values
in fresh water can range from -5 to -10‰. As most phytoplankton
obtain their carbon from the free carbon dioxide pool and this pool size
increases with decreasing temperature, temperature both directly and
indirectly contributes to the variation in $\delta^{13}C$.

Gearing *et al.* (1984) sampled different plankton sizes and found that
$\delta^{13}C$ varied with species and size, ranging from -20.3‰ for diatoms to
-22.2‰ for nanoplankton. Rau *et al.* (1990) also noted isotopic differences in both $\delta^{13}C$ and $\delta^{15}N$ with size in suspended particulate organic
matter (POM). These differences were thought to reflect microbial break-
down and fractionation of the POM, and a trophic-level effect in the
smallest size fraction. The smallest fractions were thought to consist of
nano- and picoplankton and may be a significant low trophic level component (Rau *et al.*, 1990). On the other hand, a study of size fractions of
POC in Martha's Vineyard Sound (Woods Hole, Massachusetts) found
little or no correlation of $\delta^{13}C$ versus size (S. Wainwright, personal
communication 1992). One possible explanation was a problem in obtaining pure samples of diatoms, since laboratory cultures have shown
size-related differences. Nonetheless, sampling protocols should be chosen
carefully and one must be cognizant that phytoplankton species size and
composition may affect $\delta^{13}C$.

Another variable affecting $\delta^{13}C$ in phytoplankton is the metabolic
pathway of photosynthesis. As discussed in Chapter 1, the two primary
pathways of photosynthesis utilize either RuBP carboxylase (isotopic discrimination -23 to -41‰) or PEP carboxylase (isotopic discrimination
-0.5 to -3.6‰). There is little evidence that phytoplankton have a C_4
pathway; however, they do concentrate HCO_3^-, which may eventually
result in C_4-like $\delta^{13}C$ values. Wong & Sackett (1978) found that marine
phytoplankton species will differ in their metabolic pathways, leading to a
range of $\Delta(\delta^{13}C$ algae vs. $HCO_3^-)$ of -22.1 to -35.5‰ for 17 species in
laboratory cultures. As they cautioned, this was a controlled situation and
optimal growth conditions were not achieved for all species. Within cells,
different metabolic fractions have been shown to have different $\delta^{13}C$
values. Cellulose fractions are more depleted in $\delta^{13}C$ than carbohydrates
(Degens *et al.*, 1968b), with lipids being the most depleted (Parker,
1964). Thus, variations in lipid content may account for some of the
observed isotopic differences.

Rau *et al.* (1982) synthesized the then-available information on latitudinal gradients from north to south in $\delta^{13}C$ of plankton and found a much steeper isotopic gradient in the southern hemisphere. Rau *et al.* (1990) later ascribed the poleward depletion in ^{13}C to the increasing size of the free carbon dioxide pool in seawater. This could not explain, however, the large differences between Arctic and Antarctic $\delta^{13}C$ values. Antarctic phytoplankton are also very light in ^{15}N relative to Arctic phytoplankton. This probably reflects slow growth rates in the higher concentrations of nutrients and carbon in the Antarctic. Ultimate control of $\delta^{13}C$ and $\delta^{15}N$ in the southern-most marine waters is most likely due to some environmental factor such as light or a trace element (such as iron) limitation.

The variations in $\delta^{13}C$ of POC are important when defining a pelagic food web system. For example, natural gradients among POC and phytoplankton $\delta^{13}C$ provide a powerful means of defining habitat usage and acquiring insight into the natural history of migratory fauna. Saupe *et al.* (1989) describe the gradient in $\delta^{13}C$ of zooplankton across the Alaskan Beaufort, Chukchi and Bering seas, and Schell *et al.* (1989a,b) used this gradient to identify critical feeding habitats and seasonal feeding cycles in bowhead whales (*Balaena mysticetus*). Oscillations in $\delta^{13}C$ derived from feeding in differing summer and winter habitats were recorded in the baleen plates of the whales from both the Pacific and Atlantic Oceans. These oscillations provided a means of age determination in the whales and defined regions essential for feeding (Schell, 1987).

7.5 Phytoplankton and particulate organic nitrogen (PON)

As with carbon, nitrogen isotope ratios in phytoplankton are affected by the abundance and various forms of inorganic nitrogenous nutrients. PON plays an important role in the vertical transport of material out of the euphotic zone. In terms of POM, there are two types of organic matter: rapidly sinking particles (usually made up of faecal pellets and marine snow) and slowly sinking particles, which are generally decomposed and remineralized within the euphotic zone (Saino & Hattori, 1987). C/N ratios of POM increase with depth, implying that nitrogen is more rapidly lost than carbon during degradation (Gordon, 1977; Tanoue & Handa, 1979; Saino & Hattori, 1980). It is therefore important to determine how this nitrogen is cycled, since ultimately the $\delta^{15}N$ of this nitrogen will determine the $\delta^{15}N$ of the phytoplankton (Owens, 1987; Wada & Hattori, 1991).

In the euphotic zone, the forms of inorganic nitrogen important to phytoplankton include N_2 gas, ammonia and nitrate (Wada *et al.*, 1975). Regenerated nitrogen is recycled nitrogen within the euphotic zone and typically refers to ammonia uptake and release and to a lesser extent low-molecular-weight organic nitrogen (e.g., amino acids and urea) (Dugdale

& Goering, 1967). 'New' nitrogen refers to the influx of nitrate from deeper waters, nitrogen fixation and atmospheric washout of fixed dinitrogen (Saino & Hattori, 1987). In order to maintain a nitrogen balance, the loss of nitrogen from the downward flux of particles out of the euphotic zone must be balanced by an input of this new nitrogen to the system (Eppley & Peterson, 1979).

Nitrogen fixation and atmospheric washout of fixed nitrogen appear to introduce a small fraction of new nitrogen to pelagic systems (Eppley *et al.*, 1973; Wada *et al.*, 1975; Mullin *et al.*, 1984; Altabet, 1989; see Chapter 6), although for certain species this input may be very helpful (Martinez *et al.*, 1983). This input will introduce nitrogen with a low $\delta^{15}N$, since nitrogen fixation has a small fractionation factor (for example, laboratory cultures of *Azotobacter* and *Anabaena cylindrica*, $\alpha = 0.996–1.009$; Wada *et al.*, 1975). *Trichodesmium*, a nitrogen fixer, has been found to have a lower $\delta^{15}N$ value than non-N_2 fixing phytoplankton (Owens, 1987). Oxidized nitrogen and ammonia in rainfall have also been shown to have low $\delta^{15}N$ values (Hoering, 1957; Wada *et al.*, 1975). In estuarine systems, inputs of nitrogen with low $\delta^{15}N$ material via river water become increasingly important to the overall nitrogen budget (Wada *et al.*, 1975; Mariotti *et al.*, 1984; Owens, 1985).

The most important source of nitrogen to the oceanic euphotic zone is in the form of upwelled nitrate (Altabet & McCarthy, 1985; Altabet, 1989; Liu & Kaplan, 1989). Nitrate with high $\delta^{15}N$ is left during denitrification, which has a large fractionation factor ($\alpha \leq 1.04$). Nitrate with high $\delta^{15}N$ often occurs in oxygen-depleted water (Cline & Kaplan, 1975; Liu & Kaplan, 1989). Subsequent vertical transport will introduce this enriched nitrate into the water column to be taken up by phytoplankton and in turn give higher $\delta^{15}N$ values for marine organisms (Wada *et al.*, 1975; see also Table 7.1 and Fig. 7.2). It is likely that high $\delta^{15}N$ values will be found in the biota in these areas as well. This leads to the future possibility of using nitrogen isotope ratios to follow upwelling. Values for $\delta^{15}N$ nitrate in different geographical regions can be found in Table 7.1 and Fig. 7.2.

As with POC, it is extremely difficult to obtain clean samples of phytoplankton PON for analysis. Therefore, PON samples usually include a mix of phytoplankton, detritus, microzooplankton and bacteria (Altabet & McCarthy, 1985). As stated above, nitrogen available in the euphotic zone is recycled, which may result in isotopic fractionation. The dissolved recycled nitrogen primarily is in the form of ammonia and urea, present in small quantities relative to the total PON. This pool is totally recycled and does not change the $\delta^{15}N$ of the bulk PON even if the different components differ in $\delta^{15}N$ (Altabet & McCarthy, 1985). Variations in

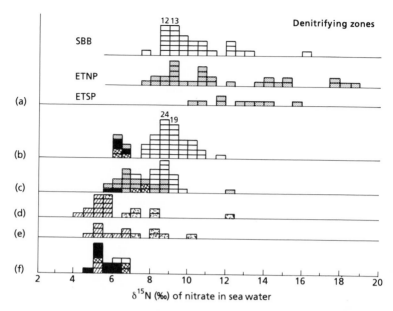

Fig. 7.2 Nitrate δ^{15}N values measured in seawater below 200 metres. (a) SBB, Santa Barbara Basin, California; ETNP, Eastern Tropical North Pacific; ETSP, Eastern Tropical South Pacific. (b) Pacific subsurface water (200–500 m); (c) Pacific intermediate water (500–1500 m); (d) Atlantic subsurface water (200–500 m); (e) Atlantic intermediate water (500–1500 m); (f) deep waters (>1500 m). ▨ = Northwestern North Pacific; ■ = Central North Pacific; ☐ = California; ▨ = Northern North Atlantic; ▨ = Eastern Pacific; tropical; ▨ = Tropical North Atlantic. (From Liu & Kaplan, 1989.)

Table 7.1 ^{15}N values for nitrate and particulate organic nitrogen (PON)

Location	δ^{15}N nitrate	δ^{15}N PON	Reference
Northeastern North Pacific	5.1–7.0		Miyake & Wada (1967)
Seawater	5.8 ± 1.6		Wada et al. (1975)
Denitrifying zones, ETNP	Up to 19		Cline & Kaplan (1975)
Denitrifying zones, ETSP	Up to 13		Liu et al. (1987)
Northern Atlantic >200 m	4.8–6.8		Liu & Kaplan (1989)
Denitrifying, ETNP 200–500 m	12–18		
ETNP >500 m	~6.5		
Central North Pacific	5–6		
North Pacific		Mean = 4.9	Wada et al. (1975)
Southern California Bight		6.5–12.1	Sweeny & Kaplan (1980b)
North Pacific		−1.7 to +9.7	Wada & Hattori (1976)
North Pacific		3.4–7.2	Miyake & Wada (1967)
Northeast Indian Ocean, surface		1.4 ± 0.8	Saino & Hattori (1980)
to 500 m		2.9–13.0	
>500 m		~13	

ETNP, Eastern Tropical North Pacific; ETSP, Eastern Tropical South Pacific.

$\delta^{15}N$ of PON within the euphotic zone appear to be due to both the influx of nitrate and a loss through sinking particles.

PON below the euphotic zone increases in $\delta^{15}N$ (Saino & Hattori, 1980; Altabet & McCarthy, 1986; Saino & Hattori, 1987; Altabet, 1989). Saino & Hattori (1987) found approximately a 6‰ enrichment in PON with depth at each station they sampled, which they ascribed to biological degradation of the sinking PON. Altabet & McCarthy (1985) hypothesized that the change with depth is due to (i) selective degradation of different chemical fractions and (ii) the sinking rates of the particles will be different, therefore more reworking will occur in slower-settling particles and the refractory components will increase in $\delta^{15}N$. Altabet (1989) proposed that the differences in $\delta^{15}N$ between the PON within and below the euphotic zone indicated the average number of trophic steps between phytoplankton and sinking particulate organic matter. He suggested that $\delta^{15}N$ values could be used as indicators of vertical change in trophic structure within the euphotic zone.

Numerous profiles of PON-$\delta^{15}N$ show a sharp decrease, then increase with depth in the euphotic zone (e.g., Saino & Hattori, 1980). The $\delta^{15}N$ minimum is usually accompanied by a decrease in nitrate concentrations. Phytoplankton have been shown to fractionate ^{15}N during assimilation of nitrate (Wada & Hattori, 1978), and this profile in $\delta^{15}N$ may result from preferential uptake of $^{14}NO_3$ by phytoplankton (Altabet et al., 1986; Liu & Kaplan, 1989). In the euphotic zone of the Sargasso Sea under stratified conditions, Altabet (1989) typically found a $\delta^{15}N$ minimum, PON maxima, and the top of the nitracline all occurring at approximately the same depth. Nitrate reduction and uptake was thought to outweigh the sinking of PON enriched in ^{15}N to produce the $\delta^{15}N$ minima. It has also been noted that suspended particulate matter in the euphotic zone has been found to usually be isotopically lighter in ^{15}N in oligotrophic than eutrophic seas (Saino & Hattori, 1980; Minagawa & Wada, 1984; Checkley & Entzeroth, 1985). Eutrophic systems are often characterized by nitrate uptake, whereas oligotrophic systems depend on recycled nitrogen uptake, especially ammonia which is generally depleted in ^{15}N (e.g., $\delta^{15}N$ of $NH_4 = -3.5‰$; Miyaki & Wada, 1967). Checkley & Entzeroth (1985) proposed that a major source of this remineralized nitrogen comes from excretion by pelagic heterotrophs (especially copepods). Since excreta was shown to be depleted in ^{15}N, sinking faeces would be enriched, leading to an overall depletion of PON within the euphotic zone of oligotrophic systems.

There are significant variations in PON seen in different sampling locations around the world. As noted above, Antarctic waters are characterized by very low $\delta^{15}N$ values in flora and fauna. The primary factor causing this variation is probably slow phytoplankton growth rates combined with a loss of ^{15}N through sinking particles. It should be noted however, that to date the ocean has been quite undersampled with respect to $NO_3^--\delta^{15}N$ and PON-$\delta^{15}N$ (Altabet & McCarthy, 1985; Liu & Kaplan, 1989). Any comprehensive survey should endeavor to include $\delta^{15}N$ measurements of nitrate and PON to develop a complete picture of the food web under study.

7.6 Marine food webs

There have been a number of studies utilizing stable isotopes to determine marine and estuarine food web structure (Table 7.2). A majority of the studies used carbon isotopes, but a few investigators have combined carbon, nitrogen and sulfur isotopes. Some investigators (e.g., Mariotti *et al.*, 1984) have found that due to the complexity of estuarine systems, using a single isotope tracer, results can be ambiguous and of little help in defining the system. The following studies are roughly divided into Antarctic/Arctic pelagic systems, offshore systems, nearshore/estuarine systems and salt-marsh systems.

7.6.1 Antarctic/arctic systems

Open-water Arctic/subarctic and Antarctic systems usually show clear trophic structures in stable isotope studies (McConnaughey & McRoy, 1979a,b; Minagawa & Wada, 1984; Wada, 1987; Wada *et al.*, 1987; 1991). In most cases there is one primary food source (phytoplankton) or at most two (phytoplankton and macrophytes) with distinct isotopic signatures that can be easily traced in the food web. McConnaughey & McRoy (1979a) found a significant enrichment in $\delta^{13}C$ with increasing trophic level in the Bering Sea, and estimated a 1.5‰ enrichment per trophic level. All animals were enriched relative to the phytoplankton ($\delta^{13}C$ = c. $-20‰$). They also noted that the benthic organisms were typically more enriched than pelagic animals. This was attributed to possible bacterial and meiofaunal reworking of the food source, or to shorter food chains with greater fractionation. The enrichment was still evident when the isotope data were corrected for lipid content. Most of the enrichment seen was thought to be due to respiration of light $^{12}CO_2$ (McConnaughey & McRoy, 1979a).

The Antarctic Ocean is characterized by phytoplankton and consumers particularly depleted $\delta^{13}C$ and $\delta^{15}N$ (Wada, 1987; Wada *et al.*, 1987a). High nitrate concentrations, high P_{CO_2} concentrations combined with low light intensities, and perhaps trace element limitations may yield slow growth rates resulting in these values. No clear pattern of trophic enrichment in $\delta^{13}C$ was noted. The isotopically light $\delta^{13}C$ and $\delta^{15}N$ values in the animals (in some cases lighter than the phytoplankton) may also reflect seasonal variations in growth conditions of the phytoplank-

Table 7.2 Marine and estuarine food web studies utilizing stable isotopes. References are listed in approximate order of discussion in the text

System	Isotopes Used	Reference
Bering Sea	Carbon	McConnaughey & McRoy (1979a)
Beaufort Sea	Carbon	Schell (1987)
Beaufort, Chuckchi, Bering Seas	Carbon	Saupe et al. (1989)
Bering Sea	Nitrogen	Minagawa & Wada (1984)
Antarctic Ocean	Carbon, nitrogen	Wada (1987)
Antarctic Ocean	Carbon, nitrogen	Wada et al. (1987a)
Barrow Strait–Lancaster Sound	Carbon, nitrogen	Hobson & Welch (1992)
Puget Sound, Washington	Carbon	Simenstad & Wissmar (1985)
Nova Scotia, Canada	Carbon	Stephenson et al. (1986)
East China Sea	Nitrogen	Minagawa and Wada (1984)
George's Bank	Carbon, nitrogen, sulfur	Fry (1988)
S. California Bight	Carbon	Rau et al. (1983)
E. Tropical Pacific	Carbon	Rau et al. (1983)
Monaco	Carbon, nitrogen	Rau et al. (1990)
Gulf of Mexico	Carbon	Fry et al. (1984)
Gulf of Mexico	Carbon, nitrogen, sulfur	Fry (1983)
Gulf of Mexico	Carbon	Thayer et al. (1983)
Nearshore Gulf of Mexico	Carbon	Fry & Parker (1979)
Sefansson Sound, Alaska	Carbon	Dunton & Schell (1987)
Tampa Bay, Florida	Carbon	Conkright & Sackett (1986)
Narragansett Bay, Rhode Island	Carbon	Gearing et al. (1984)
St. Louis Bay, Mississippi	Carbon	Hackney & Haines (1980)
Redfish Bay, Texas	Carbon	Parker (1964)
Torres Strait, Australia	Carbon	Fry et al. (1983b)
Scheldt Estuary, North Sea	Carbon	Mariotti et al. (1984)
Freshwater, Estuarine Systems	Nitrogen	Minagawa & Wada (1984)
Otcuchi River Estuary	Carbon, nitrogen	Wada (1987)
Seagrass meadow, Texas	Carbon	Fry et al. (1977)
Seagrass meadow, Texas	Carbon	Fry & Parker (1979)
Sapelo Island, Georgia	Carbon	Haines (1976a)
Sapelo Island, Georgia	Carbon	Haines & Montague (1979)
Sapelo Island, Georgia	Carbon, nitrogen, sulfur	Peterson & Howarth (1987)
Sipewissett Marsh, Massachusetts	Carbon, sulfur	Peterson et al. (1985)

ton. Again, an average 3.3‰ enrichment in $\delta^{15}N$ per trophic level was observed.

An analysis of the high arctic food web of the Barrow Strait–Lancaster Sound region by Hobson & Welch (1992) spanned five trophic levels. Higher trophic levels were characterized by little change in $\delta^{13}C$ but $\delta^{15}N$ increased an average of 3.8‰ per trophic level. This study confirmed the observed importance of arctic cod, *Boreogadus saida*, in the transfer of energy from lower trophic levels to top consumers.

7.6.2 Offshore temperate systems

Offshore ecosystems in temperate oceans also showed stable isotope enrichment in food webs (Rau *et al.*, 1983, 1990; Minagawa & Wada, 1984; Fry *et al.*, 1984; Fry, 1988). In open-water systems where the primary food source is limited to phytoplankton, a relatively clear food web can be traced using carbon and nitrogen isotope ratios. For example, both the Gulf of Mexico and George's Bank had similar $\delta^{13}C$ values (−21.7 and −21.3‰, respectively), and organisms at higher trophic levels were enriched relative to the phytoplankton (Fry *et al.*, 1984; Fry, 1988). In the Gulf of Mexico there was an increase in $\delta^{13}C$ from POC to zooplankton to benthic crustaceans. Shipboard experiments tended to confirm the 0.5–1‰ enrichment between animal and food noted by DeNiro & Epstein (1978).

Carbon isotope ratios do not always show a clear picture of food web structure even with a single primary food source. Rau *et al.* (1983) analysed whole muscle tissue from each organism in a pelagic food web. All fish isotopic compositions were enriched relative to the POC (Williams & Gordon, 1970; Rau *et al.*, 1982), although $\delta^{13}C$ enrichments did not always correspond to trophic level increases. For example, skipjack and yellowfin tuna had the same isotopic values but are known to have different trophic levels (Fry & Sherr, 1984).

Nitrogen isotopic compositions in offshore systems also showed a ^{15}N increase per trophic level, and reflected the source phytoplankton. In the East China Sea, a system dominated by nitrogen-fixing blue-green algae, trophic enrichments were lower than other systems, reflecting the initial phytoplankton depletion in ^{15}N (−0.55‰; Minagawa & Wada, 1984). Still, the average enrichment per trophic level was 3.4 ± 1.1‰.

Given the difficulties of using a single isotopic tracer, Fry (1988) illustrated the benefit of multiple isotopes in his study of George's Bank. Sulfur isotopes showed little change among all organisms ($\delta^{34}S$ = +15.6 to +17.7‰) and were similar to seawater sulfate values (+20‰). Carbon and nitrogen isotopes were heavier with increasing trophic level, although $\delta^{13}C$ increases were not as consistent as $\delta^{15}N$ increases. This was thought to be due to variations in phytoplankton isotopic values; scallop analyses showed a gradient in increasing $\delta^{13}C$ from deep to shallow water, indicating a possible gradient in phytoplankton ^{13}C, confirmed in later studies (Fry & Wainright, 1991). Nitrogen isotopes showed a consistent enrichment of 3–4‰ per trophic level and overall trophic structure estimated from $\delta^{15}N$ data agreed well with fisheries production models (Fry, 1988).

7.6.3 Nearshore/estuarine systems

In studying estuarine systems, one must focus on variables that could affect the food source at the base of the food web. These include terrestrial versus marine inputs to the system, seasonality, the importance of macrophytes and taxonomic changes in phytoplankton populations. A discussion of carbon isotopes in estuarine systems can also be found in Chapter 10. Due to the similarity of nitrogen isotopic compositions of food sources in nearshore systems, researchers in general have concentrated on carbon isotopes.

Organic inputs to estuarine systems (Fry & Sherr, 1984) can include C_3 terrestrial plant material ($\delta^{13}C$ = -23 to $-30‰$), seagrasses (-3 to $-15‰$; $-26‰$ in some species growing in low-salinity reaches), macroalgae (-8 to $-27‰$), C_3 marsh plants (-23 to $-26‰$), C_4 marsh plants (-12 to $-14‰$), benthic algae (-10 to $-20‰$) and marine phytoplankton (-18 to $-24‰$). Most systems generally have multiple inputs, making data interpretation difficult. One way to resolve this is to compare estuarine organisms with offshore organisms (Fry & Parker, 1979; Fry, 1983; Fry & Sherr, 1984). Benthic macrophytes and seagrasses are usually enriched relative to phytoplankton and epiphytes, so that as the carbon from these isotopically heavier sources is incorporated into the food chain, differences can be seen. In south Texas, Fry & Parker (1979) found that organisms collected in seagrass flats were consistently heavier than samples collected in the Gulf of Mexico (-8.3 to $-14.5‰$ vs. -15.0 to $-19.0‰$). Benthic plants were shown to be the source of the heavier $\delta^{13}C$ values.

Other estuarine systems are less clear and can have several primary producers supplying carbon to higher trophic levels. Torres Strait (Australia) is an example of this complexity (Fry et al., 1983b). Offshore benthic systems were influenced by phytoplankton, whereas inshore systems reflected ^{13}C-enriched benthic algae and seagrasses. Motile organisms did not have isotopic signatures corresponding to one primary producer and reflected access to multiple food sources. A multiple isotope study may have provided better insight into this food web system.

In systems limited to two food sources, the use of stable isotopes can provide more insight into defining food web structure. In resolving the influence of terrestrial carbon to the Gulf of Mexico, Thayer et al. (1983) found that although DOC and particulates $<0.45\,\mu m$ approached terrestrial $\delta^{13}C$ values (-24.0 and $-24.6‰$) and phytoplankton averaged $-22.7‰$, zooplankton and larval fish isotopic compositions indicated that their carbon was derived from marine phytoplankton. Macrophytes may also provide an important source of carbon to coastal ecosystems (see also Chapter 10). Stephenson et al. (1986) studied a system in Nova Scotia dominated by two species of macrophytes. The organisms sampled were lighter than the macrophytes and it was concluded that the dominant source of carbon to this food web system was marine phytoplankton.

Arctic systems are typified by more limited carbon sources and lend themselves to clearer definition with stable isotope ratios. Dunton & Schell (1987) compared seasonal feeding in sessile and motile invertebrates in Sefansson Sound near Prudhoe Bay, Alaska. Here the two sources of energy were phytoplankton and the macrophyte *Laminaria solidungula*. During the darkness of winter, only macrophyte carbon was available and the shift to macrophyte-based diets by consumers was readily evident.

In some systems there are seasonal effects on carbon isotope ratios. Since estuarine systems usually have an input of freshwater, seasonal changes in DIC $\delta^{13}C$ may be incorporated by the phytoplankton (see also Chapter 10). Conkright & Sackett (1986) found that coastal marine phytoplankton, POC, and bivalves differed in their $\delta^{13}C$ ratios between the dry and rainy seasons, with all samples lighter during the wet season. This was thought to be due to either terrestrial organics entering the food web or lighter DIC from freshwater input being incorporated by the phytoplankton. A similar situation was observed in Puget Sound, Washington (Simenstad & Wissmar, 1985). Depletions of up to 8‰ were noted in autotrophs, DOC and some herbivores in estuarine and nearshore habitats during the winter. Nearshore waters were also depleted in the winter, and it was thought that freshwater DIC influenced the isotopic values of the autotrophs. Longer lived and secondary consumers did not show a seasonal trend, due to long-term integration of isotopic values.

Other seasonal effects may be due to changes in phytoplankton populations. Gearing *et al.* (1984) noted that Narragansett Bay, Rhode Island, has distinct phytoplankton populations of differing isotopic compositions. Nanoplankton were dominant in the summer and heavier diatom blooms occurred in winter and spring. Zooplankton also showed a seasonal trend in $\delta^{13}C$ ratios and were 0.5–0.6‰ enriched relative to the phytoplankton. Larval fish did not show a seasonal trend; it was hypothesized that they were selectively using the diatoms or there was a larger fractionation between the fish and phytoplankton. All consumers and predators in the water column and benthos were enriched relative to the phytoplankton, indicating no terrestrial influence. Increasing $\delta^{13}C$ followed a trend based on the organism's presumed trophic position in the food web.

7.6.4 Salt-marsh systems

Most salt marshes of the USA are dominated by stands of *Spartina alterniflora*, which may add a significant amount of carbon to the consumers of the ecosystem (Fry & Sherr, 1984). *Spartina* $\delta^{13}C$ values generally range from −12 to −14‰, intermediate to most terrestrial C$_3$ plants and phytoplankton. When stands of marsh plants are largely monospecific, it appears that the associated invertebrates reflect that carbon isotopic composition (Haines, 1976a,b). However, physical processes within the system as well as biological processes tend to blur the distinctions. At Sapelo Island, Georgia, Haines & Montague (1979) found that sampled invertebrates tended to reflect their diet. Again, where distinct stands were evident and had differing isotopic compositions, organisms such as mud snails that stayed within a given area reflected the source carbon. More mobile species, such as foraging crabs, tended to deviate from the marsh plant $\delta^{13}C$ values and were thought to feed on a mixture of benthic diatoms and *Spartina*.

Hackney & Haines (1980) studied two marshes in St. Louis Bay, Mississippi, which had different nearly monospecific stands of marsh grass. One marsh was a C_4-dominated system with *Spartina alterniflora* ($\delta^{13}C = -12.4‰$), the other was a C_3-dominated system of *Juncus roemerianus* ($\delta^{13}C = -26.2‰$). Organisms sampled from both systems had similar carbon isotopic compositions of -20 to $-26‰$, indicating a possible mixing of material from both marshes or mixing with unsampled carbon sources. This ecosystem also had very negative values for the filter-feeding bivalves, suggesting that either a significant quantity of

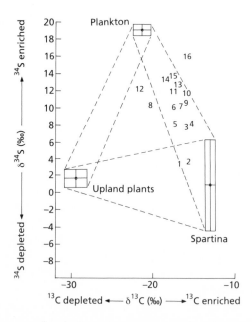

Fig. 7.3 Illustration of the utility of multiple isotope studies to resolve food web questions. Plot of $\delta^{34}S$ vs. $\delta^{13}C$ in relation to the mean of potential food sources and marsh consumers. Numbers represent consumers. (From Peterson & Howarth, 1987.)

terrestrial organic matter or very negative estuarine algae was entering the system. The authors concluded that in contrast to Georgia marshes, terrestrial C_3 plant material played a major role in this Mississippi marsh (Hackney & Haines, 1980).

Peterson & Howarth (1987) completed an extensive survey looking at organic matter flow in the salt marsh and estuarine waters of Sapelo Island, Georgia. In this study, a combination of carbon, nitrogen and sulfur isotopes were used. Sulfur isotopes were especially important, since *Spartina* stands utilize ^{34}S-depleted sulfides which can be as much as 30–40‰ lower than δ^{34}S values for sulfate. Sulfur input to terrestrial systems comes primarily from precipitation (δ^{34}S = 2–8‰) whereas marine systems utilize seawater sulfate (δ^{34}S = 21‰). Analysis of a broad range of organisms showed that the terrestrial organic matter input was not important to this system. Here, the combination of carbon and sulfur isotopes (Fig. 7.3) illustrates this pattern. The dominant source of organic matter to the systems was a mixture of algae and *Spartina*. Nitrogen isotopes tended to show a trophic level enrichment and reflected the mixed diet (Fig. 7.4). A transect study of ribbed mussels in Sippewissett Marsh, Massachusetts, came to a similar conclusion of the importance of *Spartina* in marsh food webs (Peterson *et al.*, 1985). These two studies

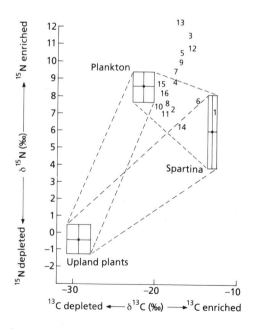

Fig. 7.4 Carbon and nitrogen isotope ratios of food sources vs. consumers for a salt marsh. Numbers represent consumers. Note the enrichment in ^{15}N for consumers vs. food sources. (From Peterson & Howarth, 1987.)

show the increased efficacy of the multiple isotope approach to resolve food web systems.

An important factor which should be considered in using isotopes is the signal-to-noise ratio of the sources, defined as the difference in the mean isotope values of the food sources versus the sum of the standard deviations of the mean values of the food sources. Since ecologists utilize isotopes in mixing models, resolution will be limited by the separation between end-member sources. This can vary with the isotope measured. For example, in the Sapelo Island system, upland C_3 plants and *Spartina* had a large ^{13}C isotopic difference (16‰) and low standard deviations, leading to a signal/noise ratio of 8.6 (Peterson & Howarth, 1987). However, ^{34}S isotope ratios were very similar and had a signal/noise ratio of 0.1. Resolution will also be limited by the sampling density; in a system with a 1‰ separation between sources, 50 samples may be necessary, whereas a system with a 10‰ separation may only need up to five samples. The multiple isotope approach can help to resolve these questions.

7.7 Summary

Stable isotope ratios offer an effective natural tracer for following energy and nutrient flows in ecosystems. Although many studies now illustrate the fundamentals of these techniques, the limitations and caveats are often not fully appreciated. These are becoming more evident. Through a synthesis of past work several generalizations can be made.

1 Carbon isotope ratios reflect the primary production important to the ecosystem energy flow. The transfer of carbon isotope ratios is essentially conservative between trophic levels although minor ^{13}C-enrichment (<1‰) may occur. This allows allocation of energy sources in the cases where major inputs are limited to two and the sources have distinctive isotope ratios.

2 An advantage of the isotope tracer approach is that organisms from all trophic levels can be analysed with relative ease, leading to quick estimates of how organic matter flows throughout an entire food web.

3 Within estuarine and coastal ecosystems carbon isotope ratios of primary producers are altered by seasonal and environmental changes. Depletion in the ^{13}C content of DIC through increased respiratory inputs or reductions in primary producer growth rates by decreasing light, nutrients or trace element concentrations will decrease the $\delta^{13}C$ of fixed carbon. Large cell sizes and fast growth rates produce ^{13}C-enriched fixed carbon.

4 Carbon isotope tracing of food webs in pelagic systems must take into account seasonal, geographical and multi-year changes in the $\delta^{13}C$ of primary producers. The $\delta^{13}C$ values of migratory fauna can reflect feeding in distant areas where isotope ratios are different from the environment in which they were sampled. Seasonal changes in lipid content will alter bulk $\delta^{13}C$ but such changes are usually small and easily corrected.

5 Nitrogen isotope ratios reflect trophic status in most ecosystem food webs and to a lesser extent, nutritional status. Food deprivation leads to enrichment of ^{15}N in body tissues. Overall $\delta^{15}N$ in fauna usually reflects the nitrogen isotope ratios in the primary producers with a gain of c. 3–4‰ per trophic level.

6 Multiple element isotope ratio studies often can provide much better insight into ecosystem processes than single element studies. Multiple source inputs can sometimes be separated with a combined tracer approach much better than by using just one isotope tracer.

CHAPTER 8

Stable isotopes in the study of marine chemosynthetic-based ecosystems

N.M. CONWAY, M.C. KENNICUTT II & C.L. VAN DOVER

8.1 Introduction

8.1.1 Background

In certain environments – deep-sea hydrothermal vents, brine seeps, shallow petroleum seeps, sewage outfall areas and some organic rich coastal sediments – microbial chemoautotrophic primary production can replace photosynthesis as the dominant source of ecosystem energy production (Jannasch & Wirsen, 1979; Rau & Hedges, 1979; Howarth & Teal, 1980; Rau, 1981a,b; Cavanaugh et al., 1981; Brooks et al., 1987; Van Dover & Fry, 1989). The term chemosynthesis, or more correctly, chemoautolithotrophy, describes the biosynthesis of organic carbon compounds from CO_2 (or other C_1 compounds such as CH_4 and CH_3OH) using energy and reducing power derived from the chemical oxidation of inorganic compounds (Fig. 8.1). Characteristic of chemosynthetic environments is the presence of high concentrations of reduced inorganic compounds, such as H_2S, S_2O_3, CH_4, H_2 and NH_4^+, which allow for chemical oxidation and support bacterial chemosynthesis (Jannasch, 1984).

While chemosynthesis has been known for over a century (Winogradsky, 1889, 1890), it is only recently that complex, chemosynthetic-based ecosystems have been discovered (see reviews of Jannasch & Mottl, 1985; Fisher, 1990). First encountered at hydrothermal vents along seafloor spreading centres (Lonsdale, 1977; Corliss & Ballard, 1977; Corliss et al., 1979), chemosynthetic communities have since been found at cold hypersaline and hydrocarbon seeps (e.g., Kennicutt et al., 1985; Kulm et al., 1986; Brooks et al., 1987), and even in shallow-water sewage outfalls (Felbeck, 1983) and reducing sediments in coastal areas (Cavanaugh, 1985). For the most part, sulfur-oxidizing chemosynthetic bacteria appear to be the major primary producers in these environments, although methanotrophic bacteria and perhaps bacteria utilizing other reduced chemicals may also be important.

Sulfur-oxidizing bacteria are capable of fixing CO_2 using the ATP and NAD(P)H obtained from the oxidation of reduced sulfur compounds such as H_2S and S_2O_3 (see reviews of Kelly, 1982; Keunen & Beudeker, 1982), while methanotrophic bacteria use methane as both a carbon and

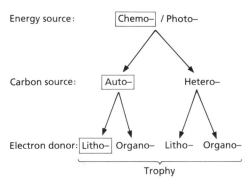

Fig. 8.1 Trophic mechanisms; boxed names indicate types of chemosynthetic bacteria.

energy source (Smith & Hoare, 1977; Colby *et al.*, 1979). Both groups of bacteria require access to oxygen in addition to the appropriate reduced inorganic fuel. Invertebrates are capable of using this bacterial production either through ingestion of free-living chemoautotrophic bacteria or via nutritional exchange with endosymbiotic chemoautotrophic bacteria (Cavanaugh *et al.*, 1981; Felbeck *et al.*, 1981; Childress *et al.*, 1986).

The last decade has produced a proliferation of research on the physiology and biochemistry of vent species and invertebrate–chemoautotroph interactions. A recent review by Fisher (1990) summarizes many of these studies; interested readers are referred to this paper. Stable isotope techniques have been an essential component of research regarding the energetic basis of chemosynthetic communities. Insight into the ecology and physiology of the dominant invertebrates found in these ecosystems can also be obtained by stable isotope analysis. Stable isotope ratios have been particularly useful in the study of marine chemosynthetic-based ecosystems because of the inaccessibility of the deep-sea environments in which they often occur. Specimens and submersible time are precious, making more traditional methods of analysis difficult to undertake. Kennicutt *et al.* (1992) have compiled and interpreted stable isotope data from hundreds of organisms from chemosynthetic environments. Stable isotope techniques have been used to address a diverse set of questions, ranging from the identification of primary nutritional sources to recognition of fossil chemosynthetic assemblages. In this review we will highlight a number of marine studies where stable isotope techniques have been applied to the study of chemosynthetic-based ecosystems to demonstrate the creative (although not always successful) scientific detective work accomplished despite the remoteness of many of the communities investigated.

8.1.2 Overview of the general processes affecting stable isotope ratios of organic matter in the marine environment

Stable isotope ratios of biosynthetically derived organic compounds are determined by the stable isotope composition of the source material utilized, availability of the source material, and kinetic fractionations during uptake, assimilation and incorporation by living organisms (Fig. 8.2). Thus, in order to use stable isotope ratios as quantitative ecological tools, these controlling factors must be fully characterized for each case studied. Due to the complexities of most environments, it is rare for the full characterization of an ecosystem to be accomplished and care must be taken when interpreting stable isotope ratios. More often, stable isotope ratios provide preliminary data on trophic relationships and ecological interactions, and qualitative data on the relative importance of organic and inorganic source materials. In order for stable isotope ratios to be interpreted, supporting data obtained from additional environmental, physiological and biochemical studies are needed. We review here some of the isotopic characteristics of carbon, nitrogen and sulfur that are relevant to a discussion of the use of stable isotopes in the study of marine chemosynthetic-based ecosystems.

Carbon

Sources.   For the most part, seawater dissolved inorganic carbon (DIC) $\delta^{13}C$ values are close to 0‰ in both surface and deep-water environments (Fig. 8.2). Hydrothermal vent effluent DIC $\delta^{13}C$ ratios may be slightly more negative (-7‰; Craig et al., 1980). Average pore-water DIC ratios are also usually close to 0‰, although the range of values in sediments can be highly variable in some settings (Brooks et al., 1984, 1987; Kennicutt et al., 1989).

Metabolic fractionations.   The predominant metabolic process affecting the $\delta^{13}C$ ratios of organic carbon in marine systems is autotrophic fixation of seawater DIC (Fig. 8.2). Large isotopic fractionations can result from these processes because the non-limiting supply of seawater bicarbonate permits substantial enzymatic discrimination against the heavier ^{13}C isotope. In certain environments, bacterial methanogenesis results in very ^{13}C-depleted biomass (Fig. 8.2).

In marine plants and algae, carbon dioxide fixation via the Calvin–Benson pathway results in the production of organic carbon having $\delta^{13}C$ ratios of -15 to -22‰ (Gearing et al., 1984). Isotopic fractionations associated with chemosynthetic enzymatic pathways are not well characterized, but studies to date suggest that reduction of DIC during

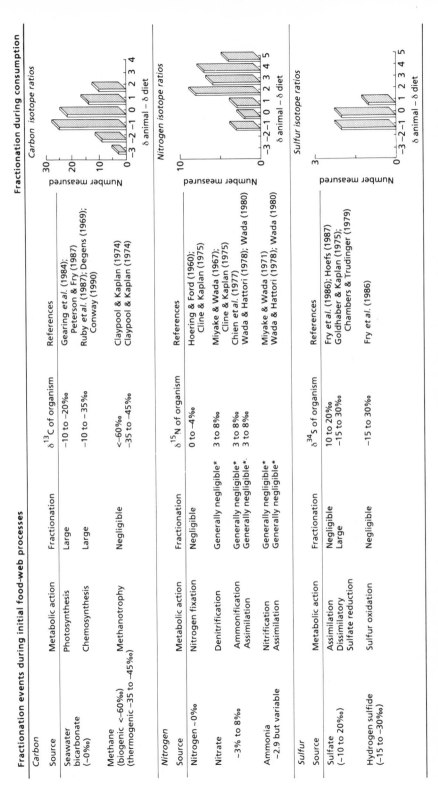

Fig. 8.2 Fractionation events during food web processes. Redrawn from Peterson & Fry (1987) and based on marine and aquatic species. *In laboratory situations where nitrogenous supplies are non-limiting, isotope fractionation can be large.

chemoautotrophy generally results in a greater utilization of the lighter isotope (i.e., more negative $\delta^{13}C$ biomass is produced than for marine photosynthetic sources; Degens, 1969; Ruby et al., 1987) although there are few experimental laboratory studies to support this hypothesis. Methanotrophic bacteria can have even more negative $\delta^{13}C$ ratios because their source of carbon may be isotopically depleted methane produced from bacterial methanogenesis ($\delta^{13}C$ = −60 to −80‰; Claypool & Kaplan, 1974; Fuchs et al., 1979) or thermogenic processes ($\delta^{13}C$ = −35 to −45‰; Claypool & Kaplan, 1974; Brooks et al., 1987; Fig. 8.2). Reports of microbial methane oxidation favouring ^{12}C during methane assimilation suggest ranges from −2.4 to −5.6‰ (Silverman & Oyama, 1968; Claypool & Kaplan, 1974) to as much as −20‰ (Zyakun et al., 1981).

In the marine environment, carbon isotope ratios are increased by c. +1‰ during most trophic interactions (Peterson & Fry, 1987; Fig. 8.2) allowing determination of the source of dietary carbon for an organism where two isotopically distinct carbon pools are available. For example, organisms with a diet based on carbon produced by marine chemoautotrophs will be expected to have $\delta^{13}C$ ratios considerably more negative than those of organisms whose diet is based on photosynthetically fixed carbon.

Nitrogen

Sources.   Atmospheric and dissolved seawater nitrogen have $\delta^{15}N$ ratios of 0‰, whereas seawater nitrate $\delta^{15}N$ is generally +3 to +6‰ (Fig. 8.2). Deep-sea particulate organic nitrogen (PON) generally has $\delta^{15}N$ ratios >+6‰ (Saino & Hattori, 1987). Molecular nitrogen (N_2) associated with thermogenic hydrocarbons can be significantly depleted in ^{15}N ($\delta^{15}N$ c. −4‰; Brooks et al., 1987).

Metabolic fractionations.   During N_2 fixation, isotope discrimination effects are small (0 to −4‰; Hoering & Ford, 1960; Cline & Kaplan, 1975). Nitrogen limitation, as with carbon, tends to minimize isotopic discrimination during the assimilation of inorganic nutrients (Fig. 8.2). During uptake and assimilation of organic nitrogen by consumers and predators, a change of c. +1 to +5‰ occurs with each trophic step (average = +3.4‰; Minawaga & Wada, 1984) due to excretion of isotopically 'light' nitrogen. As a result, animal tissue values are more positive than their dietary source $\delta^{15}N$. This increase in $\delta^{15}N$ ratios with trophic level suggests that nitrogen isotope ratios can serve as indicators of trophic level. The most positive $\delta^{15}N$ ratios are found in species at higher trophic levels of food chains (Rau, 1985), while lowest values are found in herbivores and detritivores that feed on phytoplankton and bacteria (Fry, 1988).

Sulfur

Sources. Nutritional sulfur sources available to marine organisms include seawater (δ^{34}S c. +21‰) and pore-water sulfate (δ^{34}S c. +20 to +48‰; Hartmann & Nielson, 1969), particulate organic matter (δ^{34}S c. +20‰), and biogenic reduced sulfides (δ^{34}S c. 0 to −40‰; Hartmann & Nielson, 1969). Geothermal vent sulfide deposits have slightly more positive δ^{34}S ratios than biogenic sulfides (Fry *et al.*, 1983a).

Metabolic fractionations. During assimilation of sulfate by phytoplankton, isotopic fractionation is negligible (Fig. 8.2). Marine algae have δ^{34}S ratios close to that of seawater sulfate (c. +20‰, Rees *et al.*, 1978; Fry *et al.*, 1983a; Peterson & Fry, 1987). As organic matter moves through the food chain, consumer assimilation effects are also small (Fig. 8.2). By far the most important biological process affecting the δ^{34}S ratios of sulfur-containing compounds is dissimilatory sulfate reduction by marine bacteria. Sulfate reduction can deplete the δ^{34}S ratios of source sulfides by −30 to −60‰, resulting in porewater sulfides in the 0 to −40‰ range (Goldhaber & Kaplan, 1975; Fry *et al.*, 1986). Kinetic fractionation that occurs during oxidation of these reduced sulfides is negligible, thus sulfur-oxidizing bacteria have δ^{34}S ratios close to those of marine sulfides. *A priori*, organisms that obtain their sulfur requirements from sediment or vent sulfides will have very different δ^{34}S ratios from species utilizing seawater sulfate.

8.2 The chemosynthetic basis for life at hydrothermal vents and cold water seeps: stable isotopic evidence

'*Large isotopic discrepancies between vent and pelagic organisms would clearly show that local, non-pelagic food sources are quantitatively important for vent animal growth and metabolism*' (Rau, 1985).

Input of organic matter into the deep sea is predominantly limited to the small percentage of photosynthetically derived organic matter which drifts down from the upper few hundred metres of the euphotic surface layer where light penetration permits phytoplankton growth. This food limitation in the deep sea results in low living biomass and, while the deep-sea floor can hardly be considered a desert, the prevailing fauna is often minute and inconspicuous. Thus the discovery of high living biomass in dense invertebrate assemblages clustered around hydrothermal vents along the Galapagos Rift at depths greater than 2500 m (Corliss & Ballard,

1977; Lonsdale, 1977; Corliss *et al.*, 1979) immediately raised questions regarding food resources for these species.

Hydrothermal vents occur at seafloor spreading centres (Fig. 8.3) where tectonic and volcanic activity drives hydrothermal convection cells within newly-formed ocean floor. Water vented from these areas is enriched in metals, geothermally-reduced sulfides, methane and hydrogen, and may reach extremely high temperatures (>350°C). Where hydrothermal fluids pass through unconsolidated sediments, hydrothermal petroleum formation can also occur. Tunnicliffe (1991) reviews the ecological settings of hydrothermal vents explored to date.

Diluted hydrothermal fluids, exiting at temperatures favourable to life (2–40°C), support unusually high densities of large invertebrates, including the vestimentiferan tubeworm *Riftia pachyptila* which can exceed 1 m in length (Jones, 1984). High biomass of vent-like organisms also occur at cold, hypersaline and hydrocarbon seeps such as those located along the Florida Escarpment and Louisiana Slope (Paull *et al.*, 1985; Kennicutt *et al.*, 1985; Fig. 8.3). These seep communities are not associated with hot water or magmatic sources but are associated with the expulsion of hypersaline waters or petroleum compounds along faults and fissures.

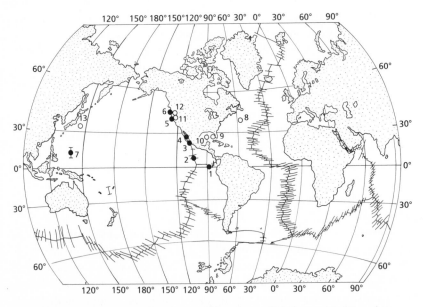

Fig. 8.3 Map showing the location of vents (●) and seeps (○) discussed in this review. Vents: 1, Galapagos Spreading Center; 2, 11°N, East Pacific Rise; 3, 21°N, East Pacific Rise; 4, Guaymas Basin; 5, Gorda Ridge; 6, Juan de Fuca Ridge; 7, Marianas Back Arc Spreading Center. Seeps: 8, Laurentian Fan; 9, Florida Escarpment; 10, Louisiana Seeps; 11, Oregon Seeps; 12, Northern California Seeps; 13, Japanese Seeps.

The slow flux of organic matter from the overlying euphotic zone was thought to be unlikely to support the high densities and fast growth rates exhibited by vent and seep species. Two alternative mechanisms were initially proposed to explain the high densities of animals found in vent regions. In one mechanism, thermal plumes rising from the vents might create inflowing bottom currents, thereby increasing the flux of photosynthetically derived organic material from the surrounding waters (Lonsdale, 1977; Enright *et al.*, 1981). This hypothesis seemed unlikely, however, as high invertebrate biomass occurs in areas where the venting water is close to ambient water temperatures, where thermal circulation would not be significant. According to the second mechanism, high concentrations of reduced compounds (such as H_2S, H_2 and CH_4) in vent waters could support local populations of chemoautotrophic bacteria as primary producers of organic compounds (Rau, 1981a,b; Ruby *et al.*, 1981).

The first biological specimens obtained from hydrothermal vents were generally damaged, and few survived for more than a few days. In order to understand the basic physiology of these species, it was useful to measure parameters that were unaffected by damage to the organism during collection and storage, such as stable isotope ratios. In classic and elegant reports, Rau & Hedges (1979) and Rau (1981a) used stable isotope compositions to test the alternative nutritional mechanisms described above. If $\delta^{13}C$ ratios of vent animals matched those of invertebrates dependent on photosynthetically derived carbon, it would be impossible to reject the first hypothesis of normal deep-sea heterotrophy. However, if $\delta^{13}C$ ratios of vent organisms differed significantly from those of typical deep-sea heterotrophs, then a local autochthonous source of organic carbon would be indicated as the most likely nutritional base for the community.

8.2.1 $\delta^{13}C$ ratios of vent and seep species

Rau & Hedges (1979) and Rau (1981a) did find that vent organisms had $\delta^{13}C$ values very different from those of non-vent deep-sea organisms (Table 8.1), suggesting that local primary producers at vents must be responsible for carbon fixation. Rau and Hedges concluded from the unusual $\delta^{13}C$ ratios they measured in vent species and reports of high concentrations of chemosynthetic bacteria in vent waters (Jannasch & Wirsen, 1979; Karl *et al.*, 1980; Ruby *et al.*, 1981) that chemosynthetic bacteria were somehow the primary source of organic carbon within the vent ecosystem. Concurrent morphological and biochemical investigations of vent organisms revealed that trophic interactions between some vent invertebrates and sulfur-oxidizing chemosynthetic bacteria occur intracellularly. Such intracellular associations between vent invertebrates and chemosynthetic bacteria were first discovered in the gutless tube-

Table 8.1 δ¹³C, δ¹⁵N and δ³⁴S stable isotope ratios of (a) hydrothermal vent and seep species and (b) non-vent organisms

Species	Site	Tissue	δ¹³C	δ¹⁵N	δ³⁴S	References
(a) HYDROTHERMAL VENT AND SEEP SPECIES						
Phylum Vestimentifera						
*Riftia pachyptila**	21°N, East Pacific Rise	Vestimentum	−10.8 to −11.0‰	+0.8 to +4.0‰		Rau (1981a,b); Rau (1985)
		Trophosome	−10.9 to −11.1‰	+1.8 to +2.0‰		
*Riftia pachyptila**	21°N, East Pacific Rise	Vestimentum	−11.7‰	+4.5‰		Van Dover & Fry (1989)
		Trophosome	−11.3‰	+3.4‰		
*Riftia pachyptila**	21°N, East Pacific Rise	Various tissues			−4.7 to 4.7‰	Fry et al. (1983a)
Oasisia alvinae	21°N, East Pacific Rise		−10.4 to −11.4‰	+2.9 to +3.4‰		Van Dover & Fry (1989)
Lamellibranchia sp.*†	Oregon Subduction Zone		−26.7 to −31.9‰			Kulm et al. (1986)
Lamellibranchia sp.*†	Louisiana Hydrocarbon Seep	Tubes and tissues	−29.8 to −57.2‰			Brooks et al. (1987)
Escarpia-like tubeworm*†	Louisiana Hydrocarbon Seep	Tubes and tissues	−21.4 to −48.6‰	+2.9‰, +5.4‰	−3.5‰	Brooks et al. (1987)
Unidentified *Vestimentiferan*†	Florida Escarpment	Not specified	−42.7 ± 0.7‰			Paull et al. (1985)
Phylum Pogonophora						
Unidentified *Pogonophoran*†	Louisiana Hydrocarbon Seep	Tubes and tissues	−30.5 to −59.3‰			Brooks et al. (1987)
Phylum Mollusca						
*Calyptogena magnifica**	Rose Garden	Mantle	−32 to −32.1‰	+2.1 to +2.8‰		Rau (1981a,b)
*Calyptogena magnifica**	Clam Acres	Mantle	−32.6‰	+4.0‰		Rau (1981a,b)
*Calyptogena magnifica**	Rose Garden	Gills	−33.2 ± 0.7‰	+1.8 ± 1.3‰		Fisher et al. (1988c)

Species	Location	Tissue				Reference
*Calyptogena magnifica**	Rose Garden	Rest	−32.6 ± 0.5‰	+4.5 ± 1.3‰		Fisher *et al.* (1988c)
*Calyptogena magnifica**	21°N, East Pacific Rise	Gill Foot			+1.5‰ −1.7 to +0.4‰	Fry *et al.* (1983a)
*Calyptogena ponderosa**†	Louisiana Hydrocarbon Seep	Not specified	−36.9 to −39.1‰	+1.1 to +7.1‰	+0.1 to +2.1‰	Brooks *et al.* (1987)
Calyptogena sp.*†	Oregon Subduction Zone	Gills 'Tissue'	−51.6‰ −35.7‰			Kulm *et al.* (1986)
Mussel (probably *Bathymodiolus thermophilus*)*	Clambake 1, Galapagos	Foot and mantle	−32.7 to −33.6‰			Rau & Hedges (1979)
*Bathymodiolus thermophilus**	Galapagos	'Tissue'	−32.8 to −33.9‰			Williams *et al.* (1981)
*Bathymodiolus thermophilus**	Rose Gardens	Gill Rest	−34.7 to −35.7‰ −33.5 to −34.4‰			Fisher *et al.* (1988a)
*Bathymodiolus thermophilus**	Rose Gardens	'Tissue'		−3.9 to +3.5‰		Fisher *et al.* (1988a)
Bathymodiolus sp.*	Mariana	Gill Muscle	−34.8‰ −32.8‰	−3.0‰ −0.5‰		Van Dover & Fry (1989)
Unidentified *Mytilid**†	Louisiana Hydrocarbon Seep	Soft parts	−50.1 to −45.5‰	+2.9 to +3.0‰	+13.4 to +7.5‰	Brooks *et al.* (1987)
Unidentified *Mytilid**†	Florida Escarpment	'Tissue'	−74.3‰			Paull *et al.* (1985)
*Alviniconcha hessleri**	Mariana	Muscle Gill	−28.0 to −29.7‰ −27.3 to −28.1‰	+3.2 to +5.2‰ +5.2 to +7.8‰		Van Dover & Fry (1989)
Pseudomiltha sp.*†	Louisiana Hydrocarbon Seep	Not specified	−30.9 to −37.7‰	−3.5 to +6.1‰	−11.5 to +1.3‰	Brooks *et al.* (1987)
Unidentified mussel*†	Louisiana Hydrocarbon Seep	Not specified	−40.1 to −57.6‰			Brooks *et al.* (1987)

continued on p. 168

Table 8.1 *Continued*

Species	Site	Tissue	$\delta^{13}C$	$\delta^{15}N$	$\delta^{34}S$	References
Solemya sp.*†	Oregon Subduction Zone	Not specified	−31.0‰			Kulm *et al.* (1986)
*Vesicomya chordata**†	Louisiana Hydrocarbon Seep	Soft parts	−36.3‰	−0.9‰		Brooks *et al.* (1987)
Phylum Annelida						
Alvinella caudata	Hanging Gardens	Branchiae, oral tentacles	−12.8‰	+3.9‰		Van Dover & Fry (1989)
Alvinella caudata	Hanging Gardens	Body wall		+6.3‰		Van Dover & Fry (1989)
Alvinella pompejana	Hanging Gardens	Branchiae, oral tentacles	−11.7‰	+4.7‰		Van Dover & Fry (1989)
Paralvinella grasslei	Hanging Gardens	Not specified	−12.8‰	+7.3‰		Van Dover & Fry (1989)
Paralvinella sp.	Mariana	Not specified	−10.2, −11.8‰	+8.4, +7.9‰		Van Dover & Fry (1989)
Phylum Arthrapoda						
Bythograea thermydon	Galapagos	Claw and leg	−13.7 to −17.6‰	+7.4 to +9.8‰		Rau (1985)
Bythograea thermydon	Galapagos	Claw muscle			−0.1 to −0.8‰	Fry *et al.* (1983b)
Neolepas zevinae	Hanging Gardens	Whole animal	−15.4, −14.7‰	+5.9, +6.5‰		Van Dover & Fry (1989)
Chorocaris sp.	Mariana	Abdominal muscle	−16.4, −16.7‰	+8.6, +8.9‰		Van Dover & Fry (1989)

(b) TYPICAL DEEP-SEA MARINE INVERTEBRATE VALUES

Phylum Chordata					
Tunicate	Near East Pacific Rise 2600 m depth	Tunic	−21.3‰	+15.7‰	Van Dover & Fry (1989)
Phylum Echinodermata					
Holothurian	Near East Pacific Rise 2600 m depth	Body wall	−17.1‰	+14.0‰	Van Dover & Fry (1989)
		Hepatopancreas	−19.3‰	+11.6‰	
Brisingid		Gonad	−18.2‰	+12.3‰	
Phylum Coelenterata					
Hydroids	Near East Pacific Rise 2600 m depth	Whole animal	−19.7‰	+14.1‰	Van Dover & Fry (1989)
Anemone		Body wall	−17.1‰	+12.9‰	
Phylum Mollusca					
Octopus	Near East Pacific Rise 2600 m depth	Tentacle	−17.0‰	+14.1‰	Van Dover & Fry (1989)
Phylum Arthropoda					
Crab	390 m Gulf of Mexico	−17.2‰			Kennicutt et al. (1985)
Crab	1390 m Gulf of Mexico	−17.4‰			Kennicutt et al. (1985)
Shrimp	1225 m Gulf of Mexico	−18.3‰			Kennicutt et al. (1985)
'Bathypelagic Crustacea'	2100 m	−18.7‰			Williams & Gordon (1970)

* Thought to contain endosymbiotic chemosynthetic bacteria.
† Seep species.

worm *R. pachyptila* (Cavanaugh *et al.*, 1981; Felbeck *et al.*, 1981). *R. pachyptila* lives in tubes attached to lavas and metal sulfides in the zone of mixing between warm-water, sulfide-rich hydrothermal effluents and oxygenated ambient seawater. High concentrations of Gram-negative prokaryotic cells reside within a special, large and highly vascularized organ called the *trophosome* (Cavanaugh *et al.*, 1981). The tubeworm host tissues are apparently nourished internally by chemosynthetic bacteria. Further studies have shown that many vent and seep tubeworms, bivalves and at least one gastropod species (= species denoted with an asterisk in Table 8.1a), also harbour chemoautotrophic bacteria and have reduced digestive systems or are totally gutless, implicating the symbionts in a nutritional role for the host. Invariably, the putative bacterial symbionts are found in specific, enlarged and vascularized tissues (gills in the case of mollusks and trophosome tissue in the case of vestimentiferan and pogonophoran tubeworms). Electron microscopy, assays for key enzymes of sulfur-based chemoautotrophy (e.g., ribulose bisphosphate carboxylase, ATP sulfurylase, APS reductase, methanol dehydrogenase), and the presence of lipopolysaccharides suggest that the symbionts are autotrophic, sulfide-oxidizing, Gram-negative bacteria in many invertebrate–symbiont associations (see reviews of Cavanaugh, 1985; Fisher, 1990).

Since the initial carbon isotope ratio reports of Rau & Hedges (1979) and Rau (1981a), there have been a number of studies of the stable isotope composition of vent and seep species, particularly sulfur-based animal-chemoautotroph symbioses (Table 8.1a). Vent tubeworms have $\delta^{13}C$ values centered around $-13‰$. In contrast, vent bivalves are consistently about $21‰$ lighter ($\delta^{13}C$ c. $34‰$). In all cases, vent symbiont species are isotopically distinct from ambient deep-sea invertebrates.

The large dichotomy in the isotopic compositions of vent tubeworms and bivalves may be a consequence of differences in the isotopic composition of the source carbon and/or differences in biochemical fractionation effects. Several lines of evidence suggest that source differences are not responsible for the $21‰$ difference between vent tubeworms and bivalves:

1 both taxa occur in similar, or even identical microhabitats;

2 radiocarbon analyses indicate that both taxa use the same DIC source (Williams *et al.*, 1981);

3 the consistency of $\delta^{13}C$ values within diverse vent tubeworm species and within diverse bivalve species argues for a large (relative to demand) isotopically-uniform source pool of inorganic carbon, namely deep-water DIC with only small, $1–2‰$ variations introduced by contributions of magmatic DIC, respired CO_2, or use of photosynthetically-derived particulate or dissolved organic carbon.

Source effects thus seem unlikely to provide an explanation for the vent tubeworm/bivalve dichotomy in carbon isotopic compositions. Carbon isotopic fractionation during CO_2 fixation in a laboratory-cultured, free-living, sulfide-oxidizing micro-organism isolated from vent water is on the order of $-20‰$ versus HCO_3^-, consistent with fractionation effects observed in vent bivalve–symbiont associations (Ruby et al., 1987), and suggesting that it is the tubeworm that expresses an 'anomalous' isotopic composition requiring explanation. Two alternative hypotheses, first suggested by Rau (1981a, 1985) have been proposed and are reviewed by Fisher (1990).

1 There may be different metabolic pathways for CO_2 fixation in bivalves versus tube-worms. Rau noted the resemblance of $\delta^{13}C$ values of tube-worms to isotopic compositions of Slack–Hatch cycle (C_4) plants, while bivalves are isotopically lighter, more like the situation in Calvin cycle (C_3) plants. Carbon isotopic differences between C_3 and C_4 plants are known to be a consequence of different metabolic pathways for CO_2 fixation (see Chapter 1).

2 A physiological CO_2-transport limitation may be imposed on the site of CO_2-fixation in the trophosome of the tubeworm. Fisher et al. (1990) provide evidence in support of this hypothesis, based on increasing $\delta^{13}C$ values with increasing size (and presumably increasing CO_2 limitation) in two vent tubeworm species, R. pachyptila and Tevnia jerichonana.

In sulfide-based symbioses, the isotopic composition of fixed carbon is dependent on enzymatic fractionation, CO_2 isotopic composition and CO_2 availability. A 12‰ (temperature-dependent) fractionation is imposed on the system in the equilibrium exchange between DIC and $CO_{2(aq)}$ at 0°C (Mook et al., 1974). The degree of carbon isotopic fractionation subsequent to the conversion of DIC to $CO_{2(aq)}$ in normal marine systems can be viewed as a continuum of possible values, from 0‰ to a maximum of 29‰ (Roeske & O'Leary, 1984). This continuum is constrained by the isotopic composition of the source DIC at the heavy end and by the fractionation effect of the initial carboxylating enzyme (RuBPCase) at the other end. Degree of expression of the enzymatic fractionation effect is dependent on the degree of CO_2 limitation in the system. The isotopic composition of fixed carbon can therefore range between 12 and 41‰ lighter than the isotopic composition of the source DIC. Assuming a large pool of isotopically uniform DIC source with a $\delta^{13}C$ near 0‰, vent tubeworms ($\delta^{13}C = -10$ to $-15‰$) and vent bivalves ($\delta^{13}C = -33$ to $-36‰$) fall near the predicted extremes. There will be a shift in tissue carbon isotopic composition in sulfide-based symbioses to match any shift in $\delta^{13}C$ composition of the DIC source, and where the DIC source has a locally variable $\delta^{13}C$ composition, we expect a consequent variability in the isotopic composition of symbiont tissues. Such

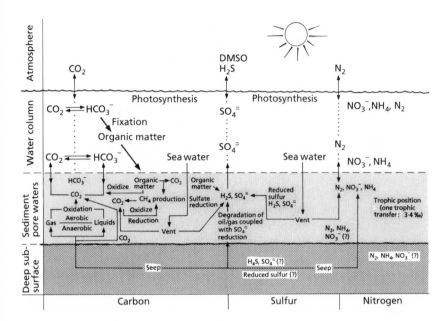

Fig. 8.4 Element cycling in vent/seep communities.

variability has been observed in sulfide-oxidizing symbiont species from non-hydrothermal areas (especially seep tubeworms) and may be a consequence of biogeochemical processes affecting the isotopic composition of local DIC reservoirs in the surrounding sedimented environment (see discussion in Brooks *et al.*, 1987).

The wide range of $\delta^{13}C$ values found in vent and seep tubeworms demonstrate some of the limitations of stable isotope ratio techniques. Without additional data on the isotopic composition and concentration of carbon and the biochemistry and physiology of vent and seep species, it is not possible to use isotope ratios to definitively demonstrate nutritional strategies. Yet, in almost every case, the stable isotope ratios of vent and seep species have been crucial in suggesting that non-photosynthetically derived organic matter is being utilized and that element cycling within the communities may be complex (Fig. 8.4).

8.2.2 $\delta^{15}N$ ratios of vent and seep species

Building on his work with carbon isotopes, Rau (1981b) analysed the nitrogen isotopic composition of some vent invertebrates (Table 8.1a), reasoning that if these vent invertebrates used nitrogen derived from deep-sea PON, their $\delta^{15}N$ values should be similar to deep-water PON sources (c. +3 to +6‰; Saino & Hattori, 1987). Rau (1981b) found that $\delta^{15}N$ values of tissues from vent vestimentiferan worms and clams were isotopically light (Table 8.1a). Subsequent analyses of additional vent and seep species (Brooks *et al.*, 1987; Fisher *et al.*, 1988a,b,c; Van Dover & Fry, 1989) reveal that the nitrogen isotopic compositions of these invertebrates are consistently lighter ($\delta^{15}N = -12$ to $+4$‰) than those of typical shallow and deep-water PON sources.

As with CO_2 fixation, expression of nitrogen fractionation effects during assimilation is dependent on substrate source and availability and differences in metabolic uptake. A number of hypotheses have been invoked to explain the unusually negative $\delta^{15}N$ ratios of vent and seep species. First, a wide variety of nitrogenous sources are available to vent and seep species from surface and deep-water input, subsurface fluids, and porewater fluids (Fig. 8.4). Secondly, the light $\delta^{15}N$ ratios may be due to the utilization of non-limiting nitrogen sources which would allow full expression of isotopic source and fractionation effects. Nitrate is especially abundant in ambient deep-sea environments ($15-45\,\mu\text{mol}\,l^{-1}$), in contrast to very low concentrations typical of surface waters. Elevated ammonia concentrations are associated with some hydrothermal effluents (Johnson *et al.*, 1988). Thirdly, N_2 fixation by free-living or symbiotic bacteria may be responsible for the negative $\delta^{15}N$ ratios observed in vent and seep symbioses. N_2 fixation has a small effect on $\delta^{15}N$ ratios (0 to -4‰), thus more negative isotope ratios could result if N_2 was the nitrogen source. Further, N_2 associated with hydrocarbon seeps has been demonstrated to be in the -4‰ range in at least one instance (Brooks *et al.*, 1987), providing the potential for more negative $\delta^{15}N$ ratios.

Because $\delta^{15}N$ ratios increase by about $+3.4$‰ with each trophic step (Minagawa & Wada, 1984; Fig. 8.2), Rau (1985) concluded that the ^{15}N-depleted $\delta^{15}N$ ratios measured in vent species suggests a local nitrogen source which has undergone little or no biological cycling. But nitrogen isotopic fractionation effects and uncertainties of the isotopic composition of inorganic nitrogenous sources make interpretation of the isotopic composition of assimilated nitrogen difficult. The ambiguity prevents determination of the relative importance of potential nitrogen sources in vent or seep species without further data on availability and isotopic composition of the sources and biochemical assessment of nitrogen metabolism within organisms.

8.2.3 *$\delta^{34}S$ ratios of vent and seep species*

The initial $\delta^{13}C$ and $\delta^{15}N$ studies of Rau and Hedges were complemented by Fry *et al.* (1983a) in a report on the $\delta^{34}S$ composition of vent organisms. Potential sources available at vents include photosynthetically-derived particulate organic matter, locally produced organic matter, vent sulfides, and porewater sulfides and thiosulfate (Fig. 8.4). Because the isotopic fractionation associated with the assimilation of sulfur compounds is thought to be small (see Section 8.1.2), the isotope ratios of vent and seep species should reflect source values. Food webs based on phytoplankton have sulfur isotopic values close to that of seawater sulfate ($\delta^{34}S$ c. +10 to +20‰; Fry *et al.*, 1983a; Peterson & Fry, 1987) while sulfuroxidizing chemoautotrophic bacteria usually have lower $\delta^{34}S$ ratios as biologically depleted or thermogenic 'light' sulfides are the source of sulfur. Fry *et al.* (1983a) found $\delta^{34}S$ ratios of −5 to +5‰ in the vent clams, tubeworms and crabs they measured (Table 8.1a), suggesting the utilization of local sulfur sources such as vent sulfides ($\delta^{34}S$ = 0 to +5‰). The sulfides are presumably used as energy sources for bacterial chemosynthesis. More recent studies (Brooks *et al.*, 1987) have extended the range of $\delta^{34}S$ ratios found in vent and seep species ($\delta^{34}S$ = −11.5 to +13.4‰; Table 8.1a).

In systems like the Galapagos vents, point sources of vent waters have been sampled, and the geothermal nature of the sulfides demonstrated (Fry *et al.*, 1983a). The variability of $\delta^{34}S$ values for species from other areas suggest that sulfur sources of different isotopic compositions may be utilized; different species do appear to preferentially utilize different sulfur species (Felbeck *et al.*, 1981; Fisher *et al.*, 1987; Stein *et al.*, 1988). Binding and dissociation of sulfate from blood components (active transport systems) may also produce isotope fractionations and variations in bacterial sulfur oxidation pathways are possible.

The sulfur-oxidizing capabilities of the symbiont-bearing tissues of some of the species examined, the small digestive systems of vent and seep species, and the presence of high levels of elemental sulfur in symbiont-containing tissues, suggest that endosymbiotic sulfur oxidizing bacteria may be the dietary source of sulfur for these animals (see reviews of Cavanaugh, 1985; Fisher, 1990). The sulfur isotope values are consistent with a diet of sulfur-oxidizing endosymbiotic bacteria.

8.3 Vent mussels: stable isotope techniques at the population level

Rose Garden (Galapagos Spreading Center, Fig. 8.3), one of the first hydrothermal vent communities discovered in 1979, has been the focus of intense biological investigations aimed at clarifying the relationship between vent organisms and their environment. This site is dominated by tubeworms, clams and mussels living in diffuse flows of warm, sulfide-enriched water. Of the three dominant, symbiont-bearing species, only the mussel, *Bathymodiolus thermophilus*, is distributed throughout the vent field (Hessler *et al.*, 1988). It is most abundant in the central clumps of tubeworms where vent flow is greatest, but is also found in peripheral areas where tubeworms and clams do not occur.

Unlike the vestimentiferan tubeworms which lack a digestive system, or the vent clams at this site which have greatly reduced guts, *B. thermo-*

philus has a functional mouth and gut (LePennec & Hily, 1984) and may supplement its nutrition from endosymbiotic bacteria by filter feeding (LePennec & Prieur, 1984). In an effort to understand the ecology of the mussels at Rose Garden, Fisher *et al.* (1988a) collected *B. thermophilus* from central and peripheral sites and subjected them to an array of chemical analyses, including stable isotope analysis. It was suspected that peripheral populations would rely more heavily on filter-feeding strategies than central populations close to the sulfur source, and that these potential differences in feeding strategy would affect animal stable isotope ratios.

Unfortunately, the results of the carbon isotope analyses shed little light on the relative importance of filter-feeding versus symbiont chemosynthesis; differences in $\delta^{13}C$ ratios between non-symbiont-bearing tissues of central and peripheral populations were negligible ($\leq 1.5\%$; Fisher *et al.*, 1988a). However, what remains indisputable is that both central and peripheral populations of *B. thermophilus* obtain most of their nutrition from non-pelagic, non-photosynthetic sources of carbon.

Fisher *et al.* (1988a) did find striking differences in nitrogen isotope ratios between central ($\delta^{15}N$ c. -3.9%) and peripheral populations of *B. thermophilus* ($\delta^{15}N$ c. $+3.5$) at Rose Garden. These differences are consistent with the hypothesis that non-vent PON with more positive $\delta^{15}N$ ratios ($> +6\%$; Saino & Hattori, 1987) is important in the nutrition of peripheral *B. thermophilus* populations. However, alternative hypotheses can also be put forward including different nitrogen source availability and the possibility of N_2 fixation by the endosymbionts of central *B. thermophilus* populations.

8.4 Vent food webs: a community approach to isotope studies

While vent communities are often spectacular showcases for symbiotic relationships between invertebrates and bacteria, a variety of non-symbiont-containing species are also abundant at vents and can contribute significantly to total vent biomass (Van Dover & Fry, 1989). Trophic relationships among the dominant non-symbiotic faunal assemblages at deep-sea vent communities are relatively obscure. These organisms include suspension-feeders (barnacles), grazers (gastropods, polychaetes) and scavengers or carnivores (decapod crustaceans), which could use chemo- and heterotrophic free-living bacteria as food sources. Van Dover & Fry (1989) used isotopic techniques to determine the extent to which these invertebrates rely on chemosynthetic production and to explore trophic relationships among a wide variety of heterotrophic and symbiont-bearing fauna at two geographically and faunistically distinct sites, a community at Hanging Gardens on the East Pacific Rise, and communities at the Marianas Spreading Center (Fig. 8.3).

At both sites, carbon and nitrogen isotopic compositions of hetero-trophic vent invertebrates are significantly different from those of non-vent deep-sea invertebrates (compare values for vent annelids and arthropods (Table 8.1a) with species listed in Table 8.1b). Surprisingly, however, many of these vent heterotrophs are enriched in ^{13}C (c. −10 to −17‰, Table 8.1a), with $\delta^{13}C$ ratios comparable to those measured in tubeworms at vents (c. −11‰). These values indicate sources with $\delta^{13}C$ values of $\geqslant -11$‰. The *a priori* expectation was that free-living chemo-synthetic bacteria comprise the base of the heterotrophic food web, and that carbon fractionation in these bacteria should result in an isotopically 'light' pool of organic carbon. The isotope ratios found suggest that there is a population of isotopically 'heavy' free-living bacteria at vents which are preyed upon by some vent heterotrophs. Predation of isotopically 'heavy' tubeworms by many of the heterotrophic vent invertebrates can be ruled out, as some species are deposit feeders and grazers which seem to feed on free-living bacteria (Van Dover & Fry, 1989). Van Dover *et al.* (1988) found similarly heavy $\delta^{13}C$ values for the shrimp *Rimicaris exo-culata* at the TAG site on the Mid-Atlantic Ridge. The stomach of *R. exoculata* contains only sulfides and bacterial cells (Van Dover *et al.*, 1988). The ^{13}C-enriched $\delta^{13}C$ ratios of both vent tubeworms and vent heterotrophs requires further study.

Van Dover & Fry (1989) also attempted to use nitrogen isotopic compositions to estimate the trophic complexity of hydrothermal vent communities. The number of trophic levels was estimated by measuring the difference in $\delta^{15}N$ ratios between symbiont-containing tissues of vent species (which were taken as representative of bacterial values) and animal values of symbiotic and non-symbiotic species. Based on the assumption that there is a systematic increase in $\delta^{15}N$ ratios with each trophic step (c. 3.4‰, DeNiro & Epstein, 1981a; Minagawa & Wada, 1984), 2.5–3.5 trophic levels were estimated for the vent sites examined. This approach also assumes that a uniform pool of nitrogen is used by microorganisms and transferred throughout the community. Microhabitat variation in vent-water chemistry (Johnson *et al.*, 1988) suggests that nitrogen pools may not in fact be uniform, and additional analyses of nitrogen availability, utilization and fractionation within vent communities are required before trophic complexity can be estimated accurately using stable isotope ratios. However, this approach has provided preliminary evidence on feeding guilds and trophic complexity at vent sites, and the stable isotope data seems to be supported by gut content analyses and observational data (Van Dover & Fry, 1989).

8.5 Fossil chemosynthetic communities: reconstruction of past environments using stable isotope techniques

Cretaceous fossil assemblages in the Canadian Arctic Archipelago have been found which are charaterized by localized abundances of bivalves and serpulid worms reminiscent of chemosynthetic communities. This discovery led Beauchamp *et al.* (1989) to examine the isotopic composition of associated authigenic sediments, in order to determine whether these fossil communities may have been based on bacterial chemosynthesis.

Carbonate $\delta^{13}C$ compositions at these fossil sites are depleted in ^{13}C ($\delta^{13}C = -25$ to $-50\text{\textperthousand}$) relative to normal marine carbonates from the same period ($\delta^{13}C = +1.1$ to $+4.0\text{\textperthousand}$; Beauchamp *et al.*, 1989). This amount of ^{13}C depletion is similar to that of methane-derived carbonate rocks from modern seep environments. If it is assumed that these isotope ratios have not been altered by diagenetic processes, it would appear that the fossil assemblages represent communities that had a significant carbon source associated with methane oxidation.

The fossil environment in the Canadian Arctic Archipelago was further characterized using $\delta^{18}O$ ratios. The temperature dependence of oxygen isotopic fractionation during equilibrium carbonate deposition can be used to estimate water temperatures at the time of deposition (Epstein *et al.*, 1953). $\delta^{18}O$ compositions of the carbonates associated with the fossil assemblages are slightly enriched in ^{18}O ($\delta^{18}O$ near $0\text{\textperthousand}$) relative to normal marine fossils from the same period (Beauchamp *et al.*, 1989). A similar enrichment is reported in modern authigenic carbonates from the Oregon Subduction Zone (Kulm *et al.*, 1986) and the North Sea (Hovland *et al.*, 1987). Because of the temperature-dependent fractionation between carbonate and seawater (Epstein *et al.*, 1953), this enrichment suggests that the fossil seep site was, if anything, cooler than the surrounding seawater, eliminating the possibility of a hydrothermally driven community. This fossil assemblage probably inhabited a cold water seep.

8.6 Methanotrophy: isotopic clues to a new symbiosis

Soon after the discovery of sulfur-based symbioses at hydrothermal vents and seeps, chemoautotrophic endosymbionts which appear to use methane as energy sources were reported in at least two seep species from the Louisiana Slope and the Florida Escarpment (Fig. 8.3; Paull *et al.*, 1985; Brooks *et al.*, 1987; Cavanaugh *et al.*, 1987), and suggested for species from the Oregon Subduction Zone (Kulm *et al.*, 1986). Utilization of methane derived from different sources will result in distinctive $\delta^{13}C$ ratios. The methane sources may be derived from methanogenic bacteria, which produce predominantly ^{13}C-depleted methane ($\delta^{13}C \leqslant -60\text{\textperthousand}$; Claypool & Kaplan, 1974; Fuchs *et al.*, 1979) with little or no higher molecular weight gases (i.e., ethane, propane), or thermogenic methane input processes which results in methane $\delta^{13}C$ ratios of $\geqslant 50‰$ (Claypool & Kaplan, 1974; Brooks et al., 1987). In addition, bacterial alteration of thermogenic methane and the mixing of different methane sources can result in a wide range of naturally occurring methane $\delta^{13}C$ ratios.

Methane-based symbiotic relationships are thought to function in a manner similar to the sulfur-oxidizing symbioses described previously in that the host provides an opportunity for the bacterial endosymbionts to obtain the gases necessary for chemoautotrophy and the endosymbionts produce organic compounds which are translocated to the animal host (Fisher et al., 1987). If methane-oxidizing symbionts incorporate methane carbon with little expression of isotopic fractionation, methanotrophic bacteria will have $\delta^{13}C$ ratios similar to the methane utilized. If the animal host is utilizing endosymbiont carbon, host $\delta^{13}C$ ratios should also reflect the isotope values of the source methane. By examining the $\delta^{13}C$ ratios of seep species (Table 8.1a), it can be seen that at one Louisiana seep site the mussel tissues $\delta^{13}C$ ($-40.6‰$) are similar to the $\delta^{13}C$ ratios of methane collected nearby ($\delta^{13}C = -41.2‰$; Brooks et al., 1987). The gases escaping at this location were shown to be thermogenic fluids reservoired 2000–3000 m deep in the subsurface (Kennicutt et al., 1988). On the other hand, the Florida escarpment mussels ($\delta^{13}C = -74.3‰$) are probably utilizing primarily biogenic methane ($\delta^{13}C < -60‰$; Claypool & Kaplan, 1974; Fuchs et al., 1979); methane is the only known primary source of carbon in near surface sediments so highly depleted in ^{13}C. Thus stable isotope ratios provide chemical evidence to suggest the utilization of endosymbiotic methanotrophic bacterial biomass for nutritional purposes in some seep species.

Although the $\delta^{13}C$ ratios of these species cannot be used to quantify the nutritional importance of the endosymbionts or to specific specific methane sources, they provide data suggestive of the potential for methane-driven chemoautotrophy at these seep ecosystems. Ultrastructural and physiological analyses of the endosymbionts of these species demonstrate the presence of cells similar to methanotrophic bacteria, enzymes involved in the oxidation of methane and methane oxidation by some species has been recorded (Childress et al., 1986; Cavanaugh et al., 1987; Cary et al., 1988) providing independent evidence of methanotrophy in these unusual animals. In addition, radiocarbon analyses (Brooks et al., 1987) support the use of methane at seep environments, and have even been used to estimate the relative importance of biogenic versus thermogenic methane utilization.

8.7 Invertebrate–chemoautotroph symbioses in non-vent/seep environments

Environmental conditions suitable for bacterial chemosynthesis are by no means limited to vent and seep environments. Free-living sulfur-oxidizing and methane-oxidizing bacteria are widespread in marine sediments where oxic and anoxic sediments co-occur. Since the initial discovery of vent symbioses in 1981, many non-vent/seep species living in organic-rich sediments (e.g., some species of Solemyid, Lucinid and Thyasirid bivalves) have been found to contain endosymbiotic sulfur-oxidizing bacterial symbionts within gill tissues (Table 8.2a). Moreover, many species which do not appear to inhabit reducing sediments, such as members of the gutless oligochaete genus *Inandrilus* (= *Phallodrilus*), some thyasirid bivalves and the lucinid bivalve *Myrtea spininfera* (Giere, 1985; Dando *et al.*, 1986; Spiro *et al.*, 1986; Giere & Langheld, 1987; Schmaljohann *et al.*, 1990) also contain endosymbiotic sulfur-oxidizing bacteria, thus this type of animal-bacteria association may in fact be widespread in the marine environment. In addition, *Siboglinum poseidoni*, a pogonophoran tube worm, has recently been reported to contain endosymbiotic methanotrophic bacteria, further extending the known niches of this type of symbiosis (Schmaljohann *et al.*, 1990). Both sulfur-oxidizing and methanotrophic symbioses can be found in the same habitat (Schmaljohann *et al.*, 1990). Some of these symbioses (e.g., *Solemya velum*) inhabit shallow-water sediments and are thus more readily accessible for study. It is perhaps ironic that it was only after the unusual vent communities were found that these more common symbioses were discovered.

For the most part, the structure and overall physiology of these non-vent/seep symbioses is similar to that of vent/seep species. Thus, some of the more accessible shallow-water symbioses have become general models of invertebrate–chemoautotroph symbioses. For example, in the case of coastal and shallow-water symbioses, it is possible to collect large numbers of samples with relative ease and preform the type of 'trial-and-error' experiments which are impossible to carry out on the few species retrieved by deep submersibles. Moreover, habitat accessibility permits more detailed characterizations of the organisms' surroundings and laboratory studies allow the biochemistry, physiology and feeding strategies of invertebrate–chemoautotroph symbioses to be investigated. Thus, carbon, nitrogen and sulfur sources can be constrained permitting a better interpretation of stable isotope data.

The stable isotope ratios of most of the non-vent/seep symbiotic species analysed suggest that, even in shallow-water environments, many invertebrate–chemoautotroph symbioses do not appear to utilize photosynthetically derived organic matter (Table 8.2a). $\delta^{13}C$, $\delta^{15}N$ and $\delta^{34}S$ ratios are all more negative than those associated with phytoplankton

Table 8.2 Stable isotope ratios of (a) non-vent/seep invertebrate–chemoautotroph symbioses in comparison with (b) non-symbiotic species inhabiting similar environments

Species	Site	Tissue	$\delta^{13}C$	$\delta^{15}N$	$\delta^{34}S$	References
(a) NON-VENT/SEEP INVERTEBRATE–CHEMOAUTOTROPH SYMBIOSES						
Phylum Pogonophora						
Siboglinum atlanticum	Southern Bay of Biscay, 1700 m	Various tissues and tube	−38.2 to −44.7‰			Southward et al. (1986)
Siboglinum atlanticum	Northern Bay of Biscay	Various tissues and tube	−42.7 to −45.8‰			Southward et al. (1986)
Siboglinum fordicum	Norwegian fjords	Various tissues and tube	−35.5‰			Southward et al. (1986)
Siboglinum ekmani	Norwegian fjords	Various tissues and tube	−45.3‰			Southward et al. (1986)
Siboglinum poseidoni	Skaggarak	Anterior / Posterior / Tube	−73.6‰ / −78.3‰ / −62.2‰			Schmaljohann et al. (1990)
Phylum Mollusca						
Solemya velum	Reducing sediments, Massachusetts, USA	Gills / Foot / 'bacterial pellet'	−32.4 to −33.9‰ / −30.9 to −32.1‰ / −31.7 to −33.6‰	+0.4 to −9.8‰ / +4.4 to −8.6‰ / −6.9 to −8.6‰	−26.7 to −28.2‰ / −29.2 to −31.1‰	Conway et al. (1989)
Solemya borealis	Reducing sediments, Massachusetts, USA	Gills / Foot	−34.6‰ / −32.0‰	−9.7‰ / −8.6‰	−15.7‰ / −32.6‰	Conway et al. (1992) Submitted
Lucinoma borealis	Devon, England (non-reducing sediments)	Gills / Remaining soft parts	−28.1 to −29‰ / −25.3 to −25.9‰			Spiro et al. (1986)
Lucinoma borealis	Norwegian fjord	Gills / Remaining soft parts	−28.8‰ / −24.1‰			Spiro et al. (1986)

Species	Location	Tissue			Reference
Thyasira flexuosa	Norwegian fjord	Gills	−29.3‰		Spiro et al. (1986)
Thyasira sarsi	Norwegian fjord	Gills	−31.0‰		Spiro et al. (1986)
		Remaining soft parts	−28.2‰		
Thyasira sarsi	Skaggarak	Gills	−39.5‰		Schmaljohann et al. (1990)
		Remaining soft parts	−37.4‰		
Myrtea spinifera	Norwegian fjord	Gills	−24.2‰		Spiro et al. (1986)
		Remaining soft parts	−23.4‰		
Codakia obicularis	Grand Bahamas	Gills	−23.9 to −28.3‰		Berg & Alatalo (1984)
	Eel grass beds	Remaining soft parts	−23.2 to −28.1‰		
Phylum Annelida					
Inandrilus leukodermatus	Calcareous sediments (non-reducing sediments)	Whole animal	−26.0‰	−2.1‰	Giere et al. (1990)
(b) NON-SYMBIOTIC SPECIES					
Phylum Mollusca					
Mya arenaria	Reducing sediments, Massachusetts, USA	Gills	−17.2, −17.5‰	+8.5‰	Conway et al. (1989)
		Foot	−17.7, −17.8‰	+8.3‰	
Tellina agilis	Reducing sediments, Massachusetts, USA	Gills	−15.4‰	+6.3, +6.5‰	Conway et al. (1989)
		Foot	−14.2, −15.2‰	+6.8, +8.2‰	
Phylum Annelida					
Polychaete	Skaggarak	Body	−20‰		Schmaljohann et al. (1990)

photosynthetic primary production (Table 8.2b). The most negative $\delta^{13}C$ ratios are found in the methanotrophic *S. poseidoni* symbiosis (Table 8.2a). The nutritional importance of the endosymbiotic bacteria varies among species. For example, the bivalve *Spisula subtruncata*, which has a normal digestive tract (Soyer *et al.*, 1987) and Lucinids (e.g., *M. spinifera*) and Thyasirids (e.g., *Thyasira flexuosa*) with small but functioning digestive systems (Spiro *et al.*, 1986; Herry & LePennec, 1987) probably utilize photosynthetically produced particulate organic compounds. In fact, $\delta^{13}C$ ratios of these species are slightly more positive (c. -23 to -31‰, Table 8.2a) than those of species known to have minimal gut digestive capabilities (e.g., *Solemya reidi*, *S. velum* and the pogonophoran tube worms, $\delta^{13}C$ c. -30 to -45‰; Table 8.2a). There appears to be a general correlation between degree of mixotrophy and more positive $\delta^{13}C$ ratios. To further complicate $\delta^{13}C$ interpretations, small pogonophoran tube worms, *Inandrilus leukodermatus*, and many of the bivalves which are now known to contain symbiotic sulfur-oxidizing bacteria are also capable of significant uptake of dissolved organic compounds (Giere *et al.*, 1984; Fiala-Médioni *et al.*, 1986). In order to use stable isotope ratios to infer the quantitative importance of endosymbiont nutrients, it is necessary to understand the potential importance of alternative modes of nutrition. Thus biochemical and physiological studies should accompany stable isotope studies where possible.

   S. velum was the first non-vent/seep species discovered to contain endosymbiotic sulfur-oxidizing autotrophic symbionts (Cavanaugh, 1983). Because there is a considerable amount of data available on both the overall physiology, biochemical composition, feeding strategies and stable isotope composition of this species, it serves as a good model of the overall organization of animal–bacteria associations. Moreover, knowledge of the overall biology of the species enables the stable isotope data for this species to be better interpreted and used in a quantitative manner.

   Analysis of the $\delta^{13}C$, $\delta^{15}N$ and $\delta^{34}S$ ratios of tissues with and without symbiotic bacteria (i.e., gill vs. foot tissue) and the analysis of an enriched bacterial fraction derived from gill sections allowed Conway *et al.* (1989) to use a simple two source mixing model to estimate the relative contribution of chemosynthetic versus photosynthetic nutrients to *S. velum*. Assuming that: (i) the values obtained for the 'bacterial pellet' were representative of endosymbionts; (ii) values obtained for other bivalve species inhabiting similar environments were representative of photosynthetically derived nutrients; and (iii) symbionts and exogenous photosynthetically produced organic matter were the only food sources for *S. velum*, then the percentage of the hosts carbon derived from these two sources could be estimated (Conway *et al.*, 1989; Fig. 8.5). From this analysis it would appear that most of the host's carbon and variable

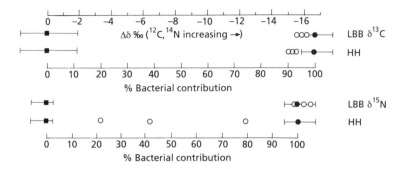

Fig. 8.5 Bacterial contribution to carbon and nitrogen nutrition in *Solemya velum* at Little Buttermilk Bay (LBB) and Hadley's Harbor (HH), calculated from a two-source mixing model. Δδ values denote isotopic differences between the foot tissue of control bivalves (■ = bacterial contribution = 0%), the bacterial pellet (● = bacterial contribution = 100%), and *S. velum* foot tissue (○). Horizontal lines represent the ranges of isotopic values recorded. (From Conway *et al.*, 1989.)

amounts of host nitrogen are derived from the bacterial endosymbionts. Also, the extremely negative $\delta^{34}S$ ratios found in *S. velum* (Table 8.2a) could only be derived from the utilization of biogenically produced H_2S.

In order to accept these quantitative estimates, we need to determine sedimentary carbon $\delta^{13}C$ and $\delta^{15}N$ signatures, rule out alternative trophic pathways for the animal host, and have supporting evidence for utilization of bacterial metabolites. The $\delta^{13}C$ and $\delta^{15}N$ ratios of the sediments at Little Buttermilk Bay, where most of the specimens were collected, have strong phytoplankton signatures (Conway *et al.*, 1989) suggesting that particulate organic matter (POM), and dissolved organic matter (DOM) in the sediments would also have photosynthetic values. The 'light' isotopic ratios of *S. velum* suggest that assimilation of POM and DOM sources by *S. velum* is probably minimal. In addition, laboratory experiments with radiolabelled algal cells (J.W.C. White, N. Conway & J. McDowell-Capuzzo, unpublished data) demonstrate negligible ingestion of POM by *S. velum*.

Analysis of the lipid composition of *S. velum*, supports the conclusions derived from the stable isotope studies. The dominant fatty acid in *S. velum* is *cis*-vaccenic acid, a fatty acid usually found in large concentrations only in bacteria (Goldfine, 1972; Fulco, 1983). *Cis*-vaccenic acid is found in all the tissues of *S. velum*, and all the lipid class pools of this species, suggesting transport and utilization of bacterial metabolites (Conway & McDowell-Capuzzo, 1991). Only low levels of plant-derived fatty acids and sterols were recorded in *S. velum*. Moreover, analysis of the $\delta^{13}C$ ratios of the fatty acids and sterols of *S. velum* reveals strong chemosynthetic $\delta^{13}C$ values (Table 8.3), suggesting that the lipids of this

Table 8.3 Stable isotope ratios of some of the biochemical components of *Solemya velum* in comparison with the symbiont-free clam *Mya arenaria* and *Thiomicrospira crunogena*, a sulfur-oxidizing chemoautotroph

Species	$\delta^{13}C$	$\delta^{15}N$	$\delta^{34}S$	References
Solemya velum				
Intact tissue	−31.5 to −33.9‰	−6.9 to −9.8‰	−28.2, −31.1‰	Conway et al. (1989)
Fatty acids	−45.4‰			Conway & McDowell-Capuzzo (1991)
Sterols	−38.5‰			Conway & McDowell-Capuzzo (1991)
Total amino acids	−30.4‰	−8.0‰		Conway & McDowell-Capuzzo (1992)
Free amino acids	−27.6‰	−9.2‰	−17.2‰	Conway & McDowell-Capuzzo (1992)
'Endosymbiont fraction'	−32.3 to −33.6‰	−6.9 to −8.0‰		Conway & McDowell-Capuzzo (1992)
Mya arenaria				
Fatty acids	−17.1 to −17.8‰			Conway & McDowell-Capuzzo (1991)
Sterols	−23.8‰			Conway & McDowell-Capuzzo (1991)
Total amino acids	−24.2‰	+9.7‰		Conway & McDowell-Capuzzo (1992)
Free amino acids	−16.6‰ −18.7‰	+7.8, +7.9‰		Conway & McDowell-Capuzzo (1992)
Thiomicrospira crunogena				
Bacteria	−27.3‰			Conway & McDowell-Capuzzo (1991)
Fatty acids	−45‰			

species are synthesized either directly by the symbionts or by the host using carbon derived from the symbionts. Similarly, the $\delta^{13}C$ ratios of the amino acids of *S. velum* reveal chemosynthetic signatures (Table 8.3) suggesting that these molecules are also synthesized from bacterial carbon and nitrogen sources. The $\delta^{34}S$ ratios of the free amino acid pool are particularly interesting ($-17.2‰$), as the free amino acids of *S. velum* consist of about 70% taurine (a sulfur-containing amino acid). The $\delta^{34}S$ ratios of the free amino acid pool thus probably represent the $\delta^{34}S$ ratios of the free taurine in this species. This demonstrates internal sulfur-isotope fractionation in *S. velum*, as the tissue $\delta^{34}S$ ratios for this species are about $-30‰$.

Overall, the stable isotope ratios of *S. velum*, when considered with the known biology of this species, can be used in quantitative manner to determine the relative utilization of different source compounds, and the results obtained with stable isotope ratios are supported by experimental studies on the physiology and biochemistry of the host.

8.8 Summary and concluding remarks

The studies outlined here have focused on some of the wide range of questions concerning chemosynthetic-based ecosystems which have been addressed using stable isotope ratios and which provide novel insights into the nature of vent and seep communities and invertebrate–chemoautotroph symbioses. Because these communities were discovered relatively recently, this research has often been preliminary in nature, yet analyses of stable isotope ratios has provided the starting point for subsequent, more detailed investigations. Stable isotope techniques have been particularly useful in the study of chemosynthesis-based ecosystems because of the following.

1 Isotopic analyses are unaffected by damage to organisms during collection. Although the first biological specimens obtained from deep-sea vents were generally damaged and often unsuitable for conventional physiological investigations, they were adequate for isotope work.

2 Preservation of material for isotope analyses requires no special equipment or procedures, an invaluable characteristic for shipboard sampling.

3 A small number of specimens can provide critical preliminary information to guide subsequent research programmes. This is particularly important in studies of deep-sea fauna, where the difficulties of sample collection makes many specimens precious commodities.

4 Isotopic data can provide synoptic measures of the nutritional and/or environmental milieu in which an organism lives.

At present, stable isotope ratios often cannot be used alone to make definitive statements regarding element cycling and trophic mechanisms in chemosynthetic-based communities until sources, environmental chemistry and animal/bacterial physiology are better understood. It is imperative that the interpretation of stable isotope ratios be confirmed by collaborative studies and that the stable isotope ratios of source materials become better characterized, as the cycling of elements within vent/seep communities can be complex (Fig. 8.4) due to multiple sources with variable isotopic compositions. In symbiont-containing systems, the stable isotope ratios of the intact symbiosis represents the end product of a complex series of chemical and biological processes that can often produce similar isotopic compositions through multiple pathways. Isotope ratios alone cannot be definitively linked to a unique process. For example, free-living chemoautotrophic bacteria will have stable isotopic compositions similar to those of symbiotic chemoautotrophic bacteria if elemental sources and source availability are similar, and a common enzymatic process is responsible. Therefore organisms capable of filter feeding can have isotopic signatures typical of chemoautotrophy without the presence of chemoautotrophic symbionts. Despite these drawbacks, stable isotope analyses are valuable ecological tools for examining the flow of nutrients and energy in a variety of marine environments, and are providing vital, and otherwise difficult to obtain, information regarding the biology and chemistry of chemosynthetic-based ecosystems.

CHAPTER 9

Physiology of isotopic fractionation in algae and cyanobacteria

R. GOERICKE, J.P. MONTOYA & B. FRY

9.1 Introduction

Carbon and nitrogen isotopes are used in tracer experiments to determine rates of photosynthesis, nutrient uptake, or respiration in aquatic algae and cyanobacteria. While tracer addition experiments are a valuable experimental approach, they also have drawbacks. For example, organisms have to be confined in incubation bottles; this confinement may adversely affect the individual organisms or the interaction of groups of organisms. Additionally, the time scale of the experiments may not be well matched to the natural time scale of the organisms. Incubation times are constrained in the case of nitrate, ammonia or phosphate uptake experiments by the turnover rates of the inorganic nutrient pools in the medium (which can range from minutes to hours); however, the relevant time scales of the organisms, the reciprocal of the growth rates, range from days to weeks.

Natural abundances of stable isotopes are increasingly being studied by ecologists in a attempt to avoid some of the problems encountered in tracer experiments. The distribution of stable isotopes in aquatic systems has been related to rates of elemental cycling or to the sources of carbon and nitrogen that are utilized by the organisms (Fry & Sherr, 1984; Peterson & Fry, 1987; Sackett, 1989). Such studies have shown that isotopic compositions of algae in natural systems are affected by three factors: the isotopic composition of inorganic nutrients, isotopic fractionation during the uptake and metabolism of those nutrients, and by fractionation during catabolic processes. The understanding of carbon and nitrogen isotopic fractionation by algae and cyanobacteria is the particular focus of this chapter.

Isotopic distributions in algae can be affected by rates of nutrient uptake, photosynthesis or respiration; in short, those processes that physiological ecologists are most interested in. Thus, it is in principle possible to use stable isotopes to study these processes. Such an approach has several advantages:

1 the processes that determine isotopic distributions act on time scales very similar to those of biological processes;

2 the isotopic composition of organic matter is usually not changed by the process of sampling and analysis;

3 the isotopic composition of organic matter, and in particular the isotopic composition of specific compounds, may be unchanged by diagenetic processes such that the isotopic signature of the source organisms may still be preserved in ancient sediments (Hayes *et al.*, 1987, 1990).

Below we will discuss the processes that affect the fractionation of carbon and nitrogen isotopes in eukaryotic algae and cyanobacteria (for brevity *algae*) and point out ways to use natural distributions of isotopes to understand the physiology of algal populations in natural systems. Unfortunately, many of the isotopically important physiological processes have not been studied in detail. At times our main motivation will be to point out gaps in our knowledge to stimulate research on those aspects of isotopic fractionation in algae. We will conclude with a model of isotopic discrimination in algae.

For the discussion of carbon uptake and fixation by algae we will rely on recent reviews, primarily those by Kerby & Raven (1985), Badger (1987) and Glover (1989); isotopic fractionation in higher plants and algae has recently been reviewed by O'Leary (1981, 1988), Raven & co-workers (Kerby & Raven, 1985; Raven, 1987; Raven *et al.*, 1987, 1990), Berry (1988), Farquhar *et al.* (1989a) and Wada & Hattori (1991). Patterns of isotopic distribution in aquatic ecosystems have been reviewed by Fry & Sherr (1984), Peterson & Fry (1987) and Wada & Hattori (1991) as well as other chapters in this book.

9.2 Modelling isotopic fractionation

9.2.1 Definitions

A variety of notations and symbols have been used to express the degree of isotopic fractionation associated with chemically or biologically mediated reactions (see the Introduction for the definitions used here). Isotopic discrimination arises when the rate constant (k) for a reaction A $\rightarrow$ B differs for two isotopic species ^mA and ^nA (for $m < n$), i.e., $^mk \neq {}^nk$. The ratio of these reaction rate constants, usually expressed as $^mk/{}^nk$ (but see Mariotti, 1981), is the isotopic discrimination factor α:

$$\alpha = {}^mk/{}^nk. \tag{9.1}$$

Values of α for reactions involving nitrogen or carbon are usually greater than 1.000 with values up to 1.040 for discrimination against the heavier isotope by 40‰. The range of variation in α for biological processes is bounded by the intrinsic fractionation effects of chemical reactions, which are on the order of 40–60‰ for carbon and about 30‰ for nitrogen (M.H. O'Leary, personal communication 1991). It is often convenient to define a per mil isotopic enrichment factor ε, which is positive for normal isotope effects in which the lighter species reacts more quickly than the isotopically heavy species, i.e., ^mA reacts faster than ^nA:

$$\varepsilon = 1000(\alpha - 1). \qquad (9.2)$$

Isotopic differences between two substances are often denoted using the delta-del notation, $\Delta\delta^{13}C$.

9.2.2 Fractionation effects

To illustrate the relationship between discrimination factors and isotopic composition expressed in δ-notation, we consider the simple uptake of a substrate (S) such as CO_2 or NH_4^+ by an organism. The isotopic system is described by the fractionation factor associated with the uptake process, ε_1, and the δ-values of the substrate pool (S) and the product (P), i.e., the substrate that was taken up by the organisms, δ_S and δ_P, respectively:

$$\delta_S \xrightarrow{\varepsilon_1} \delta_P. \qquad (9.3)$$

Assume that the substrate pool is infinitely large, such that its concentration and isotope ratio does not change significantly while the system is observed. We will also assume that the isotopic composition of the product pool does not change over time. Uptake of the substrate can be represented as:

$$S \xrightarrow{k_1} P. \qquad (9.4)$$

We assume that the reaction is first order with respect to the substrate pool such that the rate per unit biomass of the organism, ρ, depends linearly on the concentration of the substrate, i.e.,

$$\rho = k_1 S. \qquad (9.5)$$

Equation 9.5 can be written for both the light isotope m and the heavy isotope n. The ratio of the two equations gives the uptake ratio of the isotopically heavy and light substrate:

$$\frac{^n\rho}{^m\rho} = \frac{^nk_1}{^mk_1} \frac{^nS}{^mS} \qquad (9.6)$$

or rewritten using isotope ratios R_ρ, R_S (noting that R_P and R_ρ are equal at all times) and the isotopic discrimination factor α:

$$R_P = \frac{1}{\alpha_1} R_S. \qquad (9.7)$$

Equation 9.7 can be related to an expression involving δ-notation:

$$1000\left(\frac{R_P}{R_{ST}} - 1\right) = \frac{1}{\alpha_1}\left(1000\left(\frac{R_S}{R_{ST}} - 1\right) + 1000\right) - 1000 \tag{9.8}$$

which can be solved for α or for ε using Equation 9.2:

$$\alpha_1 = \frac{\delta_S + 1000}{\delta_P + 1000} \quad \text{and} \quad \varepsilon_1 = \frac{\delta_S - \delta_P}{1 + \delta_P/1000}. \tag{9.9a,b}$$

Considering the precision of most isotopic work we can neglect the denominator of Equation 9.9b and rewrite it as:

$$\delta_P = \delta_S - \varepsilon_1. \tag{9.10}$$

The above equation expresses that the isotopic composition of the substrate that is taken up is the difference of the δ-value of the substrate and the isotopic enrichment factor.

9.2.3 Steady-state fractionation models

In this chapter we will primarily be concerned with isotopic changes that occur when an inorganic substrate is taken up and metabolized by algae. We have to consider the isotopic composition of the inorganic compounds and fractionation during the uptake and subsequent metabolism. In many cases we can assume that the pool (P), which may be a whole organism or a metabolite pool, does not change its isotopic composition (δ_P) with time. This assumption is never strictly true, but the conditions are satisfied as long as changes in isotopic composition are slow compared to the rate at which material flows through the pool, in which case the system is effectively at steady state. A one-pool (P) model with a single input of a substrate S and a single output can be represented by:

$$S \xrightarrow{k_1} P \xrightarrow{k_2} \quad \text{and} \quad \delta_S \xrightarrow{\varepsilon_1} \delta_P \xrightarrow{\varepsilon_2}. \tag{9.11a,b}$$

Using Equation 9.10 we can calculate the isotopic composition of the material that enters and leaves P, i.e., δ_1 and δ_2, respectively.

$$\delta_1 = \delta_S - \varepsilon_1 \quad \text{and} \quad \delta_2 = \delta_P - \varepsilon_2. \tag{9.12a,b}$$

In the simplest case we can assume that S is infinitely large, i.e., that the isotopic composition of S is not affected by the uptake and excretion of matter by P. When the size of the pool P does not increase as a function of time, δ_1 and δ_2 must be equal. Using Equations 9.12a and 9.12b we can write the isotopic mass-balance equation

$$\delta_S - \varepsilon_1 = \delta_P - \varepsilon_2 \tag{9.13}$$

and solve it for δ_P:

$$\delta_P = \delta_S - \varepsilon_1 + \varepsilon_2. \tag{9.14}$$

This equation shows that the isotopic composition of a metabolite or elemental pool is affected by the source's isotopic composition and by isotopic fractionation during both uptake and loss of material.

Simple models of this nature have been used in diverse contexts.

1 The isotopic composition of animals has been described with a similar isotope mass-balance approach (DeNiro & Epstein, 1978). The δ-value of animals is roughly related to the isotopic composition of their diet, to the isotopic composition of their excreta, and to fractionation processes during the uptake of the diet and during excretion.

2 The isotopic composition of methane in the atmosphere has been modelled using a one-pool model with multiple inputs and outputs (Davidson *et al.*, 1987).

The assumptions made above are somewhat restrictive. Pool size often increases considerably during the observation or sampling period. Let us denote the specific growth rate of an organism by μ (day^{-1}) and the specific rate of respiration by v (day^{-1}). Thus, the specific rate of carbon uptake is $\mu + v$. The equation that describes the time-change of the pool P is:

$$\frac{dP}{dt} = (\mu + v)P - vP = \mu P. \tag{9.15}$$

When writing the isotopic mass-balance equation we have to weight the isotopic terms by the growth and respiration rates and we have to take into account the carbon that remains in the organism due to growth:

$$(\delta_S - \varepsilon_1)(\mu + v) = (\delta_P - \varepsilon_2)v + \delta_P\mu. \tag{9.16}$$

This equation can be solved for δ_P:

$$\delta_P = \delta_S - \varepsilon_1 + \varepsilon_2\frac{v}{\mu + v}. \tag{9.17}$$

In this case the isotopic effects are no longer additive, and the value of ε_2 is multiplied by the ratio of the reaction rates. Using this approach it is possible to solve more complicated models such as one-pool models with multiple inputs and outputs or multi-pool models with multiple inputs.

Inorganic carbon uptake and photosynthesis of algae provides a more complicated example. The relevant processes can be summarized in the following conceptual models that describe the flow of carbon and the isotopic system:

$$DIC_e \underset{k_{-1}}{\overset{k_1}{\leftrightarrow}} DIC_i \overset{k_2}{\to} POC \overset{k_3}{\to} \quad \text{and} \quad \delta_e \underset{\varepsilon_{-1}}{\overset{\varepsilon_1}{\leftrightarrow}} \delta_i \overset{\varepsilon_2}{\to} \delta_{POC} \overset{\varepsilon_3}{\to}. \tag{9.18a,b}$$

The parameters are: k_1, the uptake of an inorganic carbon species from the external pool of dissolved inorganic carbon (DIC$_e$); k_{-1}, the back diffusion of inorganic carbon; k_2 the incorporation of intracellular inorganic carbon (DIC_i) into particulate organic matter (POC) by photosynthesis; and k_3, dark respiration (the k_i values will be related to true physiological parameters (e.g., μ, ν, etc.) in Section 9.3.8).

Equation 9.18 can be solved for the $\delta^{13}C$ of the POC of the algae (δ_{POC}) using the methods outlined above. We will assume that DIC_e, δ_{DIC_e}, δ_{DIC_i} and δ_{POC} do not change as a function of time. We will similarly assume that DIC_i is constant. Although this assumption is not strictly true since we have allowed growth of the organisms, it is justified because the specific rate of increases of the pool is small compared to the rate at which inorganic carbon flows through it.

We first consider δ_{DIC_i}. The sum of all isotopic fluxes into DIC_i must be equal to the sum of all isotopic fluxes out of DIC_i plus the isotopic change due to biomass increase. When writing the isotopic mass balance we have to consider that DIC_i has two outputs, the back-diffusion of DIC and carbon fixation by the organism. These outputs have to be weighted by the fractions, $f = k_{-1}/k_1$ and $1 - f = k_2/k_1$, that each flux represents of the input, k_1:

$$\delta_{DIC_i} = \delta_{DIC_e} - \varepsilon_1 + f\varepsilon_{-1} + (1 - f)\varepsilon_2. \qquad (9.19)$$

We have already solved the isotopic mass-balance equation for δ_{POC} (cf. Equation 9.17). We substitute Equation 9.19 into Equation 9.17 and rearrange the terms to obtain:

$$\delta_{POC} = \delta_{DIC_e} - \varepsilon_1 + f(\varepsilon_{-1} - \varepsilon_2) + \varepsilon_3 \frac{\nu}{\mu + \nu} \qquad (9.20)$$

where μ = growth rate (day^{-1}) and ν = respiration rate (day^{-1}).

O'Leary (1981) and Berry (1988) have derived mathematically similar equations assuming that $\varepsilon_1 = \varepsilon_{-1}$ and that $\varepsilon_3 = 0$. Equation 9.20 reflects the two intuitively obvious cases (cf. O'Leary, 1981):

1 when DIC_i leaks from the organism at a rate much larger than the rate of carboxylation (i.e., when f is close to unity) fractionation of the carboxylases (ε_2) are fully expressed;

2 when only a small fraction of DIC_i leaks back to the environment and almost all the inorganic carbon that enters the cell is fixed (i.e., f is very small) observed fractionation is entirely due to the uptake of inorganic carbon, ε_1.

9.3 ^{13}C fractionation in algae

Considerable isotopic fractionation occurs as algae fix DIC. In the ocean, $\delta^{13}C$ values of plankton usually vary between -18 and $-28‰$ (Fig. 9.1); lower values of -25 to $-35‰$ are observed in the Antarctic. The isotopic composition of plankton varies somewhat as a function of latitude and

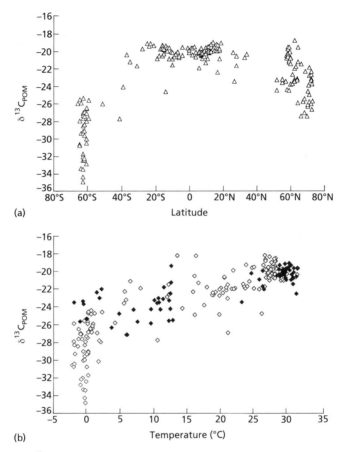

Fig. 9.1 The $\delta^{13}C$ of oceanic particulate organic matter sampled from the surface layer plotted against latitude (a) and water temperature (b). ◆ = data from the northern hemisphere; ◇ = data from the southern hemisphere. (From Goericke & Fry, in press.)

water temperature. These -18 to $-28‰$ values are actually intermediate between the value of the source carbon (DIC, about $1.5‰$ in the surface ocean) and a value of about $-40‰$ that we expect for slow-growing algae that fix CO_2 via the Calvin cycle (see Section 9.3.8). We can pose the question: Why are plankton generally $10–20‰$ enriched in $\delta^{13}C$ as opposed to the expected $-40‰$ value?

Current attempts to answer this question are focused in two directions. One focus centres on the inorganic carbon chemistry of seawater; low concentrations of free, aqueous CO_2 ($[CO_2]_{aq}$) may limit photosynthetic rates and lead to the observed enrichment of ^{13}C (Mizutani & Wada, 1982; Rau *et al.*, 1992). A contrasting view is that physiological processes within algal species, rather than external carbon chemistry, controls ^{13}C enrichment in algae. Other possible physiological explanations are the fixation of bicarbonate rather than CO_2 by β-carboxylases with small fractionation effects, in addition to fixation of CO_2 by Rubisco (Descolas-Gros & Fontugne, 1990) or the active uptake of inorganic carbon from the medium, with relatively small leakage back out of the intracellular pool (Berry, 1988). Current evidence favours the last explanation as the most likely cause of ^{13}C enrichment in phytoplankton; less is known about the other possibilities.

Below we address different views of isotopic fractionation in algae. We first review DIC chemistry relevant for algal carbon fixation (Sections 9.3.1–9.3.3), then build towards a model of isotopic fractionation by examining the roles of Rubisco and various β-carboxylases (Sections 9.3.4 and 9.3.5), and the possible importance of respiration and active carbon uptake (Sections 9.3.6 and 9.3.7). We conclude with a discussion of models of carbon isotopic fractionation in algae (Section 9.3.8).

9.3.1 DIC in aqueous media

DIC is taken up by algae in the form of CO_2 or HCO_3^-. CO_3^{2-} is thought to be biologically inactive; it is a small fraction of the total DIC at the pH of most aquatic systems (pH <8.5), and will be treated as part of the HCO_3^- pool. CO_2 is a small fraction of the total dissolved inorganic carbon (TCO_2) in seawater. At isotopic equilibrium the $\delta^{13}C$ of CO_2 (δ_{CO_2}) is about 9‰ lighter than the $\delta^{13}C$ of HCO_3^- (δ_{HCO_3}) at 25°C (Mook et al., 1974). However, these chemical and isotopic equilibria can be perturbed by biological activity.

First we will consider the effect of passive CO_2 uptake on these equilibria. Consider the following simplified chemical and isotopic system (O'Leary, 1981; Johnson, 1982):

$$[HCO_3^-] \underset{k_{-1}}{\overset{k_1}{\leftrightarrow}} [CO_2] \overset{k_2}{\to} POC \quad \text{and} \quad \delta_{HCO_3} \underset{\varepsilon_{-1}}{\overset{\varepsilon_1}{\leftrightarrow}} \delta_{CO_2} \overset{\varepsilon_2}{\to} \delta_{POC}. \quad (9.21a,b)$$

The reaction constants, k_1 and k_{-1}, are dependent on temperature, salinity and pH, with values of $k_1 = 3.7 \times 10^{-4}$ and $k_{-1} = 5.7 \times 10^{-2}$ for normal seawater (25°C, salinity of 33‰, pH of 8.2; Johnson, 1982). The relative value of these constants implies that CO_2 usually represents about 1% of the total inorganic carbon pool in seawater. The absolute value of the reaction constants implies that the chemical equilibrium between CO_2 and HCO_3^- is attained in c. 30–90 seconds, primarily depending on the pH of the medium (Johnson, 1982).

The kinetic isotopic effects on the hydration of CO_2 and the dehydration of HCO_3^- are $^{12}k_1/^{13}k_1 = \alpha_1 = 1.015$ and $^{12}k_{-1}/^{13}k_{-1} = \alpha_{-1} = 1.007$ in fresh water at 24°C (Marlier & O'Leary, 1984). Corresponding values for seawater have not been determined. The equilibrium isotopic effect, α_1 divided by α_{-1}, is about 1.008. The equilibrium fractionation factor is temperature dependent, ranging from 8.4‰ at 30°C to 12.0‰ at 0°C (Mook *et al.*, 1974). Assuming that the $\delta^{13}C$ of DIC in surface seawater is 1.5‰, the $\delta^{13}C$ of CO_2, at a temperature of 20°C, will be -8‰ at isotopic equilibrium.

Attainment of the isotopic equilibrium is impeded by the slow reaction kinetics of HCO_3^- and CO_2. Thus, biological activity may shift this equilibrium in the bulk medium when the uptake of inorganic carbon by algae is high relative to the dissociation rates of HCO_3^-. We can calculate the values of the chemical and isotopic equilibrium in the presence of photosynthetic uptake of CO_2 (pp = $\mu \cdot POC$). We assume that $[TCO_2]$ is constant and that a chemical equilibrium has been established, i.e., $d[CO_2]/dt = 0$. The assumption of constant $[TCO_2]$ is not strictly correct and will introduce a small error, but it simplifies the derivation. We calculate the steady state concentrations of CO_2 and $[HCO_3^-]$ from:

$$\frac{d[CO_2]}{dt} = 0 = k_1[HCO_3^-] - k_{-1}[CO_2] - pp \tag{9.22}$$

and

$$[TCO_2] = [HCO_3^-] + [CO_2]. \tag{9.23}$$

Equation 9.22 solved for the concentrations of CO_2 and HCO_3^- normalized by $[TCO_2]$, κ'_{-1} and κ'_1, respectively, is

$$\kappa'_{-1} = \frac{[CO_2]_{eq}}{[TCO_2]_{eq}} = \frac{k_1 - pp/[TCO_2]}{k_1 + k_{-1}} \tag{9.24a}$$

and

$$\kappa'_1 = \frac{[HCO_3^-]_{eq}}{[TCO_2]_{eq}} = \frac{k_{-1} + pp/[TCO_2]}{k_1 + k_{-1}}. \tag{9.24b}$$

Using Equations 9.21a,b and 9.24a,b we can solve the isotopic mass-balance equation for δ_{CO_2}:

$$\delta_{CO_2} = \frac{(\delta_{HCO_3^-} - \varepsilon_1)k_1\kappa'_1 - \varepsilon_{-1}k_{-1}\kappa'_{-1} - \varepsilon_{k_2}\,pp/[TCO_2]}{k_{-1}\kappa'_{-1} + k_2/[TCO_2]}. \tag{9.25}$$

δ_{CO_2} was calculated for varying concentrations of phytoplankton growing at a rate of 1 day^{-1} and varying values of ε_2, and plotted against the ratio of algal biomass to $[TCO_2]$ (Fig. 9.2). The results indicate that biological activity does not change the isotopic equilibrium appreciably in a closed system under normal conditions when algal biomass is less than $0.1\,mol\text{-}Cl^{-1}$ in seawater. Expected perturbations of the isotopic equilibrium are less than ± 0.2‰ since rates of carbon uptake are usually

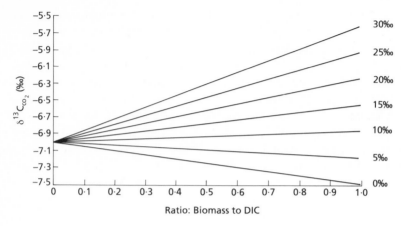

Fig. 9.2 The $\delta^{13}C$ of $[CO_2]_{aq}$ in seawater calculated from Equation 9.31 for varying ratios of phytoplankton biomass and $[TCO_2]$ and various carboxylation fractionation factors ε_2, ranging from 0 to 30‰. Most natural waters have biomass/$[TCO_2]$ ratios <0.1. When $\varepsilon_2 = 0$‰ only fractionation due to the dehydration of bicarbonate is observed.

orders of magnitude lower than rates of hydration and dehydration. Similar calculations for the uptake of bicarbonate would show that significant isotopic disequilibrium will not occur at pH values in the range 7 to 9 because the concentrations of CO_2 and CO_3^{2-} are too small relative to the concentration of HCO_3^- to affect the isotopic composition of the HCO_3^- pool.

9.3.2 Diffusion of CO_2 and HCO_3^- in unstirred layers

Chemical and isotopic disequilibria can arise in the unstirred layers adjacent to algal cell surfaces. The presence of such unstirred layers can influence the isotopic composition of the inorganic carbon that crosses the cell membrane. Conceptually, unstirred layers can be defined as those regions in the vicinity of a surface where the movement of solutes relative to the surface is dominated by molecular diffusion as opposed to turbulence. Figure 9.3 summarizes the physical and chemical processes that occur in the unstirred layer of algae that take up inorganic carbon. Inorganic carbon diffuses to the cell surface in the form of CO_2 and HCO_3^- and is taken up passively as CO_2 or is taken up actively as CO_2 or HCO_3^-. The isotope effect associated with diffusion of inorganic carbon in fresh water at 25°C is 0.7‰ (O'Leary, 1984).

The isotopic and chemical state of the inorganic carbon system in unstirred layers can be determined using a diffusion-reaction model. Such a calculation is beyond the scope of this chapter; however, we can estimate the importance of these processes using a rough calculation for a 10 μm diameter algal cell in seawater. Inorganic carbon uptake per cell (growth

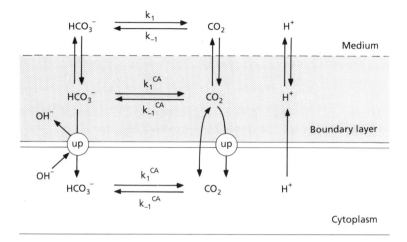

Fig. 9.3 A summary of the boundary layer processes discussed in the text.

rate × carbon per cell = 1×10^{-16} mol-C s^{-1}), dehydration of HCO$_3^-$ in a volume of seawater equivalent to the cell volume ($k_1 \times$ [HCO$_3^-$] × cell volume = 3×10^{-16} mol-C s^{-1}) and diffusion of CO$_2$ through an unstirred layer equivalent to the cell diameter (cell surface area × diffusivity of CO$_2$ × concentration difference of 5 μmol-C/unstirred layer thickness = 4 × 10^{-16} mol-C s^{-1}) are all of the same order of magnitude. This indicates that CO$_2$ uptake of larger phytoplankters (>10 μm diameter) and particularly macroalgae and aquatic plants can be diffusion limited and that the CO$_2$ in the vicinity of the cell can be significantly enriched in ^{13}C versus the dissolved CO$_2$ in the bulk medium when the carboxylation reactions discriminate against ^{13}C. On the other hand, marine algae that take up HCO$_3^-$ actively are not likely to be limited by HCO$_3^-$ availability, nor will the uptake process alter the isotopic distribution of the inorganic carbon in the unstirred layer significantly because of the high bicarbonate concentrations in seawater.

The effect of unstirred layer thickness on the δ^{13}C of fresh water and marine algae has been documented numerous times by comparing algal species grown in low- and high-turbulence environments. Degens *et al.* (1968a) noted that the δ^{13}C of phytoplankton was highest in cultures that were violently aerated with CO$_2$. This effect was interpreted in terms of 'CO$_2$ diffusion limited' photosynthesis. Raven *et al.* (1982) and Osmond *et al.* (1981) studied the covariation of water flow and δ^{13}C of aquatic macroalgae growing in streams with different flow environments. A significant correlation between water flow (i.e., turbulence) and δ^{13}C of the macroalgae was observed with high δ^{13}C occurring under low flow conditions. However, Cooper & McRoy (1988) did not observe such a correlation between $\delta^{13}C$ and turbulence for the marine macrophytes *Phyllospadix* spp. growing in tide pools and in the surge zone. Raven *et al.* (1987) pointed out that an effect of the flow environment on the $\delta^{13}C$ of aquatic macroalgae should only be expected for species that rely on the diffusive entry of CO_2. Thus, turbulence effects on $\delta^{13}C$ should not be expected for species which have an active inorganic carbon uptake system (see Section 9.3.7), as may be the case for *Phyllospadix* spp.

9.3.3 Carbonic anhydrase and the uptake of inorganic carbon

A potentially rate-limiting step associated with reactions involving CO_2 at physiological pH is the slow hydration–dehydration reaction of CO_2. In most organisms this reaction is catalysed by an enzyme, carbonic anhydrase (Aizawa & Miyachi, 1986), which has one of the highest turnover numbers of any known enzyme (Paneth & O'Leary, 1985). It is often present in the cytoplasm of algal cells and in the stroma of chloroplasts at concentrations sufficient to enhance HCO_3^- dehydration rates by orders of magnitude (Aizawa & Miyachi, 1986). Carbonic anhydrase does not perturb the chemical equilibrium between CO_2 and HCO_3^-; this is similarly true for the isotopic pairs $^{12}CO_2$-$^{12}HCO_3^-$ and $^{13}CO_2$-$^{13}HCO_3^-$. It follows that carbonic anhydrase does not perturb the isotopic equilibrium between CO_2 and HCO_3^-. The kinetic fractionation factor associated with the carbonic anhydrase catalysed dehydration of HCO_3^- is 10‰ at 25°C (Paneth & O'Leary, 1985), and the fractionation associated with the reverse reaction can be calculated from the difference between the forward reaction fractionation effect and the equilibrium fractionation effect; the value is 1‰. For carbonic anhydrase catalysed reactions we can assume that the CO_2–HCO_3^- system is always at chemical and isotopic equilibrium with an equilibrium fractionation factor given by the equations of Mook *et al.* (1974).

Carbonic anhydrase activity has been reported for a variety of Chlorophyceae on the cell surface or in the periplasmic space (Fig. 9.4), the space between the cell wall and the plasmalemma (Aizawa & Miyachi, 1986). This extracellular carbonic anhydrase activity was enhanced in Chlorophyceae that were grown at 'low' CO_2 concentrations, 0.03% vs. 3–5% v/v (Aizawa & Miyachi, 1986). Extracellular carbonic anhydrase activity is usually absent in other groups of eukaryotic microalgae (Aizawa & Miyachi, 1986; Patel & Merrett, 1986; Dixon *et al.*, 1987; Munoz & Merrett, 1989); extracellular carbonic anhydrase activity in marine algae has only been reported by Burns & Beardall (1987). Conventional carbonic anhydrase assays do not detect extracellular carbonic anhydrase activity in cyanobacteria (Aizawa & Miyachi, 1986; Tu *et al.*, 1987; but see also below).

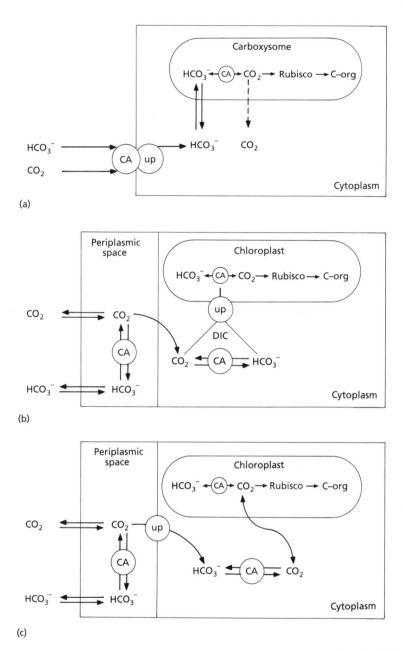

Fig. 9.4 Various inorganic carbon uptake systems, as proposed by different authors. Active uptake of inorganic carbon is indicated by 'up' enclosed in a circle. CA, carbonic anhydrase. Details of the figure are discussed in the text. (a) the dissolved inorganic carbon uptake system of cyanobacteria (From Kaplan *et al.*, 1989). (b) and (c) two possible uptake mechanisms that have been proposed for different green algae (From Badger, 1987).

9.3.4   Fractionation associated with Rubisco

The enzyme ribulose 1,5-bisphosphate carboxylase-oxygenase (Rubisco) is responsible for all net carbon fixation in algae. In eukaryotes the enzyme is located in the stroma of chloroplasts; it is also found in the pyrenoids of Chlorophyceae. In cyanobacteria Rubisco is located in the cytoplasm and in carboxysomes (Badger, 1987; Glover, 1989). Rubisco uses CO_2 and ribulose 1,5-bisphosphate (RuBP) as a substrate for its carboxylation function and oxygen and RuBP to produce 2-phosphoglycolate and 3-phosphoglycerate when it functions as an oxygenase. The oxygenase function of Rubisco is responsible for photorespiration. The carboxylase and oxygenase functions compete, and their relative activities depend on the ratio of the concentrations of CO_2 and O_2 (Jordan & Ogren, 1981). The affinities of the enzyme for CO_2 and O_2 are expressed as apparent half-saturation constants, K_c and K_o, respectively. The substrate specificity factor, V_cK_o/V_oK_c, is calculated from the half-saturation constants and the maximal velocities of the two reactions, V_c and V_o, respectively. Jordan & Ogren (1981) have shown that substrate affinities and specificity factors vary significantly among higher plants, green algae, cyanobacteria and photosynthetic bacteria (Table 9.1). The specificity for CO_2 is highest in the higher plants and lowest in cyanobacteria and photosynthetic bacteria, e.g., *Rhodospirillum rubrum* (Jordan & Ogren, 1981). The only marine organism analysed is the cyanobacterium *Synechococcus* spp.; its Rubisco has properties very similar to Rubisco from freshwater cyano-

Table 9.1 Rubisco affinity factors for CO_2 (K_c) and O_2 (K_o), the specificity factor, $S = \{(V_cK_o)/(V_oK_c)\}$, and carbon isotope fractionations (ε) associated with the carboxylation of RuBP by Rubisco. The kinetic data are from Jordan & Ogren (1981)

Group	S	K_c	K_o	ε (‰)
C_3 plants	80	13	530	29*
C_4 plants	80	25	725	–
Green algae	59	31	520	–
Cyanobacteria	48	113	1100	21.5†
Rhodospirillum rubrum‡	15	89	406	19–23§

* Rubisco isolated from spinach (Roeske & O'Leary, 1984; Guy *et al.*, 1987).
† Rubisco isolated from *Anacystis nidulans* (Guy *et al.*, 1987).
‡ At least two types of Rubisco are found in photosynthetic bacteria, a low-molecular-weight protein that is found in *Rhodospirillum rubrum* and a high-molecular-weight protein that is found together with the low-molecular-weight protein in *Rhodopseudomonas sphaeroides* (Jordan & Ogren, 1981). We have only listed the characteristics of the *R. rubrum* Rubisco because the fractionation factor for the high-molecular-weight protein has not been determined.
§ Roeske & O'Leary (1985); J.A. Berry *et al.* (unpublished data) noticed a dependence on Mg^{2+} concentrations.

bacteria (Andrews & Abel, 1981). Kinetic data for Rubisco from other groups of algae are unfortunately not available.

It is conceivable that the variation of isotopic fractionation factors of Rubisco from different groups of algae and cyanobacteria is similar to the variation of the kinetic constants (Roeske & O'Leary, 1985). Indeed, isotopic fractionation factors for Rubisco isolated from higher plants and photosynthetic bacteria (Table 9.1) differ by about 10‰. The fractionation factors listed in Table 9.1 are only those that were measured using high-precision methods, i.e., the specific label method (Roeske & O'Leary, 1984) or the substrate depletion method (Schmidt et al., 1978). Earlier measurements based on the combustion of carboxylation reaction products (e.g., Estep et al., 1978a,b) often do not agree with the more recent measurements. The green algae are the only group of eukaryotic algae for which the Rubisco kinetic constants were determined; however, the isotopic fractionation factor for Rubisco from green algae has not been determined. It is tempting to speculate that the isotopic fractionation factor of Rubisco from green algae as well as other eukaryotic algae is intermediate between the factors for higher plants and cyanobacteria. Thus it is not possible to assume that the isotopic fractionation factors of Rubisco from algae and higher plants have the same value, 29‰; a value between 20 and 30‰ may be more appropriate.

9.3.5 β-Carboxylases

The carbon isotopic composition of algae can also be influenced by other carboxylation reactions when these use bicarbonate as a substrate or have fractionation factors that differ from the fractionation factors of Rubisco. The β-carboxylation of either phosphoenolpyruvate (PEP) or pyruvate are the most important carboxylation reactions next to the carboxylation of RuBP (Kerby & Raven, 1985; Glover, 1989). However, β-carboxylation does not represent net fixation of carbon because it is linked to decar-boxylations (Fig. 9.5). In the light β-carboxylation of PEP or pyruvate usually serves to replenish the metabolites of the tricarboxylic acid cycle to produce carbon skeletons for a variety of biosynthetic reactions (Fig. 9.5) such as amino acid synthesis (Glover, 1989; Raven, 1990). Two enzymes are primarily responsible for the β-carboxylation of phosphoen-olpyruvate to oxaloacetate in microalgae: PEP-carboxylase (PEPC) and PEP-carboxykinase (PEPCK). PEPC uses HCO_3^- as an inorganic carbon substrate (Cooper et al., 1968; Cooper & Wood, 1971) and PEPCK uses CO_2 as a substrate (Cooper et al., 1968; Johnston & Raven, 1989). In cyanobacteria and Chlorophyceae only PEPC is present; either PEPC or PEPCK are found in Dinophyceae, Crysophyceae, Cryptophyceae and Prymnesiophyceae (Glover, 1989; Descolas-Gros & Fontugne, 1990). A third β-carboxylase, pyruvate carboxylase is only found in some

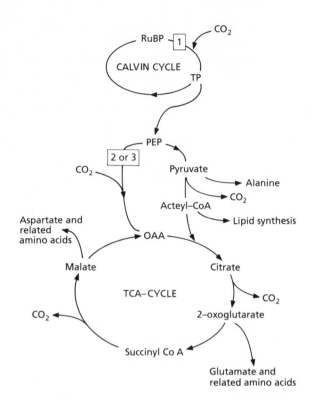

Fig. 9.5 The two important biochemical sites of carboxylations in algae. Net carbon fixation only occurs via the Calvin cycle (1, rubisco). 3-Phosphoglycerate is shuttled out of the chloroplast and is converted via some biochemical intermediates into phosphoenolpyruvate (PEP). PEP is the substrate of either PEPC (2) or PEPCK (3), producing oxaloacetate (OAA), or it is decarboxylated to produce acetyl-coenzyme A. OAA is only synthesized and supplied to the TCA-cycle when carbon skeletons are drawn from the TCA-cycle for biochemical syntheses. For every β-carboxylation decarboxylations must occur, resulting in no net fixation of carbon.

Dinophyceae (Appleby *et al.*, 1980). In planktonic diatoms either PEPC or PEPCK are present (Glover, 1989; Zimba *et al.*, 1990); benthic diatoms are characterized by PEPC only. Descolas-Gros & Fontugne (1988, 1990) generally detected PEPCK in natural phytoplankton communities; PEPC activity was usually lower than PEPCK activity or undetectable.

PEPC is the enzyme responsible for the initial fixation of carbon in terrestrial Hatch–Slack Cycle (C_4) plants that usually have $\delta^{13}C$ values of -11 to $-15‰$. A consequence of the C_4 biochemistry is the high activity of PEPC relative to Rubisco, which is usually taken to be indicative of a C_4 biochemistry in higher plants (Edwards & Huber, 1981; Osmond & Holtum, 1981; Beardall, 1989). Similar measurements in algae demonstrated very high PEPC or PEPCK activities in some Pheophyceae and

Bacillariophyceae, particularly benthic species (see review by Morris, 1980). It was concluded based on these data that some algae have a C_4 biochemistry, i.e., inorganic carbon is initially incorporated into oxalo-acetate by a β-carboxylase (Beardall *et al.*, 1976). However, it was shown later for a variety of species, using short-term (<10 seconds) $^{14}CO_2$ incubations, that label is initially incorporated into 3-PGA, the product of the carboxylation of RuBP. In those cases when high labelling of tricar-boxylic acid cycle end products was observed, label was homogeneously distributed in the C_4 acids, consistent with primary fixation of inorganic carbon via carboxylation of RuBP (Kerby & Raven, 1985; Beardall, 1989; Glover, 1989; Zimba *et al.*, 1990). The consensus at the present time is that marine microalgae have a typical C_3 biochemistry although many have 'C_4 gas exchange' characteristics due to active inorganic carbon uptake mechanisms (see Section 9.3.7).

High *in vitro* β-carboxylase-Rubisco activity ratios *per se* are not inconsistent with a C_3-biochemistry in microalgae. Extremely high *in vivo* rates of β-carboxylation have been observed transiently in nitrogen-starved algal cultures that had been enriched with ammonia (Guy *et al.*, 1989; Vanlerberghe *et al.*, 1990). Over short periods of time mol-specific rates of ammonia assimilation can be as high as steady-state mol-specific rates of photosynthesis when nitrogenous nutrients are supplied intermittently. Under these conditions β-carboxylation serves to replenish TCA-cycle intermediates that are used as carbon skeletons in the course of ammonia assimilation (Fig. 9.5). It is possible that high concentrations of β-carboxylases are generally present in marine microalgae for the rapid but intermittent assimilation of ammonia pulses (see McCarthy & Goldman, 1979). Nonetheless, the available evidence indicates that β-carboxylases are only responsible for a small fraction of the carbon fixation when averaged over the generation time of an alga. Beardall (1989) estimated that β-carboxylation cannot account for more than 25% of the net carbon fixation in microalgae (see Fig. 9.5), assuming balanced growth and that all fixed carbon is shuttled through the TCA cycle. Vanlerberghe *et al.* (1990) calculated the expected rate of β-carboxylation for the green algae *Selenastrum minutum* from an analysis of its amino acid composition. The estimated value of 0.3 mol carbon fixed via β-carboxylation per mole nitrogen assimilated agreed with the measured rate of 0.3 mol carbon. Assuming a molar carbon/nitrogen ratio of 6.6, this rate implies that only 4.5% of the cellular carbon was derived from β-carboxylation.

High β-carboxylase activities have also been observed in many macro-algae, particularly brown algae (Kremer, 1981). The consensus at the present time is that most macroalgae have the basic C_3 biochemistry, but may have inorganic carbon uptake mechanisms which lead to 'C_4-like' gas exchange characteristics (Bowes, 1985; Prins & Elzenga, 1989; Reiskind *et al.*, 1989; Raven *et al.*, 1990). A low level CAM-activity has been observed in the brown alga, *Ascophyllum nodosum* (Johnston & Raven, 1986) that is characterized by high rates of β-carboxylations in the dark, pronounced pH changes over the diel light cycle, and concomitant malate concentration changes. Nonetheless, the C_4 gas exchange characteristics of this algae are likely to be due to a DIC concentrating mechanism (Johnston & Raven, 1987; see Section 9.3.7). Reiskind *et al.* (1988) have demonstrated high activities of PEPCK and rapid ^{14}C-labelling of C_4-acids in the green algae, *Udotea flabellum*. These results, in conjunction with PEPCK-inhibitor studies, suggest an important role of PEPCK in this species, possibly similar to a C_4-like photosynthetic system (Reiskind *et al.*, 1988, 1989). Nonetheless, it still remains to be demonstrated directly that *U. flabellum* has a C_4-like biochemistry.

The kinetic fractionation factor of PEPC is 2‰ for the carboxylation of PEP using HCO_3^- as a substrate at a pH of 7.5; the fractionation factor is possibly pH dependent since values of 0.9‰ and -2.7‰ were observed at a pH of 9.0 and 10.0, respectively (O'Leary *et al.*, 1981). The fractionation factor of PEPCK isolated from *Chloris gayana* for the carboxylation of PEP using CO_2 as a substrate ranged from 24 to 40‰, depending on the assay conditions (Arnelle & O'Leary, 1992). *In vivo*, the value is likely to be in the 30–40‰ range (M. O'Leary, personal communication 1991). Fractionation factors for other β-carboxylases are not known at the present time. The large fractionation factor of PEPCK implies that the $\delta^{13}C$ of algae using this enzyme will not be affected by varying rates of β-carboxylation since the fractionation factors of PEPCK (20–40‰) and Rubisco (29‰) are similar. However, fixation of bicarbonate by PEPC will lead to the enrichment of ^{13}C in algae in proportion to the ratio of the *in vivo* activities of Rubisco and the β-carboxylases.

For example, the effect of high β-carboxylase activity on the isotopic fractionation during carbon fixation was studied in the green alga *Selenastrum minutum* which uses PEPC (Guy *et al.*, 1989). High rates of β-carboxylation were induced in nitrogen starved algae by pulsing the cultures with ammonia. Whereas rates of β-carboxylation were virtually undetectable in controls, rates of β-carboxylation accounted for up to 70% of total carbon fixation while the pulse of ammonia was assimilated. Such high rates of β-carboxylation inhibit Rubisco activity (Fig. 9.5) due to RuBP depletion since 3-PGA is drawn out of the chloroplast and is converted to PEP, the substrate of β-carboxylations (Elrifi & Turpin, 1986; Elrifi *et al.*, 1988). Guy *et al.* (1989) reported carbon isotopic discriminations relative to the dissolved CO_2 in the medium of 29‰ in the controls and 2–20‰ in cultures with high rates of β-carboxylations. These experiments dramatically illustrated the possible effects of β-carboxylations on the $\delta^{13}C$ of algae. However, as indicated above, such large effects should not be observed when discriminations are averaged over the generation time of an alga, and such effects should be absent when PEPCK is the primary enzyme responsible for β-carboxylations.

Descolas-Gros & Fontugne (1985) measured the activities of β-carboxylases and Rubisco and the $\delta^{13}C_{POC}$ of the diatom *Skeletonema costatum* grown in a batch culture. The ratio of PEPCK and Rubisco activity, measured on crude algal extracts, was less than 0.01 (Fig. 3b of Descolas-Gros & Fontugne, 1985). The $\delta^{13}C$ of the diatom culture changed from −21‰ at day 1 to −13‰ at day 4, as the culture reached stationary growth. Descolas-Gros & Fontugne (1985) argued that this isotopic shift was caused by an increasing importance of β-carboxylation. However, it is likely that the observed $\delta^{13}C_{POC}$ shifts during growth represented the depletion of ^{12}C from the medium as it was taken up by the algae (Johnston & Raven, 1992) rather than the effect of β-carboxylation on the isotopic composition of the algae because the algae were grown at high cell densities and large pH changes were observed during growth.

Falkowski (1991) reported $\delta^{13}C$ variations of algal cultures ranging from −29.7‰ to −5.5‰. Approximately 8‰ of the 25‰ variation could be attributed to changing concentrations of dissolved CO_2; the extremely high values near −5‰ were attributed to active uptake of inorganic carbon. Falkowski (1991) argued that the residual variation was due to varying β-carboxylation/Rubisco-carboxylation ratios, which were not measured in the study.

A positive correlation between chlorophyll-normalized PEPC activity and $\delta^{13}C$ has been reported for POM collected off Portugal (Descolas-Gros & Fontugne, 1985) and a positive correlation has also been found between the ratio of β-carboxylase to Rubisco activity and $\delta^{13}C$ for a subset of data from the Southern Ocean (Fontugne *et al.*, 1991). These reports are difficult to evaluate because only the regression coefficient but no data were given in the first study. In the second study no ratios of enzymatic activities for the individual β-carboxylases were given. No consistent variations of β-carboxylase activities or of the ratio of β-carboxylase to Rubisco activity and $\delta^{13}C$ were reported by Descolas-Gros and Fontugne for the Mediterranean and the St Lawrence Estuary (Descolas-Gros & Fontugne, 1990; Fontugne *et al.*, 1991). These initial results indicate that the effect of β-carboxylations on the $\delta^{13}C$ of algae has to be explored in more detail, since varying rates of β-carboxylation may be responsible for some of the observed variation of $\delta^{13}C_{POC}$ in the natural environment.

9.3.6 Respiration

Three pathways are primarily responsible for the oxidation of organic carbon in algae: the TCA cycle, glycolysis and the oxidative pentose phosphate (OPP) pathway (Raven, 1990). All three pathways have a dual function: to provide ATP and/or reductant by oxidating respiratory substrates and to provide carbon skeletons for diverse biosynthetic pathways. The above respiratory pathways are subsumed under the name 'dark respiration' to differentiate these from other decarboxylations or oxygen consuming processes, such as photorespiration (see above) that only occur in the light. Dark respiration occurs to a limited extent in the light, mostly to provide carbon skeletons for biosynthesis. Rates of dark respiration in the dark scale with rates of photosynthesis, save for a minimum rate, termed basal metabolism. Dark respiratory rates usually range from 5 to 15% of the light-saturated rate of photosynthesis, and extreme values range from about 1 to 50% (Raven, 1990).

Respiratory processes can, in theory, markedly affect isotopic compositions of algae. For example, large fractionations are known to occur in many enzymatically mediated decarboxylation reactions (Table 9.2), and if isotopically light carboxyl-derived CO_2 were lost from cells, remaining algal carbon could become strongly enriched in ^{13}C. This view from the enzymatic level is, however, not complete. Carbon entering catabolic pathways is often completely metabolized to CO_2, rather than being partially degraded via decarboxylation reactions. When substrates are quantitatively converted to CO_2, no fractionation can result. The eventual complete catabolism of cellular substrates appears to be the rule rather than the exception in overall cell metabolism, and consequently isotopic changes due to respiration are thought to be small (O'Leary, 1981; Raven, 1990).

A review of the literature shows that isotopic fractionation during respiration in algae is generally poorly studied and difficult to document unless *in vitro* systems are used. The effect of respiration on the isotopic composition of an organism can be appreciable, even in the absence of fractionation during respiration, when the $\delta^{13}C$ of the respiratory sub-

Table 9.2 Isotope effects associated with decarboxylations. The enzymes were isolated from a variety of heterotrophic species

Enzyme	ε (‰)	Reference
Malic enzyme	30	Hermes *et al.* (1982)
Glucose-6-phosphate dehydrogenase	17	Hermes *et al.* (1982)
NADP isocitrate dehydrogenase	28–40	Grissom & Cleland (1988)
6-Phosphogluconate dehydrogenase	9.6	Rendina *et al.* (1984)
Pyruvate decarboxylase	8.3 (10.1–2.2)	O'Leary (1976), Jordan *et al.* (1978)
Arginine decarboxylase	15	O'Leary (1988b)
Glutamate decarboxylase	18	O'Leary (1988b)
Histidine decarboxylase	31	O'Leary (1988b)
Aspartate β-decarboxylase	10	O'Leary (1988b)

strate differs significantly from the $\delta^{13}C$ of the whole organism. This effect was described by Guy et al. (1989) for a green algae: CO_2 respired in the dark was enriched with ^{13}C relative to the organisms by 8‰. Starch isolated from the algae, the probable respiratory substrate, was similarly enriched by 7‰ relative to the whole organism. It cannot be inferred from the data presented if the isotopically heavy starch was an experimental artifact – preferential synthesis of starch in a nitrogen-starved culture after the $\delta^{13}C_{DIC}$ had shifted – or if this reflects the usual isotopic composition of starch in this alga.

Degens et al. (1968a) attempted to measure the effect of respiration on the $\delta^{13}C$ of microalgae by incubating the algae for 10–20 days in the dark. Although the $\delta^{13}C$ of the green alga *Dunaliella* decreased by 6‰ over a 20-day incubation, $\delta^{13}C$ changes in the diatom *Skeletonema* were erratic over 12-day incubation periods. It is questionable if such long-term experiments actually measure the isotopic effects of dark respiration, which normally occurs for about 12 hours at night. Raven (1990) concluded that in general respiratory substrates should be isotopically similar to the whole cell and that isotopic changes due to respiration are expected to be small.

9.3.7 Active uptake of inorganic carbon

Many algae and cyanobacteria take up DIC actively, raising intracellular DIC 5–1000-fold over extracellular concentrations (Badger, 1987). Active inorganic carbon uptake mechanisms can affect the $\delta^{13}C$ of algae via the species of inorganic carbon that is taken up (HCO_3^- vs. CO_2) and via its effect on the leakiness of the cell membrane. The $\delta^{13}C$ of an organism can potentially increase by 10–12‰ when HCO_3^- is taken up instead of CO_2, and the $\delta^{13}C$ of cells that do not leak CO_2 to the external environment will approach values of the inorganic carbon that is transported. However, for most algae it is not known which inorganic carbon species is transported across the cell membrane; the transport of CO_2, HCO_3^-, or both simultaneously has been described. It is also not known if isotopic discrimination occurs during the active uptake of inorganic carbon. It is usually assumed that this effect is negligible (Kerby & Raven, 1985; Berry, 1988).

A physiological consequence of elevated intracellular DIC is that the affinity of algae to CO_2 will be much higher than the *in vitro* affinity of Rubisco to CO_2, particularly in the case of cyanobacteria whose Rubisco has a low affinity to CO_2 (Table 9.1). The second consequence is the inhibition of the oxygenase function of Rubisco by CO_2 (Glover, 1989). These physiological characteristics – low CO_2 compensation points and virtually no photorespiration – are very similar to those of plants with a C_4 biochemistry, and have been dubbed 'C_4 gas exchange' characteristics

(Johnston & Raven, 1989). Extensive reviews of the inorganic carbon uptake mechanism in algae and cyanobacteria and its effects on carbon isotopes have been published by Raven (1970), Kerby & Raven (1985), Badger (1987), and in two symposium volumes (Lucas & Berry, 1985; Coleman, 1991).

The uptake mechanism in eukaryotic algae and cyanobacteria is generally thought to be located in the cytoplasmic membrane and/or in the inner chloroplast envelope in the case of eukaryotes (Badger, 1987). The uptake is an energy-dependent process, and most likely ATP is used as a source of energy (Badger, 1987). Energy consumption due to inorganic carbon uptake can be a significant fraction of the total energy expenditure during photosynthesis (Kerby & Raven, 1985). In the case of HCO_3^- transport the ionic balance has to be maintained. A variety of mechanisms which will maintain the ionic balance have been suggested for different organisms (Badger, 1987).

A consequence of high intracellular DIC is the leakage of CO_2 from the cell back into the medium. This leakage can be a significant fraction (50%) of the inorganic carbon taken up (Raven, 1985; Badger et al., 1985). Common to most algal species studied is that the inorganic carbon uptake mechanism is inducible. It is usually observed that cells grown at about 0.03% CO_2 in air are able to transport inorganic carbon actively, whereas cells grown at 2–10% CO_2 in air are not able to transport inorganic carbon. It seems likely that active inorganic carbon transport occurs in some algae under natural conditions (pH = 7–9; DIC = 200–2000 μmol-Cl^{-1}). However, the characteristics of the inorganic carbon uptake mechanisms differ widely between groups of microalgae. Cyanobacteria, green algae and a variety of marine micro- and macroalgae have been studied.

Cyanobacteria can increase intracellular concentrations of DIC more than 1000-fold over extracellular concentrations (Badger, 1987). In some cyanobacteria a carbonic anhydrase-like front end is associated with the uptake mechanism (Fig. 9.4a), so that both CO_2 and HCO_3^- are used as substrates, regardless of the species of inorganic carbon actually transported (Badger & Price, 1989; Price & Badger, 1989a,b). Kaplan et al. (1989, 1991) recently proposed a role for carboxysomes in inorganic carbon uptake and carboxylation in cyanobacteria (Fig. 9.4a). The carboxysome is a polyhedral body in cyanobacteria that contains most of the Rubisco and possibly all their carbonic anhydrase (Badger, 1987). The location of cyanobacterial carbonic anhydrase in carboxysomes may lead to the inability of conventional carbonic anhydrase assays to detect its activity. Kaplan et al. (1989) suggested that both CO_2 and HCO_3^- are taken up actively by the cells, but that only HCO_3^- is released into the cytoplasm. HCO_3^- will accumulate in the cytoplasm without significant conversion to CO_2 since carbonic anhydrase is thought to be absent. HCO_3^- will diffuse into the carboxysome where CO_2 is generated in the immediate vicinity of Rubisco due to the presence of carbonic anhydrase. CO_2 is consumed by Rubisco before it can diffuse out of the carboxysome (Kaplan *et al.*, 1991). However, major aspects of this hypothesis still await experimental verification. A testable consequence of this hypothesis is that cyanobacteria are expected to have a very high $\delta^{13}C$ due to their use of HCO_3^- and their efficient utilization of the CO_2 generated in the carboxysome.

The other group of microalgae that has been studied extensively are the Chlorophyceae. A variety of different mechanisms have been described; Figure 9.4b,c show some possible models. Most Chlorophyceae grown under low CO_2 conditions have free carbonic anhydrase activity in the periplasmic space (Aizawa & Miyachi, 1986). Usually, both CO_2 and HCO_3^- can be taken up actively (Badger, 1987). Recently two different uptake systems, both inducible at low concentrations of inorganic carbon, have been described for *Scenedesmus* (Thielmann *et al.*, 1990): a CO_2 uptake system that operates at a low pH (5–8) and a HCO_3^- uptake system, possibly an ATPase linked HCO_3^- transporter that operates at a high pH (7–11). Generally, Chlorophyceae concentrate intracellular DIC only by a factor of 5–20 relative to extracellular DIC (Badger, 1987). This internal pool of inorganic carbon is not readily exchangeable with the external medium; it was shown in an elegant experiment that this pool has a half-time of depletion of 2.5 to 7 minutes in *Dunaliella salina* (Zenvirth & Kaplan, 1981). This result indicates that the permeability of the cell membrane to CO_2 is lower than expected based on experiments with artificial lipid bilayers (Gutknecht *et al.*, 1977). Recently it was shown that Chlorophycean chloroplasts are also capable of inorganic carbon uptake (Moroney *et al.*, 1987; Goyal & Tolbert, 1989).

Most marine microalgae studied can take up inorganic carbon actively (Colman & Gehl, 1983; Patel & Merrett, 1986; Burns & Beardall, 1987; Dixon & Merrett, 1988; Munoz & Merrett, 1989). It was shown for *Pheodactylum tricornutum* that HCO_3^- is transported actively across the cell membrane (Patel & Merrett, 1986; Dixon & Merrett, 1988). It is likely that HCO_3^- is taken up actively in other marine microalgae as well, given the high pH of normal seawater (8.2) and the absence of extracellular carbonic anhydrase. Most marine microalgae do not concentrate inorganic carbon intracellularly relative to concentrations in seawater, as do Chlorophyceae (Patel & Merrett, 1986; Munoz & Merrett, 1989; Dixon *et al.*, 1989). Burns & Beardall (1987), however, observed a fivefold concentration in four species.

A variety of inorganic carbon uptake mechanisms have been described for macroalgae. Evidence for the use of HCO_3^- exists for many macroalgae, particularly marine ones (Sand-Jensen, 1987). Other macroalgae, such as the red alga *Chondrus crispus* simply enhance the passive diffusion of CO_2 into their thalli by extracellular carbonic anhydrase (Smith & Bidwell, 1989). Macroalgae that do not have active inorganic carbon uptake systems are often able to extract CO_2 directly from the atmosphere or from sediments via their roots (Sand-Jensen, 1987). Unstirred layer effects are easily noticed in macrophytes due to their large size; in the light, a uniform alkaline boundary layer can be observed for *Hydrodictyon*, indicating the active uptake of HCO_3^- (Walker, 1985). Some species of aquatic macroalgae pump protons actively into the boundary layer, thus decreasing the pH, to shift the $CO_2-HCO_3^-$ equilibrium towards CO_2 (Prins & Elzenga, 1989).

The effect of inorganic carbon uptake on isotopic discrimination during photosynthesis in the green algae *Chlamydomonas reinhardtii* has been documented by Sharkey & Berry (1985). A culture of the algae was grown at a high concentration of CO_2 in air. The culture was enclosed in an experimental vessel and bubbled with an air mixture containing 0.33% CO_2. Isotopic fractionation during the experiment was monitored by sampling the air in the culture vessel and comparing its $\delta^{13}C$ with the $\delta^{13}C$ of the CO_2 in the inlet air. Carbon isotopic discrimination in these high-CO_2 adapted cells, relative to the $\delta^{13}C$ in the inlet air, was about 25–29‰ (Fig. 9.6), values close to the 29‰ kinetic fractionation factor of Rubisco (Roeske & O'Leary, 1984). When the CO_2 concentration in the air mixture was decreased to 0.02% CO_2, photosynthesis dropped immediately by a factor of four, but the isotopic discrimination remained high, at

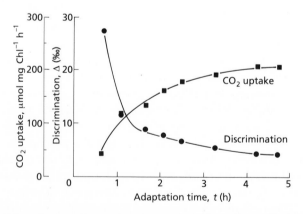

Fig. 9.6 Time changes of inorganic carbon concentration, photosynthesis and isotopic discrimination for a culture of *Chlamydomonas reinhardtii* during the adaptation from high dissolved inorganic carbon (DIC) to low DIC. Note the large and rapid change of isotopic fractionation. (From Berry, 1988.)

29‰. However, after 1–2 hours, CO_2 uptake had recovered to levels obtained under high CO_2 but isotopic discrimination had decreased by about 10‰ (Fig. 9.6). After continued growth isotopic discrimination was 4‰. Sharkey & Berry (1985) interpreted the initial decline of carbon fixation as CO_2 diffusion-limited photosynthesis. The low concentration of CO_2 induced the capacity for active inorganic carbon uptake over the next 3 hours, and carbon fixation returned to its original levels. The observed isotopic discrimination under low CO_2 was analysed using a simple carbon-flux and isotopic model, assuming that only HCO_3^- was taken up during the low-CO_2 phase of the experiment (Berry, 1988):

$$\mathrm{DIC_e} \underset{k_{-1}}{\overset{k_1}{\leftrightarrow}} \mathrm{DIC_i} \overset{k_2}{\to} \quad \text{and} \quad \delta_e \underset{\varepsilon_{-1}}{\overset{\varepsilon_1}{\leftrightarrow}} \delta_i \overset{\varepsilon_2}{\to} \tag{9.26a,b}$$

for $\varepsilon_1 = \varepsilon_{-1}$ and $\delta_e = \delta_{HCO_3}$. This model is a simplified version of a model derived above (Equation 9.18a,b). It can be solved for the $\delta^{13}C$ of the organic carbon fixed, δ_i minus ε_2:

$$\delta_i - \varepsilon_2 = \delta_{DIC_e} - \varepsilon_1 \frac{k_{-1} - k_1}{k_1} + \varepsilon_2 \frac{k_{-1}}{k_1}. \tag{9.27}$$

Sharkey & Berry (1985) assumed that isotopic fractionation during the uptake and backdiffusion of CO_2 is negligible, i.e., the second term of Equation 9.27 is close to zero. They calculated that the ratio of carbon uptake (k_1) to backdiffusion of CO_2 (k_{-1}), i.e., the leakiness of the cell membrane, was about 0.5. This experiment demonstrated that active carbon uptake can increase the $\delta^{13}C$ of algae significantly.

Beardall et al. (1982) studied the effects of CO_2 supply and nitrogen limitation on the $\delta^{13}C$ of the green algae Chlorella emersonii grown in continuous cultures at high and low concentrations of CO_2 in air (0.03 and 5%) and high and low concentrations of nitrogen in the medium (4.0 and 0.004 mmol-N l^{-1}). The $\delta^{13}C$ of the algae relative to the $\delta^{13}C$ of the air, the $\Delta\delta^{13}C$, was -25‰ for high CO_2–high nitrogen grown cells. The $\Delta\delta^{13}C$ of low CO_2–low nitrogen and low CO_2–high nitrogen treatments was similar, about 7–8‰. Ancillary data – CO_2 compensation points and intracellular versus extracellular concentrations of inorganic carbon – indicated that CO_2 was actively taken up in the low CO_2 treatments, regardless of nitrogen treatment. The small discrimination values for the low CO_2 treatments are consistent with an intracellular enrichment of CO_2, derived from HCO_3^- which prevented the full expression of the fractionation factor of Rubisco. The surprising result of these experiments was that high CO_2–low nitrogen treated algae took inorganic carbon up actively and had a $\Delta\delta^{13}C$ of -12.6‰. Beardall et al. (1982) suggested that nitrogen-use efficiency will be increased when inorganic carbon is concentrated intracellularly because high intracellular concentrations of DIC will inhibit the oxygenase function of Rubisco, i.e., photorespiration, and the subsequent deamination of glycine as part of the photorespiratory cycle. Although most of the ammonia generated during the deamination of glycine is reassimilated, some of it may diffuse out of the cells, thus decreasing nitrogen-use efficiency (Beardall, 1989). It is unknown to what extent other microalgae use a similar strategy of active carbon uptake to prevent photorespiration and thus increase nitrogen-use efficiency.

9.3.8 Synthesis

A conceptual model is needed to relate algal $\delta^{13}C$ and environmental and physiological parameters. It is not possible at the present time to propose a conceptual model which describes the active uptake of inorganic carbon since most relevant parameters are ill-constrained. Thus, we will only consider the diffusive uptake of CO_2 with subsequent fixation via Rubisco and β-carboxylases. We use Equation 9.20 as a starting point:

$$\delta_{POC} = \delta_{CO_{2_c}} - \varepsilon_1 + f(\varepsilon_{-1} - \varepsilon_2) + \varepsilon_3 \frac{v}{\mu + v}.$$

As a first step we parameterize Equation 9.20 for the case of the passive uptake of CO_2. A comparison of isotopic effects for processes that affect isotopic fractionation in algae (Table 9.3) indicates that isotopic fractionation associated with the carboxylation reactions (ε_2) is the dominant term in Equation 9.20. ε_2 is the combined fractionation due to Rubisco and β-carboxylase (PEPC, PEPCK, etc.) carboxylations. The value of ε_2 will depend on the ratio of the two carboxylation reactions and on the substrates used. We will assume that the fractionation factor of Rubisco, $\varepsilon_{Rubisco}$, is 29‰, and that all β-carboxylations are mediated by PEPC

Table 9.3 The dominant processes associated with ^{13}C-fractionation in algae

Reaction	Substrate	ε (‰)
Diffusion of CO_2	CO_2, HCO_3^-	<0.7
Dehydration of HCO_3^-	HCO_3^-	8–12
Passive DIC uptake	CO_2	<0.7 ?
Active DIC uptake	CO_2, HCO_3^-	Small ?
Backdiffusion of CO_2	CO_2	<0.7 ?
Rubisco carboxylation	CO_2	20–29
β-carboxylation PEPC	CO_2	2.0
β-carboxylation PEPCK	HCO_3^-	20–40
Respiration	(CO_2)	Small ?
Excretion of DOC	–	?

DIC, dissolved inorganic carbon; DOC, dissolved organic carbon.

whose substrate is HCO_3^-, with a fractionation factor, ε_{PEPC}, of 2‰. Since the two carboxylases use different substrates we have to calculate an 'apparent fractionation factor' for PEPC, which is the fractionation relative to the $\delta^{13}C$ of CO_2, the carbon species which crosses the cell membrane. We assume that the intracellular chemical equilibrium between HCO_3^- and CO_2 is mediated by carbonic anhydrase, such that the $\delta^{13}C$ of CO_2 differs by 9‰ from the $\delta^{13}C$ of HCO_3^-. This value of 9‰ has to be subtracted from ε_{PEPC} bringing its value to $-7‰$. To calculate the value of ε_2 the values of $\varepsilon_{Rubisco}$ and ε_{PEPC} have to be added after normalizing each factor by the fraction that each carboxylation represents of the total carboxylation:

$$\varepsilon_2 = (1 - c)\varepsilon_{Rubisco} + c\varepsilon_{PEPC}. \tag{9.28}$$

The factor c is the ratio of the β-carboxylations to total carboxylations. At the present time it is not possible to determine the value of c experimentally since it is questionable that assays of carboxylation enzymes reflect *in vivo* rates of carboxylations accurately (see above). Above we have argued that the maximum value of c cannot be larger than 0.25 in 'C_3-algae', likely values are in the range of 0.02 to 0.10 (Vanlerberghe *et al.*, 1990). Thus, the lowest possible value of ε_2 is 20.0‰ when c equals 0.25, likely values of ε_2 range from 25.4‰ ($c = 0.10$) to 28.3‰ ($c = 0.02$). The possible effect of PEPCK on the value of ε_2 is small since its substrate is CO_2 and its fractionation factor is likely to be similar to $\varepsilon_{Rubisco}$.

The value of the factor f in Equation 9.20 that describes the leakiness of the cell membrane to intracellular CO_2 determines to what degree ε_2 will affect the isotopic composition of algae. The value of f has only been measured twice in cells actively taking up inorganic carbon (Zenvirth & Kaplan, 1981; Badger & Andrews, 1982); values of about 0.5 were determined. The leakiness of cells which acquire inorganic carbon by passive diffusion of CO_2 has not been determined. We will parameterize f using quantities which can be measured directly. Recall that f was defined as the ratio of the diffusive flux of CO_2 out of the cell to the rate of gross inorganic carbon uptake, $f = k_{-1}/k_1$. This ratio can be rewritten in terms of k_2, which can be measured using short term ^{14}C-uptake experiments:

$$f = 1 - \frac{k_2}{k_1}. \tag{9.29}$$

The rate of photosynthesis must be expressed in units of moles carbon per unit cell surface area per unit time for dimensional consistency with the diffusive fluxes that depend on cell surface area. Photosynthesis is equal to the growth rate, μ, times the carbon content per cell. Carbon content per cell can be expressed in terms of carbon per unit cell volume, γ, times cell volume. The cell volume for a spherical cell is given by $4/3\pi r^3$ and the cell surface by $4\pi r^2$, where r is the cell's radius. Combining these equations we obtain an expression for photosynthesis as a function of growth rate, μ, cell carbon content per unit volume, γ, and cell radius, r:

$$k_2 = \frac{1}{3}\gamma\mu r. \tag{9.30}$$

The parameter k_1 describes the gross flux of CO_2 into the cell. Whereas the net diffusion of CO_2 into the cell is controlled by the CO_2 concentration gradient and the boundary layer thickness (b, which includes the unstirred layer, the cell membrane and the intracellular distance from the cell membrane to the carboxylation sites) the gross flux of CO_2 into the cell is controlled by the concentration of CO_2 in the external medium. Thus, this flux should scale with the extracellular concentration of $CO_2([CO_2]_{aq})$, the diffusivity of CO_2 in the medium, the cell wall, and in the cytoplasm (D_{CO_2}), and with the boundary layer thickness. We will parameterize k_1 with an analogue of Fick's diffusion equation:

$$k_1 = J_D = D\frac{[CO_2]_{aq}}{b}. \tag{9.31}$$

This parameterization is not very realistic since it is very difficult to calculate a mean diffusivity for CO_2 or determine the boundary layer thickness experimentally. However, Equation 9.31 reflects qualitatively the effect of unstirred layer thickness, which will influence the value of b and the effect of $[CO_2]_{aq}$. Equations 9.30 and 9.31 can be substituted into Equation 9.29 to obtain:

$$f = 1 - \frac{\gamma}{3D}\frac{\mu r b}{[CO_2]_{aq}}. \tag{9.32}$$

Equation 9.32 can be substituted into Equation 9.20:

$$\delta_{POC} = \delta_{CO_{2_e}} - \varepsilon_1 + \left(1 - \frac{\gamma}{3D}\frac{\mu r b}{[CO_2]_{aq}}\right)(\varepsilon_{-1} - \varepsilon_2) + \varepsilon_3\frac{v}{\mu + v}. \tag{9.33}$$

However, as a first approximation we can assume that fractionation effects associated with the diffusion of CO_2 into and out of the cell (ε_1, ε_{-1}) and respiration (ε_3) are zero. Thus Equation 9.33 reduces to:

$$\delta_{POC} = \delta_{CO_{2_e}} - \left(1 - \frac{\gamma}{3D}\frac{\mu r b}{[CO_2]_{aq}}\right)\varepsilon_2. \tag{9.34}$$

This equation suggests that there may be considerable isotopic variation among phytoplankton within a given water parcel of uniform $[CO_2]_{aq}$. In the absence of active inorganic carbon uptake small cells with slow growth rates should be most depleted in ^{13}C, while large rapidly growing cells should be most enriched in ^{13}C. Equation 9.34 also predicts that algal δ^{13}C is expected to decrease for higher values of $[CO_2]_{aq}$; note however, that the relationship between δ^{13}C and $[CO_2]_{aq}$ is non-linear (Fig. 9.7).

The potential importance of the parameters used in Equation 9.34 has been pointed out in numerous studies. The possible effects of inorganic carbon concentration and growth rate on phytoplankton δ^{13}C were recognized in early studies by Deuser *et al.* (1968) and Mitzutani & Wada (1982). Recently Rau *et al.* (1989, 1992) proposed that fractionation in open ocean phytoplankton communities varies with the concentration of aqueous CO_2. Although the latitudinal variation of $\delta^{13}C_{POM}$ between the Antarctic and the equator correlates well with P_{CO_2} (Rau *et al.*, 1989), the large variations of observed $\delta^{13}C_{POM}$ at any one latitude and the asymmetric latitudinal distribution of $\delta^{13}C_{POM}$ (Fig. 9.1) indicate that P_{CO_2} cannot be the sole cause of the observed distribution of $\delta^{13}C_{POM}$ in the open ocean. Fry & Wainright (1991) observed a strong correlation between rates of biomass increase and phytoplankton δ^{13}C in shipboard cultures of natural algal populations (Fig. 9.8), thus supporting the predicted effect of growth rate on algal δ^{13}C. Recently Fogel *et al.* (1991) reported large isotopic variations of estuarine $\delta^{13}C_{POM}$, in spite of small variations of $[CO_2]_{aq}$ and $\delta^{13}C_{[CO_2]_{aq}}$. Fogel *et al.* (1991) argued, using simple models of isotopic fractionation (Farquhar *et al.*, 1982; Sharkey &

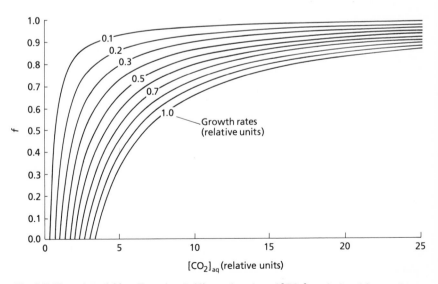

Fig. 9.7 The value of f (see Equation 9.32) as a function of $[CO_2]_{aq}$ calculated for varying growth rates μm. $[CO_2]_{aq}$ and μm are given as arbitrary units only.

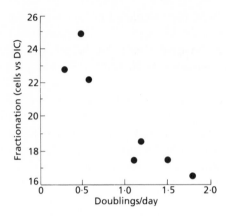

Fig. 9.8 Carbon isotopic discrimination factors decrease with specific rates of growth for natural phytoplankton populations from George's Bank grown in seawater enrichment cultures. (From Fry & Wainright, 1991.)

Berry, 1985), that the observed patterns can only be explained by variations of the factor f (see Equation 9.20), either due to varying rates of cellmembrane leakiness in the case of passive uptake of CO_2 or due to varying proportions of active inorganic carbon uptake. The predicted large effect of cell size (see Equation 9.32) on phytoplankton $\delta^{13}C$ has possibly been observed by Fry & Wainright (1992), who noted that large diatoms on George's Bank are isotopically heavier than POM which is dominated by small phytoplankton.

Models for the active uptake of CO_2 (as opposed to HCO_3^-) still have to use Equations 9.20 and 9.29 as starting points; however, we would have to substitute an uptake function for k_1. Unfortunately, very little is known about the kinetics of inorganic carbon uptake mechanisms. Nonetheless, we still expect that the uptake rate is dependent on $[CO_2]_{aq}$. The $\delta^{13}C$ of the algae would still be dependent on the parameters of Equation 9.20, i.e., cell carbon content, cell size and growth rate. In the case of HCO_3^- uptake by marine algae it is quite possible that the uptake mechanism operates at substrate saturation, which would simplify the modelling. However, apparent fractionation associated with the backdiffusion of CO_2 will be an important parameter because in this case it has to be calculated relative to the $\delta^{13}C$ of HCO_3^-, the species crossing the cell membrane.

Clearly, our understanding of the basic processes of inorganic carbon uptake and carboxylation are inadequate at the present time for a quantitative analysis of carbon fractionation in phytoplankton. The open questions are the importance of β-carboxylases, the value of Rubisco fractionation factors for the different algal groups, changes of cell membrane permeability associated with the induction of the inorganic carbon uptake mechanism, and the controls of inorganic carbon uptake mechanisms in natural algal populations. To approach these questions in the field, it may be necessary to determine the isotopic composition of individual groups of algae using biomarkers (e.g., sterols, carotenoids or fatty acids) that are distinct for different groups of algae.

9.4 ^{15}N fractionation in algae

In contrast to our extensive knowledge of the biochemistry of carbon isotopic fractionation associated with photosynthesis and respiration, our understanding of the mechanisms of nitrogen isotopic fractionation in algae is sketchy at the present. To date, much of the research on nitrogen isotopic fractionation by algae has been designed to further our knowledge of the factors that affect the δ^{15}N of bulk particulate nitrogen in aquatic ecosystems. Given this focus on isotopic fractionation at the whole-organism level, relatively little effort has been devoted to a detailed examination of the biochemical pathways that could lead to isotopic fractionation of nitrogen within algal cells.

9.4.1 Fractionation associated with nitrogen fixation

Dinitrogen fixation by bacteria and cyanobacteria is accompanied by relatively little isotopic fractionation. In the earliest quantitative study of biologically mediated isotopic fractionation, Hoering & Ford (1960) found no significant fractionation ($\alpha = 1.000$) during nitrogen fixation by four strains of *Azotobacter*. More recently, Delwiche & Steyn (1970) measured small but significant values of α ($1.0031 - 1.0054$) for nitrogen fixation by *A. vinelandii*, one of the species used by Hoering & Ford (1960). Similar small fractionation factors have been measured for nitrogen fixation by cyanobacteria (Table 9.4).

Since N_2 dissolved in water has an isotopic composition very similar to that of atmospheric dinitrogen (Benson & Parker, 1961; Miyake & Wada, 1967; Cline & Kaplan, 1975), the very small fractionation associated with nitrogen fixation implies that nitrogen-fixing cyanobacteria should have a δ^{15}N of approximately 0‰. The δ^{15}N of *Trichodesmium* sp. ranges between -2.1 and $+0.05$‰ (Wada & Hattori, 1976; Wada, 1980), which suggests that nitrogen fixation makes a substantial contribution to the overall nitrogen ratio of this marine cyanobacterium.

At the ecosystem level, low δ^{15}N values in bulk particulate nitrogen and zooplankton collected in the East China Sea and North Pacific Central Gyre have been interpreted as evidence that nitrogen fixation is an important source of new nitrogen in those regions (Wada & Hattori, 1976; Mullin *et al.*, 1984). Alternatively, such low δ^{15}N values may reflect the preferential export of ^{15}N from the upper water column as a result

Table 9.4 Isotopic fractionation factors for transformations of nitrogen mediated by algae. For each species, the range or mean ($\pm$SD) value of ε is given, along with the number of estimates summarized (n)

Process	ε (‰)	n	Reference
N$_2$-fixation			
Anabaena sp.	$+1.3 \pm 2.0$	no data	Wada (1978)
Anabaena cylindrica	-1.0 to 8.0	no data	Minagawa & Wada (1978)*
Anabaena sp.	$+2.4 \pm 0.10$	3	Macko *et al.* (1987)
NO$_3^-$ uptake			
Pheodactylum tricornutum	0.7 to 23.0	7	Wada & Hattori (1978)*
Chaetoceros sp.	0.9, 4.5	2	Wada & Hattori (1978)*
	0.0 to 2.5	no data	Wada (1980)
Skeletonema costatum	9.0 ± 2.1	8	Montoya (1990)†
Thalassiosira weissflogii	12.1 ± 2.9	3	Montoya (1990)†
Isochrysis galbana	3.2 ± 1.0	4	Montoya (1990)†
Chroomonas salina	2.2 ± 0.9	3	Montoya (1990)†
Dunaliella tertiolecta	3.4 ± 1.9	8	Montoya (1990)†
Pavlova lutheri	0.9 ± 3.5	6	Montoya (1990)†
Auke Bay plankton	4.0	1	Goering *et al.* (1990)‡
Chesapeake Bay plankton	7.0 ± 1.7	19§	Horrigan *et al.* (1990)§
NO$_2^-$ uptake			
Pheodactylum tricornutum	0.7	1	Wada & Hattori (1978)*
NH$_4^+$ uptake			
Chaetoceros sp.	$-9.7, -5.3$	2	Wada & Hattori (1978)*
Delaware Bay plankton	9.1 ± 0.8	8 ‖	Cifuentes *et al.* (1989)
Chesapeake Bay plankton	6.5, 8.0	2	Montoya *et al.* (1991)¶

* Experiments done in batch culture under a variety of growth conditions.
† Fractionation factors measured in continuous cultures at N different steady-state growth rates.
‡ Calculated from temporal changes in the δ^{15}N of particulate nitrogen in Auke Bay, Alaska.
§ Mean $\pm$ SE calculated from spatial variations in the δ^{15}N and concentration of NO$_3^-$ in Chesapeake Bay using a least-squares technique ($n = 19$).
‖ Mean $\pm$ SE ($n = 8$) calculated from spatial variations in the δ^{15}N and concentration of NH$_4^+$ in Delaware Bay using a least-squares regression.
¶ Estimated from temporal changes in the δ^{15}N and concentration of NH$_4^+$ in Chesapeake Bay.

of the isotopic fractionation associated with zooplankton feeding and excretion (Checkley & Entzeroth, 1985; Checkley & Miller, 1989; Altabet & Small, 1990; Montoya *et al.*, 1992). Such processes are likely to be important in deep, oligotrophic waters where recycled nitrogen (e.g., excreted NH$_4^+$) makes a significant contribution to the phytoplankton nitrogen ratio and where sinking particles are unlikely to be mixed back up into the euphotic zone. The latter explanation is consistent with the available data on the rate of nitrogen fixation by *Trichodesmium* which indicate that usually nitrogen fixation makes a minor contribution to the planktonic nitrogen budget (McCarthy & Carpenter, 1983); however, certain parts of the Indian Ocean and other localized areas are a possible exception (Carpenter & Capone, 1992).

9.4.2 Fractionation during the uptake of inorganic nitrogen

Algae are able to use NH_4^+, NO_2^- and NO_3^- as sources of inorganic nitrogen. An extensive body of research has been devoted to the kinetics of nutrient uptake by a wide variety of algae, but the isotopic fractionation associated with the uptake and assimilation of inorganic nitrogen has not received much attention. The magnitude of α for the uptake of NO_2^-, NO_3^- and NH_4^+ has been measured for phytoplankton growing in light-limited batch culture (Wada & Hattori, 1978; Wada, 1980), and for natural assemblages of estuarine (Cifuentes *et al.*, 1989; Montoya *et al.*, 1991), coastal (Goering *et al.*, 1990) and oceanic (Altabet & McCarthy, 1986) phytoplankton. These studies have documented rather wide variations in the magnitude of α for the uptake of each form of inorganic nitrogen (Table 9.4). In the laboratory, some of this plasticity appears to result from variations in culture conditions. For example, different degrees of aeration and mechanical mixing led to marked differences in the measured value of α for uptake of NO_3^- by the diatom *Pheodactylum tricornutum* (Wada & Hattori, 1978; Wada, 1980). In addition, the magnitude of α for uptake of either NO_3^- or NN_4^+ by the diatoms *P. tricornutum* and *Chaetoceros* sp. varied inversely with growth rate in light-limited cultures (Wada & Hattori, 1978; Wada, 1980). Wada & Hattori (1978) explained these results in terms of a two-step model of isotopic fractionation: (i) a transport step resulting in movement of inorganic nitrogen across the plasmalemma accompanied by little or no isotopic discrimination; and (ii) an internal assimilatory step (e.g., reduction of NO_3^- to the oxidation state of amino nitrogen) which fractionates strongly. Wada & Hattori (1978) argued that the rate of transport across the cell membrane should be rate limiting at high growth rates, while the assimilation of nitrogen into amino acids should be limiting at low growth rates. The different degrees of fractionation associated with these two steps would then lead to an inverse relationship between growth rate and the overall fractionation factor for production of organic matter from inorganic nitrogen.

One difficulty in applying this conceptual model of isotopic fractionation to natural systems is that the isotopic fractionation associated with intracellular processes (e.g., NO_3^- reduction) will lead to measurable differences between the $\delta^{15}N$ of dissolved inorganic nitrogen (DIN) and algae only if there is a significant efflux of DIN out of the cell. In other words, the fractionation associated with NO_3^- reduction will not be observable if all of the NO_3^- transported into the cell is assimilated. Since the transport of NO_3^- across the plasmalemma and its subsequent intracellular reduction both require the expenditure of ATP (Wheeler, 1983), excretion of DIN is energetically costly and is not commonly observed in algae.

An alternative possibility is that significant fractionation occurs during the initial transport of NO_3^- into the cell, and that fractionation associated with intracellular assimilation of NO_3^- is masked by the absence of significant exchange between internal and external pools of DIN. According to this idea, the inverse relationship between α and growth rate observed in batch culture may simply reflect differences in growth conditions (e.g., the energy charge of light-limited algae) rather than a shift in the site of isotopic fractionation. Recent measurements of the isotopic fractionation associated with uptake of NO_3^- by phytoplankton grown in continuous culture (Montoya, 1990) are consistent with this hypothesis. This experimental system allowed the determination of α at a variety of different growth rates under otherwise invariant culture conditions. A total of six species were used in this study (two diatoms, two haptophytes, one chlorophyte and one cryptophyte), none of which showed significant variation in α as a function of growth rate. Instead, the two diatoms had a fractionation factor significantly higher than that for the four flagellates used in the study (Table 9.4), which suggests that diffusion through the unstirred layer adjacent to the cell surface may make a significant contribution to the overall magnitude of α for NO_3^- uptake. Similarly, Wada & Hattori (1978) have found that α for NO_3^- uptake is significantly higher in unstirred than in well-mixed cultures of the diatom *P. tricornutum*.

Taken together, the available data indicate that the magnitude of α for uptake of DIN may vary significantly with growth conditions and between taxa. Further study of the patterns of nitrogen isotopic fractionation by algae under controlled conditions in the laboratory is clearly needed before the mechanism of isotopic fractionation can be elucidated. However, in nature, conditions that allow the expression of fractionation effects are rarely observed since nitrogenous nutrients are usually depleted in oceanic and limnic surface waters. Exceptions are bloom conditions, possibly upwelling areas, and the oceanic subarctic Pacific and the Antarctic where nitrogenous nutrients are not depleted in surface waters during the growth season.

9.4.3 Directions for future research

Our understanding of the factors that control the $\delta^{15}N$ of algae is constrained both by the relatively small body of research done to date and by the variety of experimental techniques used in studying isotopic fractionation by algae. Since data obtained using different experimental approaches are not directly comparable, we are far from having a model of nitrogen isotopic fractionation that would facilitate the use of algal $\delta^{15}N$ values as direct indicators of the physiological state or growth history of the algae. Clearly, further study of the interactions between environmental (e.g., light intensity, substrate availability) and physiological factors are needed before such a model can be constructed.

The intracellular fractionation of nitrogen by algae has received very little attention, though an understanding of the effects of different biosynthetic pathways on the $\delta^{15}N$ of algal cells is essential in any attempt to derive physiological information from natural patterns of variation in $\delta^{15}N$. Macko *et al.* (1987) have found that protein nitrogen isolated from the cyanobacterium *Anabaena* sp. is enriched in ^{15}N by roughly 3‰ relative to the whole cell. Individual amino acids isolated from *Anabaena* have a wide range of $\delta^{15}N$ values which presumably reflect both the nitrogen source and the isotopic fractionation associated with different biosynthetic pathways within the cell (Macko *et al.*, 1987). The measurement of isotopic fractionation factors for the individual reactions that constitute these pathways is an important, but largely unexplored, area of study (Macko *et al.*, 1986; Bada *et al.*, 1989).

CHAPTER 10

The use of stable carbon isotopes to study microbial processes in estuaries

R.B. COFFIN, L.A. CIFUENTES & P.M. ELDERIDGE

10.1 Introduction

Elemental cycles in estuaries are complex, owing to physical, geological, chemical and biological interactions that are specific to each estuarine system. Over the last 30 years, the stable isotopes of carbon, nitrogen, hydrogen, sulfur and oxygen have been used to differentiate and trace components of these respective elemental cycles. Essentially, the specific signature provided by the stable isotope ratio of that element is used to examine its sources and fates. This approach is uniquely suited to studies of estuarine systems.

Because of the broad range of isotopic values for carbon sources in estuarine and coastal waters, stable carbon isotopes have been used extensively to examine diverse aspects of carbon cycling in estuaries. For example, there are multiple possible sources of particulate organic carbon (POC) in estuaries, such as riverine input, salt-marsh macrophytes, phytoplankton, seagrasses and benthic algae. In various studies, the major source has been identified on the basis of $\delta^{13}C$ measurements. Alternatively, $\delta^{13}C$ ratios have been used to study the fate of organic matter in estuaries. These studies focus on either the trophic assimilation of carbon in microbial and higher animal foodchains, or on the physical transport of carbon into sediments and coastal waters. Using a different approach, researchers have applied property–salinity models to $\delta^{13}C$ measurements of dissolved inorganic carbon (DIC). Deviations from values predicted by conservative mixing have been used to infer the predominance of autotrophic versus heterotrophic activity. Finally, stable carbon isotopes have been used to describe the spatial and seasonal dynamics of inorganic and organic carbon pools.

Numerous studies have examined carbon cycling by large heterotrophic organisms. Until recently, stable isotopes have not been used to examine carbon cycling by bacteria. The few published and unpublished isotopic studies of bacteria have shown that bacteria can alter isotope ratios of both inorganic and organic pools. An important next step in isotope biogeochemistry should be to better understand how bacteria alter the isotope ratio of other carbon pools. Therefore, the purpose of this chapter is to: (i) briefly review the stable carbon isotope literature emphasizing information that pertains to estuarine carbon sources for bacteria; (ii) present our most recent carbon isotope data on bacterial carbon sources in different estuaries; (iii) document potential isotopic relationships among DIC, dissolved and particulate organic carbon, and bacteria with data from the Perdido Estuary, Florida; and (iv) present a model that predicts the relative importance of respiration and atmospheric mixing on $\delta^{13}C$ of DIC in estuaries.

10.2 Estuarine carbon pools

10.2.1 POC

Because of large differences in $\delta^{13}C$ of organic matter from marine, terrestrial and marsh environments, stable isotopes provide a useful tool to trace the transport of material in the water column and sediments and follow the trophic assimilation of carbon. Most of the POC in estuaries is derived from photosynthesis. The $\delta^{13}C$ of photosynthetic organisms is, in part, a function of the CO_2 fixation pathway, Calvin–Benson pathway (C_3), Slack–Hatch pathway (C_4) or crassulacean acid metabolism (CAM), as outlined in Chapter 1. Generally, organic matter from terrestrial sources (C_3 plants) is relatively depleted in ^{13}C, aquatic macrophytes and C_4 marsh grasses are enriched in ^{13}C, and phytoplankton are intermediate (Fry & Sherr, 1984). The isotope ratio of DIC (see below), the concentration and speciation of DIC, and physiological differences among primary producers also influence stable carbon isotope values (Fogel & Cifuentes, 1993).

Organic matter from primary production in marine, freshwater, marsh and terrestrial ecosystems has a $\delta^{13}C$ range of -35 to $-5‰$ (Fig. 10.1).

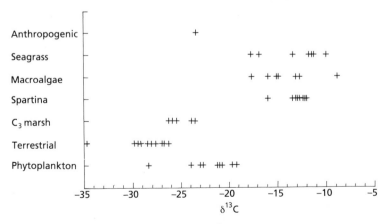

Fig. 10.1 Ranges of stable carbon isotope ratios ($\delta^{13}C$) of potential carbon sources for bacteria.

According to these data, terrestrial carbon sources can usually be resolved from aquatic carbon sources. Moreover, phytoplankton sources can be distinguished from marsh and terrestrial carbon sources. In contrast, C_3 marsh and upland plants cannot be differentiated by $\delta^{13}C$ alone. The same holds for C_4 marsh plants, seagrasses and macroalgae. For these cases, it may be necessary to use additional geochemical tracers to distinguish carbon sources with overlapping carbon isotope values (e.g., Peterson *et al.*, 1985).

In estuaries, the $\delta^{13}C$ of POC range from approximately -33 to $-18‰$ (Fig. 10.2). Generally, reported values indicate that the primary sources of particulate organic matter are terrestrial organic matter and phytoplankton. The relative importance of these two main sources depends on sample location and hydrodynamics of the estuary. Seagrasses, macroalgae and C_4 marsh plants are not usually major contributors to estuarine POC. There may be several reasons for this observation. First, the majority of studies have been conducted in estuaries where upland plants and phytoplankton are the major source. It is also possible, in some systems, that seagrasses, macroalgae and C_4 marsh plants (i.e., isotopically heavy carbon) are mixed with a significant fraction of terrestrially derived C_3 plant material (i.e., isotopically light carbon). This mixture would be confused with planktonic organic matter. Finally, the relative importance of carbon sources in providing nutrition to heterotrophic organisms will most likely be site specific. For example, in

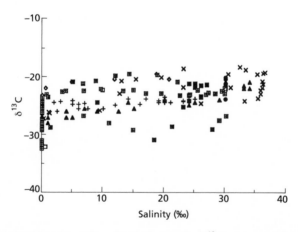

Fig. 10.2 Compiled data for stable carbon isotope ratios ($\delta^{13}C$) of suspended particulate matter vs. salinity (‰). $\times$ = Cai *et al.* (1988); $\diamond$ = Cifuentes *et al.* (1988); $\bullet$ = Gearing *et al.* (1977); $\blacklozenge$ = Laane *et al.* (1990); $\blacksquare$ = LeBlanc *et al.* (1989); $\bigcirc$ = Spies *et al.* (1989); $\blacktriangle$ = Spiker & Schemel (1979); $\bowtie$ = Wada *et al.* (1987b); $\boxdot$ = Simenstad & Wissmar (1985); $\triangle$ = Tan & Strain (1979); + = Lucotte (1989); $\square$ = Hedges *et al.* (1986); $\blacksquare$ = Fogel & Cifuentes (1991).

Baratarian Basin, Louisiana, the abundance of *Spartina* detritus in POC is inversely related to the size of the aquatic environment (Delaune & Lindau, 1987). Thus seagrasses, marsh grasses and macroalgae can be important sources of organic matter on the fringes of an estuary, but the contribution of these sources is largely diluted in the main stem of larger estuarine systems.

Among estuarine carbon sources, POC from terrestrial primary producers has the most consistent $\delta^{13}C$ values. Here, the $\delta^{13}C$ of POC that is attributed to terrestrial inputs generally range from -31 to $-26‰$ (Tan & Strain, 1979; Hackney & Haines, 1980; Simenstad & Wissmar, 1985; Cai *et al.*, 1988; Hedges *et al.*, 1988; Lucotte, 1989; LeBlanc *et al.*, 1989). *Juncus*, a freshwater marsh plant that is dominant in estuarine headwaters, is relatively enriched in ^{13}C with reported $\delta^{13}C$ values of $-26‰$ (Hackney & Haines, 1980). Because the $\delta^{13}C$ of *Juncus* is similar to ranges of terrestrial organic matter, organic matter from *Juncus* marshes may not be distinguished from terrestrial sources.

POC that is primarily composed of phytoplankton has a wide range of $\delta^{13}C$ values, from -44 to $-20‰$, a result of differences in the isotopic ratio and concentration of CO_2, and in physiology among genera of phytoplankton (Wada *et al.*, 1987; Tan & Strain, 1988; Goering *et al.*, 1990; Mook & Tan, 1991; Fogel *et al.*, 1992). The range in carbon isotope values for phytoplankton and phytoplankton-dominated POC is larger in the upper estuary compared with the seaward end of the estuary. Initially, this larger range in $\delta^{13}C$ of POC in the freshwater region of estuaries could be explained solely by the corresponding variation in the $\delta^{13}C$ of DIC (discussed below). Throughout the Delaware Estuary, however, changes in $\delta^{13}C$ of POC could not be accounted for by variations in the isotope ratio of DIC alone (Fogel *et al.*, 1992). Generally, the isotope ratio of phytoplankton is a function of the enzymatic fractionation of $^{13}CO_2$ and $^{12}CO_2$ during carbon fixation. Fogel *et al.* (1992) suggested that another important factor controlling the $\delta^{13}C$ of phytoplankton is the concentration of CO_2. During rapid growth periods CO_2 concentrations may limit phytoplankton production. Under these conditions, the diffusion of CO_2 across the cell membrane results in decreased isotopic fractionation. Another possibility, however, is that when CO_2 becomes limiting phytoplankton switch to a bicarbonate carbon source. Both factors can contribute to the wide range in $\delta^{13}C$ that is measured in phytoplankton.

The marsh grass, *Spartina*, is a dominant source of estuarine carbon in marshes associated with many East Coast estuaries. A review of studies examining the fate of particulate carbon from salt marshes in estuaries is presented in Chapter 7. The importance of *Spartina* to secondary production in the estuarine water column has been widely debated. Interpretation of results of stable carbon isotope studies in salt marsh estuaries led to the conclusion *Spartina* does not contribute to POC in the estuary (Haines, 1976a; Hackney & Haines, 1980). In contrast, more recent studies indicate that *Spartina* production can be a significant factor in the secondary production of the estuarine water column via the detrital food chain (Peterson & Howarth, 1987; Deegan *et al.*, 1990). The divergent interpretation of the importance of *Spartina*-derived plant detritus in estuaries points to the need to understand physical processes in estuaries (Delaune & Lindau, 1987, as mentioned above), which undoubtedly control the amount of organic matter from *Spartina* that is available to bacteria.

Aquatic seagrasses and macroalgae are another highly productive component of many estuarine and coastal ecosystems and therefore are a potential source of POC for estuarine foodchains. The range of stable carbon isotope values of seagrasses is quite broad, from -24 to $-3‰$ for 47 species representing 12 genera (McMillan *et al.*, 1980). In another study, $\delta^{13}C$ values as low as $-50‰$ have been reported in rapidly flowing spring waters (Osmond *et al.*, 1981). The reasons for the large variation have been widely investigated. Early studies attributed isotopically enriched $\delta^{13}C$ to C_4 metabolism (Benedict & Scott, 1976; Doohan & Newcomb, 1976; Benedict, 1978; Faganeli *et al.*, 1986). Other work proposed that variations in stable carbon isotope values are influenced by isotopic variation of inorganic carbon (Smith *et al.*, 1976; McMillan *et al.*, 1980; Osmond *et al.*, 1981; McMillian & Smith, 1982), differential uptake among species of bicarbonate and CO_2 (Andrews & Abel, 1979; Benedict *et al.*, 1980; Faganeli *et al.*, 1986), temperature (Degens *et al.*, 1968a; Wong & Sackett, 1978), light intensity (Wefer & Killingley, 1986; Cooper & DeNiro, 1988), and water mixing affecting CO_2 diffusion and subsequently enzymatic fractionation as a function of concentration (Osmond *et al.*, 1981; Raven *et al.*, 1982).

Several studies present strong evidence for the importance of seagrasses and macroalgae to aquatic food webs. The incorporation of carbon from seagrasses, however, appears to be a function of feeding mechanisms. In many ecosystems, pathways involving POC as an intermediary are not important mechanisms for assimilating carbon from seagrasses and macroalgae. For example, in an Arctic kelp community over 50% of the carbon demand for herbivores is met by *Laminaria*, while suspension feeders such as hydroids, soft corals and bryozoans obtain the greatest proportion of their carbon demand from phytoplankton (Dunton *et al.*, 1987). In the Hood Canal, Washington, $\delta^{13}C$ of POC along the estuarine salinity gradient are consistent with phytoplankton or terrestrial input as the major source of carbon, rather than the dense eelgrass or macroalgae beds (Simenstad & Wissmar, 1985). While the isotopic values of suspension feeders (*Mytilus* and *Balanus*) are consistent with that of POC, the $\delta^{13}C$ of predators indicate that macroalgae and eelgrass contribute a large proportion of the carbon flux to higher trophic levels (Thayer *et al.*, 1978a; Simenstad & Wissmar, 1985). However, other researchers report that detritus from macrophytes form the major pathway for assimilating carbon into foodchains (McConnaughey & McRoy, 1979b; Nichols *et al.*, 1985). In the Izembek Lagoon, Alaska, eelgrass contributes a significant amount to total carbon fluxes and is a significant fraction of the carbon balance for suspension feeders (McConnaughey & McRoy, 1979b). The consensus of these reports indicates that particulate organic matter from seagrasses may not be a significant source of organic matter for bacteria, although dissolved organic carbon (DOC) from seagrasses may be an important source of substrate for bacteria.

10.2.2 DOC

There are few reports of $\delta^{13}C$ for DOC because of methodological difficulties in concentrating enough material for isotopic analysis. With recent technical advances (Peterson *et al.*, in press; L.A. Cifuentes & R.B. Coffin, unpublished data) more studies of stable isotopes in dissolved carbon pools will be forthcoming. Published $\delta^{13}C$ values of DOC range from -28.7 to $-16.3‰$ (Fig. 10.3; Jeffrey, 1969; Shultz, 1974; Eadie *et al.*, 1978; Kerr & Quinn, 1980; R.B. Coffin & L.A. Cifuentes, unpublished data). As expected, the compilation of data suggests that both marine and terrestrial sources contribute to DOC pools in estuarine waters. Terrestrial sources are noted primarily in samples from rivers

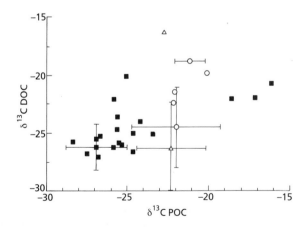

Fig. 10.3 Stable carbon isotope ratios ($\delta^{13}C$) of dissolved organic carbon (DOC) vs. suspended particulate matter. Included are individual values for river samples (■) and ranges for estuarine (△) and marine (○) samples.

whereas phytoplankton have more of a contribution in bays and open ocean waters. A comparison of $\delta^{13}C$ of POC and DOC demonstrate that the values are commonly uncoupled, especially in rivers (Fig. 10.3).

10.2.3 DIC

Primary producers in seawater fix inorganic carbon as CO_2 and/or HCO_3^- (Fogel & Cifuentes, 1993). The isotope ratio of DIC is an important consideration for tracing carbon sources because this ratio strongly influences the value for the primary producer (as was discussed above). The $\delta^{13}C$ of DIC in aquatic ecosystems can vary spatially and seasonally within an estuary. Stable isotope ratios of DIC in estuaries range from -29 to $+2\permil$ (Fig. 10.4; Spiker & Schemel, 1979; Osmond et al., 1981; Tan & Strain, 1983, 1988; Benninger & Martens, 1983; Mook & Tan, 1991; Fogel et al., 1992; L.A. Cifuentes & R.B. Coffin, unpublished data). Values in fresh waters are generally more negative and variable than in seawater, while the $\delta^{13}C$ of DIC in the marine end-member of estuaries is consistently around 0 to $+3\permil$ (Spiker & Schemel, 1979; Benninger & Martens, 1983). The trends for $\delta^{13}C$ of DIC in estuaries plotted against salinity are generally similar, with the exception of the Perdido Estuary, Florida (Fig. 10.4). Unusually light or heavy isotopic ratios of DIC may result from net-heterotrophic or net-autotrophic processes, respectively. The effect of respiration on $\delta^{13}C$ of DIC was presented in a study of macrophytes in a tropical estuary (Zieman et al., 1984). *Thalassia* and *Halodule* were more isotopically negative in poorly

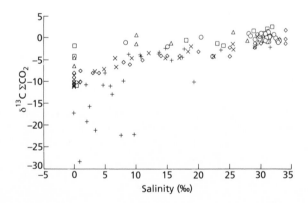

Fig. 10.4 Compiled data for stable carbon isotope ratios ($\delta^{13}C$) of dissolved inorganic carbon (DIC) vs. salinity ($\permil$). $\square$ = Gulf of St Lawrence (Tan & Strain, 1988); $\diamondsuit$ = San Francisco Bay (Spiker & Schemel, 1979); $\triangle$ = Hood Canal, Washington (Simenstad & Wissmar, 1985); $\bigcirc$ = St. Margaret's Bay, Nova Scotia (Tan & Walton, 1975); $\times$ = Delaware Estuary (Fogel & Cifuentes, 1993); + = Perdido Estuary, Florida (L.A. Cifuentes & R.B. Coffin, unpublished).

flushed coastal zones suggesting that remineralization may provide significant amounts of DIC that is depleted in ^{13}C for macrophyte production. The control of heterotrophic processes on the δ^{13}C of DIC in the Perdido Estuary are discussed below. Primary production can also alter δ^{13}C of DIC, resulting in more positive ratios. For example, in rapidly flowing freshwater streams the δ^{13}C of DIC varied from -16 to $-21‰$, while in still waters values increased to $-5‰$ (Osmond et al., 1981). This result suggested that with carbon limitation, enzymatic discrimination against ^{13}C during photosynthesis is diminished.

10.3 Isotopic measurements of bacteria in estuaries

Recently, stable isotopes have been used to examine sources of carbon used by bacterioplankton (Coffin et al., 1989, 1990). To measure the isotopic composition of bacteria, it was necessary to separate bacteria or a component of bacteria from all other particles. Initially, bioassay incubations were used to determine the source of organic matter on which bacteria were potentially growing (Coffin et al., 1989). This approach consisted of filtering water through 0.2 μm filters to remove all particulate material and then inoculating the filtered water with a 1% addition of water containing indigenous bacteria. Samples were incubated until enough bacterial carbon biomass was produced for a stable isotopic analysis (48–72 hours). In preliminary experiments testing this approach, bacteria grown on a variety of carbon sources had δ^{13}C values that were similar to each source. In the field, a survey of isotope ratios of bacteria grown in bioassay experiments from a northeastern Massachusetts salt-marsh estuary suggested that DOC leached from *Spartina* grasses is an important source of organic matter to bacteria. Although phytoplankton or other particulate sources did not appear to be an important source of organic matter for bacteria in this system, the bioassay may underestimate these sources as a result of the filtration.

In a related study, the *in situ* growth substrate for bacteria was determined by extracting nucleic acids from <1.0 μm particles that were concentrated from 50 to 100 l seawater samples (Coffin et al., 1990). With this approach, nucleic acids are used as a biomarker for bacteria. Studies demonstrated that nucleic acids have δ^{13}C values similar to the whole bacteria (Blair et al., 1987; Coffin et al., 1990). Furthermore, the nucleic acids can be traced to the organism from which they are extracted with 16S rRNA probes (Giovannoni et al., 1988, 1990; Coffin et al., 1990; Delong, 1993), to insure that the extracted nucleic acids are from bacteria. This method has recently been tested in several different estuarine and coastal environments, and in some cases the bioassay method has been compared to the nucleic acid extraction (Table 10.1). Although a large range in δ^{13}C was measured in bacterial nucleic acids, the values were

Table 10.1 Stable carbon ($\delta^{13}C$) and nitrogen isotope ($\delta^{15}N$) ratios values in bacteria measured using the bioassay (Coffin *et al.*, 1990) and nucleic acid extraction (Coffin *et al.*, 1990)

Site	$\delta^{13}C_{na}$ (‰)	$\delta^{13}C_{ba}$ (‰)	$\delta^{15}N_{na}$ (‰)	$\delta^{15}N_{ba}$ (‰)
Prince William Sound, Alaska				
Disk Island Beach	−19.9	−26.5	8.5	8.0
Alyeska Drainage	−27.0	−24.9	5.0	9.0
Disk Island High Tide	n.d.	−20.6	n.d.	8.5
Guayas Estuary, Ecuador				
Churute River	n.d.	−25.3	n.d.	6.1
Gulf of Guayaquil	n.d.	−24.0	n.d.	9.3
Shrimp Pond	n.d.	−23.1	n.d.	4.9
Parker Estuary, Massachusetts				
Upper Estuary	−23.7	−23.3	n.d.	n.d.
Mid Estuary	−21.4	−19.6	n.d.	n.d.
Perdido Estuary, Florida				
Eleven Mile Creek	−27.5	−31.9	n.d.	n.d.
Perdido River	−26.8	−26.4	n.d.	n.d.
Lower Estuary	−27.9	−29.9	n.d.	n.d.
Santa Rosa Sound, Florida				
GBERL	−21.9	−23.3	n.d.	n.d.
Range Point	−20.7	−21.0	n.d.	n.d.
Laguna Madre, Texas				
Laguna Madre	−15.3	n.d.	3.1	n.d.
Beach	−17.2	n.d.	5.8	n.d.
Baffin Bay	−20.1	n.d.	4.1	n.d.
Compuerta Pass	−20.2	n.d.	2.7	n.d.
Gulf of Mexico*	−24.1 (5)	−23.8 (5)	9.0 (5)	9.1 (5)

* Numbers in parentheses next to isotope values represent replicate samples.
ba, bacterial bioassay; na, nucleic acid; n.d., not determined.

similar to the isotope values of expected sources of organic matter in the environments from which the cells were collected (values ranged from −27.9 to −15.3‰). Cells enriched in ^{13}C, −15.3‰, were observed in waters overlying seagrass beds from Laguna Madre, Texas. Isotopically light cells were measured in waters that receive high carbon loadings from pulp mills in the Perdido Estuary and in waters that receive carbon from mangrove marshes and sewage wastes in the Guayas Estuary, Ecuador (see Table 10.1). Bacteria with $\delta^{13}C$ values similar to phytoplankton were collected in Santa Rosa Sound, Florida and Prince William Sound, Alaska.

It is instructive to compare $\delta^{13}C$ of bacteria grown in bioassay experiments with that of nucleic acids (Table 10.1). In the Perdido Estuary (Fig. 10.5), samples were taken from three locations, 2 km downstream at the mouth of a tributary (Eleven Mile Creek) that carries effluent from a local pulp mill to the Perdido Estuary, and the third, further downstream in the lower estuary (see Coffin *et al.*, 1990). At the mouth of Eleven Mile Creek and in the lower estuary, $\delta^{13}C$ from nucleic acid extractions and bioassay incubations were remarkably different. Nucleic acids were as much as 3‰ lighter than the bioassays. We attribute these differences to multiple substrates being assimilated by bacteria. The sources of organic matter that are available for bacteria in this system are discussed in Section 10.4. Another comparison that resulted in different $\delta^{13}C$ values for nucleic acids and bioassay experiments was in interstitial waters from oil contaminated beaches in Prince William Sound. Water pumped from wells on the beach during flooding tides contained bacteria with $\delta^{13}C$ of −19.9‰, suggesting that phytoplankton were an important source of organic matter. After flooding the oil contaminated beaches, bacteria grown in water from the well using the bioassay approach had isotope ratios around −27‰, suggesting the oil became an important nutrient source. For many other sample sites the carbon isotope values from the two techniques were similar (Table 10.1). Both techniques provide useful analytical capabilities; however, results from each must be interpreted

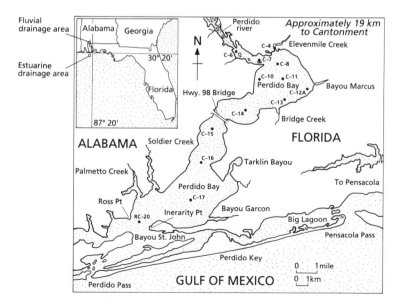

Fig. 10.5 Map of the Perdido Estuary, Florida, showing its location on the US Gulf coast. Station locations are indicated on the map.

within the limitations of the methods. The bioassay approach provides information on some of the potential sources of organic matter for bacteria. Nucleic acids indicate the organic matter that is used by bacteria *in situ*. A combination of the methods may resolve multiple substrate sources.

10.4 The Perdido Estuary: a case study

In the following discussion, we will use data from the Perdido Estuary, Florida, to discuss patterns between the isotope ratios of CO_2, organic matter and bacteria. This discussion will be followed by a simple model that investigates the potential influence of bacterial respiration and atmospheric exchange on the $\delta^{13}C$ of CO_2.

As discussed earlier, pulp mill effluent drains into the Perdido Estuary through Eleven Mile Creek (Fig. 10.5). Initially, we characterized suspended particulate matter (SPM) in pulp mill effluent and in the Perdido River with carbon and nitrogen isotopes ($\delta^{15}N$). These sources represent the major allochthonous sources of organic matter to the system. In turn, we characterized two seaward stations in the Perdido Estuary (C-17 and RC-20), which represent autochthonous sources of organic matter. To assess the extent of pulp mill contamination in the estuary itself, we then measured the $\delta^{13}C$ and $\delta^{15}N$ of SPM collected between the Perdido River and Eleven Mile Creek inputs and the high salinity stations. Surprisingly, many of those SPM samples had $\delta^{13}C$ values that were more negative than values predicted by mixing of the characterized sources (Fig. 10.6).

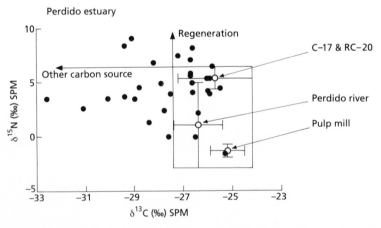

Fig. 10.6 Stable nitrogen isotope ratios ($\delta^{15}N$) vs. stable carbon isotope ratios ($\delta^{13}C$) of suspended particulate matter (SPM) in the Perdido Estuary, Florida. Ranges for pulp mill effluent, the Perdido River and stations C-17 and RC-20 are designated by error bars. The box indicates the expected domain for mixtures of these three sources. Data outside the box imply other sources.

It follows from this data that an additional, uncharacterized carbon source was important in the Perdido Estuary.

At first glance, the light isotope ratios could be attributed to another terrestrial C_3 source. Algae, however, can have isotope ratios similar to terrestrial C_3 plants if grown on high concentrations of CO_2, which is isotopically negative. To test this possibility, we measured the $\delta^{13}C$ of DIC on two occasions, April and May 1991. Most of the values at low salinites (<10‰) were less than −10‰ with values reaching −28‰ (see Fig. 10.1). Furthermore, and as discussed earlier, the data taken in May were isotopically more negative compared with that from April. Algae growing on such isotopically light DIC could certainly be the source of SPM with the lighter isotope ratios shown in Fig. 10.6.

Following the arguments in Fogel et al. (1992), the isotope ratio of carbon fixed by autotrophic processes is influenced not only by the isotope ratio of DIC, but by the isotopic discrimination between DIC and fixed carbon. The latter can be a function of the concentration and speciation of DIC. In April, there was a strong correlation between $\delta^{13}C$ of SPM and that of DIC ($r = 0.97$, Fig. 10.7a). The isotopic discrimination between SPM and DIC, however, was not constant, but increased from fresh to saline waters. If algae were the major contributor to SPM and autotrophic processes resulted in the correlation shown in Fig. 10.7a, then the availability of CO_2 to algae should have decreased from saline waters to fresh waters resulting in less isotopic discrimination. Because fresh waters generally have higher concentrations of $CO_{2(aq)}$, this correlation more likely resulted from mixing of fresh water containing terrestrially derived SPM and isotopically light CO_2 with coastal water containing marine derived SPM and CO_2 with $\delta^{13}C$ values near 0.

Heterotrophic processes can also lead to a correlation between the $\delta^{13}C$ of DIC and SPM. Contributing factors include the isotope ratio of the respired substrate, the isotopic discrimination between substrate and respired CO_2, and the size of the DIC pool to which the respired CO_2 is added. The most negative $\delta^{13}C$ for DIC were measured in the May sampling period, suggesting heterotrophic processes were important at that time. There was not, however, a strong linear relationship between the $\delta^{13}C$ of SPM and that of DIC ($r = 0.75$, Fig. 10.7b). If heterotrophic processes influenced the $\delta^{13}C$ of DIC in May, then the organic matter respired by bacteria would have been sufficiently negative and abundant to alter the isotope ratio of the DIC pool. The $\delta^{13}C$ of DOC, which integrates over the labile and refractory components, was light during the May sampling period (Fig. 10.8a) with a mean $\delta^{13}C$ of −7.3 ± 0.5‰. Although there was not a linear relationship between SPM and DOC (Fig. 10.8b), the mean $\delta^{13}C$ for SPM was −28.6 ± 0.7‰, or only 1.3‰ more negative than that for DOC. Since the isotope data we have for

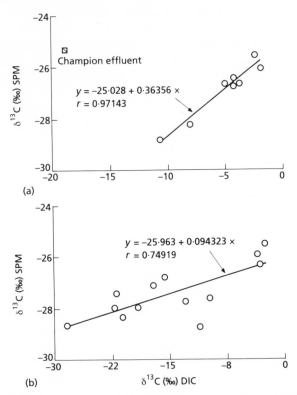

Fig. 10.7 Stable carbon isotope ratios ($\delta^{13}C$) of dissolved inorganic carbon (DIC) vs. $\delta^{13}C$ of suspended particulate matter (SPM) in the Perdido Estuary, Florida, in (a) April 1991 and (b) May 1991. Linear regressions of the data are included.

DOC in pulp mill effluent is in the range of −25‰, or heavier than the DOC in the estuary, the pulp mill could not be the major source in the system. It is certainly possible that plankton are a major source of DOC in this system because carbohydrates and proteins are generally isotopically heavier than other planktonic carbon pools (Deines, 1980). The Perdido River could also be an important source of DOC in the system, samples will examine the $\delta^{13}C$ of DOC in the Perdido River in the future.

Bacteria grown in bioassay experiments (Coffin *et al.*, 1989) can also be used as an indirect measure of the isotopic ratio of the DOC that is assimilated. The $\delta^{13}C$ of bacterial bioassays from the Perdido Estuary ranged from −24.1 to −31.9‰ (Table 10.2). The mean $\delta^{13}C$ was −27.2 ± 2.5‰. We have no isotopic data for DOC that directly correspond to the bacterial bioassay data. The mean $\delta^{13}C$ for DOC in May (−27.3 ± 0.5‰) though was similar to that of the bacterial bioassays. This link, however,

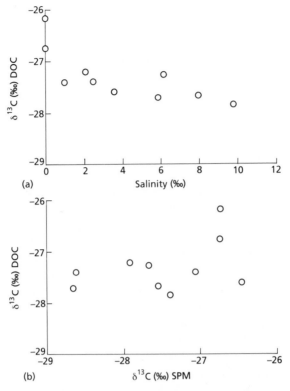

Fig. 10.8 Stable carbon isotope ratios ($\delta^{13}C$) of dissolved organic carbon (DOC) vs. (a) salinity (‰) and (b) suspended particulate matter (SPM) in the Perdido Estuary, Florida, during May 1991.

Table 10.2 $\delta^{13}C$ for bacteria grown in bioassay experiments (see Coffin *et al.*, 1989) with water from the Perdido Estuary

Station	Date	$\delta^{13}C$ (‰)
C-4	June 1989	−31.9
	January 1991	−24.6
	March 1991	−25.1
C-6	June 1989	−26.4
	January 1991	−26.3
	March 1991	−28.5
C-8	March 1991	−26.4
C-17	June 1989	−29.9
	January 1991	−24.1
	March 1991	−29.0

may be established more strongly when we have parallel $\delta^{13}C$ data for bacteria and DOC.

10.5 Modelling $\delta^{13}C$ of CO_2 in estuaries

10.5.1 Description of the model

The case study described above demonstrates the strong influence that heterotrophic processes may exert on $\delta^{13}C$ of DIC and POC in estuaries. Physical and chemical controls on the CO_2 system in estuaries determine the extent to which autotrophic assimilation or heterotrophic respiration can alter the DIC concentration or its isotope ratio. A factor that was not discussed in the treatment of Fogel et al. (1992) is the importance of atmospheric exchange in buffering the effect of either removal of CO_2 by autotrophic processes or of addition of CO_2 by heterotrophic processes. Whitfield & Turner (1986) examined the physical chemistry of CO_2 systems in estuaries. According to these authors, the chemistry of CO_2 in estuaries is probably intermediate between extremes predicted by open- and closed-system calculations. Their open-system calculations assumed that P_{CO_2} was in equilibrium with the atmosphere and that carbonate alkalinity was conservative, whereas their closed-system calculations assumed that both ΣCO_2 and carbonate alkalinity was conservative. Their intermediate case incorporated an exchange term across the air–sea boundary, which allowed for limited exchange of CO_2. We have adapted the equations of Whitfield & Turner (1986) to study the relative importance of respiration and atmospheric exchange on isotope ratios of DIC in estuaries.

For simplicity we assumed steady state and used simple, one-dimensional, advection diffusion equations for the two stable isotopes of carbon in DIC:

$$0 = -\mu\frac{d[DI^{12}C]}{dx} + K_{eddy}\frac{d^2[DI^{12}C]}{dx^2} + EX^{12}C + RI^{12}C \qquad (10.1)$$

and

$$0 = -\mu\frac{d[DI^{13}C]}{dx} + K_{eddy}\frac{d^2[DI^{13}C]}{dx^2} + EX^{13}C + RI^{13}C. \qquad (10.2)$$

Here, DIC is equivalent to ΣCO_2 (mmol l^{-1} C), μ (m s^{-1}) is velocity, K_{eddy} (m^2 s^{-1}) is the eddy diffusion coefficient, EXC (mol s^{-1}) is CO_2 exchanged across the air–sea interface, and RIC (mol s^{-1}) is respiration. The superscripts denote the particular isotope. The total DIC (i.e., [DIC] = [DI^{12}C] + [DI^{13}C]) is the measured parameter. Since DI^{13}C is approximately 1% of the total DIC, we assumed [DI^{12}C] [DIC] without much loss in accuracy. In turn, [DI^{13}C] was approximated from the concentration and isotope ratio of DIC with:

$$\delta^{13}C = \left[\frac{\left(\frac{^{13}C}{^{12}C}\right)_{DIC}}{\left(\frac{^{13}C}{^{12}C}\right)_{STD}} - 1 \right] \times 1000. \tag{10.3}$$

In Equation 10.1, we used an air–sea exchange term that was similar to that used by Whitfield & Turner (1986):

$$\frac{DK_0 \times 10^4}{zh}[CO_{2(a)} - CO_{2(w)}] \tag{10.4}$$

where D ($cm^2 s^{-1}$) is the exchange coefficient for CO_2, z (cm) is the boundary layer thickness, K_0 ($mol\,cm^{-3}\,atm^{-1}$) is the CO_2 solubility, h (m) is estuarine depth, $CO_{2(a)}$ is the partial pressure in the air (atm) and $CO_{2(w)}$ is the partial pressure in the water (atm). Multiplication by 10^4 was needed to maintain the proper units. In Equation 10.2 we used an exchange coefficient that was related to the molecular diffusivities of ^{12}C and ^{13}C, and the equilibrium fractionation factor between $CO_{2(a)}$ and $CO_{2(w)}$ (see Wanninkhof, 1985).

Having incorporated the respective exchange coefficients into Equations 10.1 and 10.2, we then transformed them in terms of salinity with the following (e.g., Fox & Wofsy, 1983):

$$0 = -u\frac{dS}{dx} + K_{eddy}\frac{d^2S}{dx^2}. \tag{10.5}$$

Following this transformation, Equations 10.1 and 10.2 were converted into sets of first-order differential equations. These were solved by Runga–Kutta–Fehlberg integrations and standard shooting techniques on a Macintosh IIci with MATLAB™.

10.5.2 Model results

We will focus on the influence of respiration and atmospheric exchange on the isotope ratio of DIC in an estuary. First, we compare isotope profiles of DIC versus salinity for an open system (P_{CO_2} is in equilibrium with the atmosphere and carbonate alkalinity is conservative), a closed system (CO_2 and carbonate alkalinity are conservative), and a system with limited atmospheric exchange (Fig. 10.9a). In turn, we discuss the effect of respired CO_2 of different isotope ratios in both the closed and limited atmospheric exchange systems (Fig. 10.10a,b).

The parameters used in the model are found in Table 10.3. The open-system calculation resulted in isotope ratios in the range of 0‰ except at very low salinities where the value decreased to <-1‰ (Fig. 10.9a). This result was quite different from the closed-system calculation, which

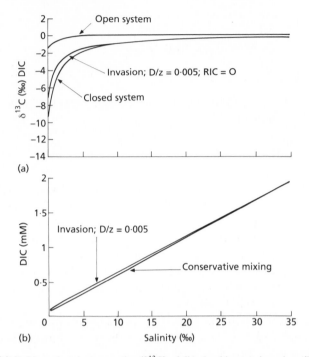

Fig. 10.9 (a) Stable carbon isotope ratios (δ^{13}C) of dissolved inorganic carbon (DIC) vs. salinity (‰) from model calculations. Solutions to the closed, open and limited atmospheric exchange ($D/z = 0.005$ cm s^{-1}) systems are shown. (b) DIC concentration vs. salinity (‰) from model calculations. Solutions to the closed and limited atmospheric exchange systems are shown. (See text and Table 10.3 for details.)

exhibited the non-linear behaviour observed often in estuaries (see Fig. 10.9a and Fig. 10.1). As expected, limited atmospheric exchange, $D/z = 0.005$ cm s^{-1} (a value suggested by Whitfield & Turner, 1986) resulted in an isotope profile intermediate between that of the open and closed systems, but more similar to the latter (Fig. 10.9a). Moreover, the inclusion of atmospheric exchange resulted in a slight increase in DIC concentration at low salinites (Fig. 10.9b). The relationship between DIC concentration and salinity, however, remained quite linear.

When respiration occurred without atmospheric exchange, and the isotope ratio of respired CO_2 varied from -10 to -26‰, the DIC in the estuary became significantly lighter (Fig. 10.10a) compared with the closed system calculation. If atmospheric exchange, $D/z = 0.005$ cm s^{-1}, was included in the calculations, the effect of respiration was unimportant at lower salinities (Fig. 10.10b). Specifically, below salinities of 10‰, the curves were intermediate between the closed system case and the case of limited atmospheric exchange without respiration (compare Figs 10.9a

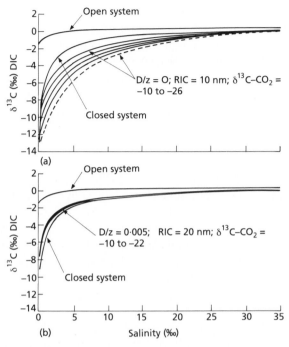

Fig. 10.10 (a) Stable carbon isotope ratios ($\delta^{13}C$) of dissolved inorganic carbon (DIC) vs. salinity (‰) from model calculations. Solutions to a closed system with respiration ($10\,\mathrm{mmol}\,l^{-1}\,s^{-1}$) of varying isotope ratio (-10, -14, -18, -22‰) are shown. (b) Stable carbon isotope ratios ($\delta^{13}C$) of DIC vs. salinity (‰) from model calculation. Solutions to a limited atmospheric exchange ($D/z = 0.005\,\mathrm{cm}\,s^{-1}$) system with respiration ($10\,\mathrm{mmol}\,l^{-1}\,s^{-1}$) of varying isotopic ratio (-10, -14, -18, -22‰) are shown. (See text and Table 10.3 for details.)

Table 10.3 Parameters used in the model of $\delta^{13}C$ of dissolved inorganic carbon (DIC) in estuaries. See text for details

Parameter	Estuary	Fresh water	Seawater
μm	$0.005\,\mathrm{m}\,s^{-1}$	–	–
K_{eddy}	$100\,\mathrm{m}^2\,s^{-1}$	–	–
D/z	$0.005\,\mathrm{cm}\,s^{-1}$	–	–
h	$10\,\mathrm{m}$	–	–
RIC	$10\,\mathrm{nmol}\,l^{-1}\,s^{-1}$	–	–
$\delta^{13}C$ of atmospheric CO_2 (‰)	-8	–	–
S (‰)		0	35
T (°C)		15	15
pH		7.0	8.2
Carbonate alkalinity		$0.058\,\mathrm{mmol}\,l^{-1}$	$2.093\,\mathrm{mmol}\,l^{-1}$
DIC		$0.072\,\mathrm{mmol}\,l^{-1}$	$1.944\,\mathrm{mmol}\,l^{-1}$
$\delta^{13}C$ of DIC (‰)		-10	0

and 10.10b). At greater salinities, respiration resulted in slightly more negative values, in spite of atmospheric exchange. Thus, the extent to which a net heterotrophic system will result in non-conservative, isotopic behaviour of DIC is not only dependent on the rate and isotope ratio of respiration, but on the extent of atmospheric exchange. We recognize, however, that in this model, we assumed steady state and isotopic equilibrium within the CO_2 system, assumptions which may not be warranted in every estuary.

10.6 Summary

Although stable isotopes have been used extensively in estuarine studies for over 30 years, the application of this technique to bacterial processes has only taken place in the last 5 years. Recent studies (Coffin *et al.*, 1989, 1990) have demonstrated that bacterial carbon sources can be traced into the bacterial assemblage. This development makes it possible to include bacterial carbon pools and bacterial dynamics in studies of estuarine carbon cycling. It will eventually lead to a better understanding of the influence of heterotrophic processes on the isotope ratio of DIC, and consequently organic carbon pools in estuaries. We suggest that field data, which includes all the carbon pools, coupled with modelling efforts can be used to study the extent of eutrophication in estuarine systems.

CHAPTER 11

Compound-specific approaches using stable isotopes

S. A. MACKO

11.1 Introduction

Stable isotopic determinations made on previously discussed bulk materials are, in fact, the weighted averages of the isotopic compositions of mixtures of hundreds to thousands of chemical compounds, each of which having its own isotopic abundances. The relative contribution of each of these materials to the isotopic content of the bulk material could theoretically be quantified through mass-balance or isotopic-mixing equations which have also been previously discussed. The stable isotope analysis of individual molecular components holds great potential as a method of tracing the source, biochemistry, diagenesis or indigeneity of a material. This final chapter endeavours to present a perspective on recent research on isotope analyses of individual compounds, or compound specific isotope analyses. The compounds studied to date include hydrocarbons, tetrapyrroles (chlorophyll derivatives), fatty acids, carbohydrates and amino acids. Carbon isotope analyses of components of petroleum and hydrocarbon extracts of sediments have indicated the preservation of original source materials. Isotope analysis of individual amino acids has been useful in detailing indigeneity of organic matter in meteorites and fossils, and helping to understand diagenesis. Inscribed in the isotopic signature is an indication of the biosynthetic pathway used in the formation of the compound. The transfer of nitrogen and carbon within the organism forming the component is thus able to be better understood. These pathways in turn imprint the signature of the organism in the rocks and sediments from which the compounds can later be isolated.

11.2 Molecular level isotope analysis

Over the years, numerous attempts have been made to isolate individual molecular components using liquid or gas chromatographic (GC) techniques in order to better interpret or trace the source of history of an organic material. The possibility of comparative biochemistry in modern or fossil organisms has been suggested through the assessment of the isotopic differences between compounds of a family of components. Such differences are the result of enzymatic fractionation effects during synthesis or metabolism of the compound; an example of such an effect has been clearly seen using the enzyme transaminase, with nitrogen isotopic fractionations being observed in acetyl-glucosamine and in the amino acids asparagine and glutamine (and others) in both cultured and natural populations of organisms (Macko *et al.*, 1986, 1987, 1991). Isotopic compositions of individual hydrocarbons have the potential for establishing sources for the materials, bacterial or otherwise, and have been useful in correlation techniques both in the petroleum industry and in pollution assessment. Individual carbohydrate isotope compositions also show great potential in metabolic and diagenetic studies. Depletions in the carbon isotopic compositions of the products of reactions permit calculations to be done which quantify use and production of new organic materials and resolve them from native materials, even though the chemical compositions of the substances are identical. In addition to establishing separation techniques to provide sufficient material for stable isotope analysis (usually milligram quantities for liquid chromatographic separations), the analytical scheme must produce little or no isotopic fractionation of the compound. Large differences in isotopic composition of a single compound exist across the chromatographic peak, owing to chromatographic isotopic fractionation effects (Macko *et al.*, 1987). Further, addition of carbon (or nitrogen) to an isolated component must be minimized and corrected for in the calculation of the authentic isotopic composition of the compound. Certain compounds may require unique separation schemes owing to close similarities in chemical structure. However, the studies to date have yielded important information regarding the source and history of the compounds characterized. Through recent technological advancements, gas chromatographic effluents can be combusted and the resulting carbon dioxide directly introduced into a stable isotope ratio mass spectrometer (IRMS). This modification, GC/IRMS, allows for rapid analysis of the carbon isotopes on components in a mixture, and with increased sensitivity, on the order of 0.5 nmol of each compound (Freedman *et al.*, 1988).

11.3 Individual amino acid isotope analysis

The foundation for the recent advancements in the use of a combination of molecular techniques in conjunction with stable carbon isotopic compositions may be ascribed to the initial work by Abelson & Hoering (1961) on algal cultures. Using a liquid chromatographic separation procedure for the isolation of sufficient material for stable carbon isotope analyses of individual amino acids and their carboxyl groups, these authors strengthened our understanding of natural biosynthetic pathways and the effects of decarboxylation during the diagenesis of organic matter. Studies on nitrogen metabolism and biosynthesis have also been attempted using stable isotopic compositions of individual amino acids (Schoenheimer &

Rittenberg, 1939; Gaebler *et al.*, 1963, 1966) using similar separation approaches and subsequent isotope analysis.

Amino acid isotopic compositions have also continued to be investigated, with pathways for nitrogen incorporation and intermolecular transfer (Macko *et al.*, 1987) and comparative biochemistry in fossil materials (Hare *et al.*, 1991) being elucidated. Nitrogen isotope abundances in the individual amino acids appear to be related to kinetic isotope effects associated with transamination reactions during synthesis (Macko *et al.*, 1986). Because the process of amino acid racemization has little apparent effect on the isotopic compositions of stereoisomers (Engel & Macko, 1986) the possibility of an absolute criterion for determining the indigeneity of amino acids in fossil or extraterrestrial materials has been suggested (Engel & Macko, 1984; Serban *et al.*, 1988). Unlike hydrocarbons, the application of GC/IRMS to amino acid analysis is complicated by the fact that amino acids are non-volatile, multifunctional molecules that require derivatization prior to gas chromatographic analysis. Although the derivatization procedure introduces additional carbon and an apparent fractionation during the esterification and acylation (Silfer *et al.*, 1991), carbon isotopic compositions of individual amino acids and/or their stereoisomers can be computed through analysis of standards prepared in a similar fashion, using mass-balance equations which incorporate fractionations inherent to the derivitization procedure. Extreme enrichments in amino acids and similar enrichments in both stereoisomers of the same amino acid from a carbonaceous chondrite analysed using GC/IRMS, the Murchison meteorite, have confirmed the extraterrestrial origin of those components and supported the lack of contamination by terrestrial compounds in the absolute concentrations and stereoisomer relationships (Engel *et al.*, 1990). Such a study required that the technology for the isotope analysis have the sensitivity for determinations on only fractions of nanomoles of the individual stereoisomers. Applications of this technology (GC/IRMS) to interpret fossil organic matter, to establish its indigeneity or to suggest modern biosynthetic relationships have recently been explored (Macko & Engel, 1991). Few studies have analysed the isotopic compositions of chemical components other than hydrocarbons or amino acids.

11.4 Isotopic compositions of lipid components

Because the techniques involved are extremely labour-intensive, only a limited number of reports utilizing molecular/stable isotopic characterizations have appeared in the literature. The greatest application for this technique has been in the analysis of lipids, with sources of petroleum derived materials (Welte, 1969; Gilmour *et al.*, 1984; Summons & Powell, 1986) or foodchain origins of hydrocarbon natural products determined through $\delta^{13}C$ isotope analysis of individual alkanes (Des Marais *et al.*, 1980). Further refinements in the isotopic characterization of potential biological sources of petroleum-related hydrocarbons are presently being attempted through direct interfacing of the gas chromatograph through a combustion furnace to an isotope ratio mass spectrometer or GC/IRMS, requiring much less of a compound for analysis (Freedman *et al.*, 1988). Petroleum-derived or related components appear to show a relationship to the original biological metabolism prior to deposition (Hayes *et al.*, 1989; Kennicutt & Brooks, 1990; Freeman *et al.*, 1990; Hayes *et al.*, 1990, Bjorøy *et al.*, 1991). Extraterrestrial origins of certain hydrocarbon components which were extracted from the Murchison meteorite were analysed using this new technology (Franchi *et al.*, 1989). Origins of sedimentary lipids derived from tree waxes were also documented using GC/IRMS (Rieley *et al.*, 1991). Lipids that originated from tree waxes were clearly more depleted than those derived from phytoplankton, and thus resolution of chemically indistinguishable sources is now possible.

Gas chromatographic based systems are presently constrained by column and component resolution considerations and the volatility of the components investigated. For this reason, hydrocarbons appear to be ideal components for direct GC/IRMS analysis, owing to their excellent resolution using capillary gas chromatography and little need for derivatization to increase volatility (Fig. 11.1a). The isotope ratio of the carbon (as carbon dioxide, mass 45/mass 44, or 2/1) changes dramatically across a single peak (Fig. 11.1b). As a result, integration of the entire peak along with background and column bleed corrections need to be incorporated in the data analysis; there are the corrections for the addition of carbon to the actual peak. Hydrocarbon components from fossil fuels have isotopic compositions which are readily resolvable from those of 'natural' lipids, and thus have potential in the analysis and tracking of pollutants.

A more recent development in the field of lipid-related materials has been the utilization of a separate isolation scheme using liquid chromatography for the purification of chlorophylls, chlorophyll-derived pigments and other tetrapyrroles, followed by isotopic characterization (Hayes *et al.*, 1987; Boreham *et al.*, 1989; Ocampo *et al.*, 1989; Bidigare *et al.*, 1990; Kennicutt *et al.*, 1992). This separation allows for the determination of both carbon and nitrogen isotopic compositions of the pigments. Analyses of this sort enable a strictly biochemical basis for interpretations of inputs and preservation of primary production in sediments to be addressed in both modern and ancient depositional environments.

Few attempts at the assessment of the isotopic compositions of long chain fatty acids exist (Parker, 1962; Vogler & Hayes, 1980). Those studies observed depletions in the individual fatty acids which were con-

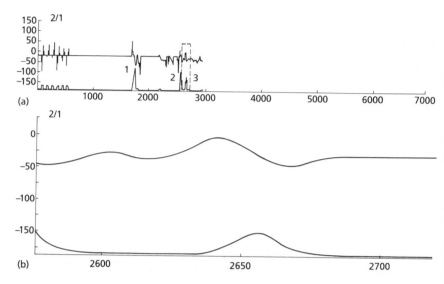

Fig. 11.1 (a) GC/IRMS gas chromatogram and isotope ratio trace for a separation of hydrocarbons. Peak 1, naphthalene; peak 2, 2-methylnaphthalene; peak 3, 1-methylnaphthalene. Top curve is the instantaneous isotope ratio of the carbon dioxide (mass 45/mass 44 or 2/1). The gas chromatographic separation (lower curve) is seen by the mass 44 ion trace. (b) Expanded 2/1 trace of peak 3, 1-methylnaphthalene, from the above separation. Note the fluctuation in the isotope ratio across the peak.

sistent with fractionations associated with lipid synthesis, and could yield clues as to the origins of long-chain hydrocarbons in sediments and fossil fuels, as well as the sources of the fatty acids themselves. Recently, analysis of fatty acids using GC/IRMS technology has been attempted through analysis of the methyl esters of the fatty acids (Abrajano *et al.*, 1992).

11.5 Isotope analyses of carbohydrates

The largest reservoir of carbon which can be chemically characterized is that of carbohydrates. Degens *et al.* (1968a,b), through isolation and isotope analysis of major biochemical fractions of phytoplankton which included different groupings of carbohydrates, concluded that the most labile materials were carbohydrates and proteins. Utilizing an ion exchange technique, Shimmelmann & DeNiro (1986) isolated *N*-acetyl-D-glucosamine, the monomeric unit of chitin. The study of this monosaccharide has suggested a use in palaeo-environmental and palaeoclimatic reconstruction by the determination of $\delta^{13}C$ on a single marine organism-derived compound (Shimmelmann *et al.*, 1986). By using one sugar from a single oceanic source, the $\delta^{13}C$ might better reflect oceanic changes in palaeoclimate. Recent developments in high-pressure liquid chromato-graphy techniques and stable isotope analysis enabled Macko *et al.* (1990) to study the isotopic compositions of carbohydrates isolated from indi-vidual organisms. In general, individual sugar isolates from organisms or polymers including glucans, galactans and chitins all have carbon isotopic compositions similar to but, typically, depleted by approximately 2‰, from the organism for which they were isolated. The monosaccharide essentially maintains the isotopic signal of the organism and reflects the primary source of carbon entering the organism, i.e., from the Calvin cycle (C_3 pathway), Hatch–Slack cycle metabolism (C_4 pathway) or from marine bicarbonate. Carbohydrates may be derivatized to acetates and analysed for their carbon stable isotopic compositions using GC/IRMS following a correction for the isotopic effect of derivatization and the addition of the acetate carbon (Macko *et al.*, 1992).

The similarity in isotopic composition of the monosaccharide to the organism is not maintained in the nitrogen isotopic compositions of the *N*-acetylglucosamine isolated from chitin. An approximately 9‰ deple-tion is observed in samples of the polymer and pure compound relative to the organism (Schimmelmann & DeNiro, 1986; Macko *et al.*, 1990). As sources of nitrogen for growth and metabolism, amino acids donate nitrogen for cellular biosynthesis. The isotope ratios of the carbohydrate which receives a donated nitrogen reflects the action of enzymatic isotopic fractionation which occurs during the transfer of nitrogen. Aspartic acid has been observed to be depleted by up to 9‰ in ^{15}N relative to glutamic acid in the transamination of glutamic acid to aspartic acid (Macko *et al.*, 1986). Transaminase enzymes are likely influential in the transfer of nitrogen from glutamine to fructose in the initial synthesis of glucosamine, the precursor of *N*-acetylglucosamine. The consequences of not con-sidering the above depletions in the nitrogen isotopic signature, in environments impacted by inputs of chitin, could include inappropriate assignment of sources of organic nitrogen. Marine environments receive large amounts of chitin from the exoskeletons of marine invertebrates. An incorrect assessment of the amount of nitrogen from more δ^{15}N-enriched sources could be possible under these circumstances.

In an initial investigation on a commercial peat, it was observed that the hexose mannose, isolated through chromatography, was isotopically similar to the whole peat and extracted carbohydrate (Macko *et al.*, 1991). However, xylose showed a distinct depletion in ^{13}C by over 7‰ when compared with other carbohydrates or the bulk carbon of the peat. This fractionation is consistent with a direction which indicated produc-tion of new material (Macko *et al.*, 1986, 1987). Similar isotopic trends have been noted in *Sphagnum*, with certain sugars remaining constant (rhamnose, arabinose) whereas other sugars (xylose, galactose and glu-

Table 11.1 *Sphagnum* and associated peats: isotopic compositions ($\delta^{13}C$ expressed as ‰) of isolated individual monosaccharides. (From Macko *et al.*, 1991)

Monosaccharide	*Sphagnum*	Peat 15 cm	Peat 30 cm
Rhamnose	−29.03	−29.75	−29.15
Arabinose	−27.15	−27.36	−27.55
Xylose	−26.38	−27.11	−28.17
Mannose	−26.88	n.d.	n.d.
Galactose	−26.24	−29.42	−29.22
Glucose	−26.13	−30.70	−29.28

n.d., not determined.

cose) become increasingly depleted in ^{13}C with depth (Table 11.1). In general, plant carbohydrates are isotopically similar to plant bulk carbon; isotopic depletions indicate that these lighter sugars are being newly produced, and when compared with the changes in the chemical compositions, gives an indication of the lability and rates of utilization and turnover. The constancy of the arabinose and rhamnose isotopic compositions with the relative increases in the molar concentrations of these materials (three- to fourfold) may reflect the amount of the plant mass which has been lost during decomposition reactions.

11.6 Conclusions

The potential for application of GC/IRMS analysis to any multitude of environmental, ecological or biochemical research areas is only beginning to be realized. Extension of compound-specific isotope analytical data derived from modern organisms and settings to yield interpretations of ancient depositional environments certainly appears possible. Further application of GC/IRMS approaches to understand the cycling of carbon and nitrogen, the identification and alteration of pollutants, or resolve metabolic relationships between compounds in living or extinct organisms are all within the scope of future research.

References

Abelson P.H. (1954) Organic constituents of fossils. *Carnegie Institution of Washington Yearbook*, **53**, 97–101.

Abelson P.H. & Hoering T.C. (1961) Carbon isotope fractionation in formation of amino acids by photosynthetic organisms. *Proceedings of the National Academy of Sciences USA*, **47**, 623–632.

Aber J.D., Nadelhoffer K.J., Steudler P.A. & Melillo J.M. (1989) Nitrogen saturation in forest ecosystems. *BioScience*, **39**, 378–386.

Aber J.D., Melillo J.M., Nadelhoffer K.J., Pastor J. & Boone R. (1991) Factors controlling nitrogen cycling and nitrogen saturation in northern temperate forest ecosystems. *Ecological Applications*, **1**, 303–315.

Abrajano T.A., Fang J., Comet P.A. & Brooks J. (1992) Compound-specific carbon analysis of fatty acids. 203rd National Mtg. Chemical Society, San Francisco, Abstract 104.

Aizawa K. & Miyachi S. (1986) Carbonic anhydrase and CO_2 concentrating mechanisms in microalgae and cyanobacteria. *FEMS Microbiological Reviews*, **39**, 215–233.

Ajie H.O., Hauschka P.V., Kaplan I.R. & Sobel H. (1991) Comparison of bone collagen and osteocalcin for determination of radiocarbon ages and paleodietary reconstruction. *Earth and Planetary Science Letters*, **107**, 380–388.

Allison G.B. & Hughes M.W. (1983) The use of natural tracers as indicators of soil-water movement in a temperate semi-acid region. *Journal of Hydrology*, **60**, 157–173.

Allison G.B., Barnes C.J. & Hughes M.W. (1983) The distribution of deuterium and ^{18}O in dry soils. *Journal of Hydrology*, **64**, 377–397.

Allison P.A. & Briggs D.E.G. (1991) Taphonomy of nonmineralized tissues. In: Allison P.A. & Briggs D.E.G. (eds) *Taphonomy: Releasing the Data Locked in the Fossil Record*. Plenum Press, New York, *Topics in Geobiology*, **9**, 25–70.

Altabet M.A. (1988) Variations in nitrogen isotopic composition between sinking and suspended particles: implications for nitrogen cycling and transformation in the open ocean. *Deep-Sea Research*, **35**, 535–554.

Altabet M.A. (1989) A time-series study of the vertical structure of nitrogen and particle dynamics in the Sargasso Sea. *Limnology and Oceanography*, **34**, 1185–1201.

Altabet M.A. & Deuser W.G. (1985) Seasonal variations in ^{15}N natural abundance in particles sinking to the deep Sargasso Sea. *Nature*, **315**, 218–219.

Altabet M.A. & McCarthy J.J. (1985) Temporal and spatial variations in the natural undance of ^{15}N in PON from a Warm-Core-Ring. *Deep-Sea Research*, **37**, 755–772.

Altabet M.A. & Small L. (1990) Nitrogen isotopic ratios in fecal pellets produced by marine zooplankton. *Geochimica et Cosmochimica Acta*, **54**, 155–163.

Altabet M.A. & McCarthy J.J. (1986) Vertical patterns in ^{15}N natural abundance in PON from the surface waters of several warm core rings and the Sargasso Sea. *Journal of Marine Research*, **44**, 185–201.

Altabet M.A., Robinson A.R. & Walstad L.J. (1986) A model for the vertical flux of nitrogen in the upper ocean: simulating the alteration of isotopic ratios. *Journal of Marine Research*, **44**, 203–225.

Aly A.I.M. (1975) N-15-Untersuchungen zur anthropogenen Störung des natürlichen Stickstoffzyklus. PhD Thesis, Aachen University, Germany.

Aly A.I.M., Mohamed M.A. & Hallaba E. (1981) Mass spectrometric determination of the nitrogen-15 content of different Egyptian fertilizers. *Journal of Radioanalytical Chemistry*, **67**, 55–60.

Aly A.I.M., Mohamed M.A. & Hallaba E. (1982) Natural variations of ^{15}N-content of nitrate in ground and surface waters and total nitrogen of soil in the Wadi El-Natrun area in Egypt. In: Schmidt H.-L., Forstel H. & Heinzinger K. (eds) *Stable Isotopes*, pp. 475–481. Elsevier, Amsterdam.

Amarger A., Mariotti A. & Mariotti F. (1977) Essai d'estimation du taux d'azote fixe symbiotiquement chez le lupin par le tracage isotopique naturel (^{15}N). *Comptes Rendus de l'Académie des Sciences de Paris, Ser D*, **284**, 2179–2182.

Amberger A. & Schmidt H.-L. (1987) Natürliche Isotopengehalte von Nitrat als indikaten für dessen Herkunft. *Geochimica et Cosmochimica Acta*, **51**, 2699–2705.

Ambrose S.H. (1990) Preparation and characterization of bone and tooth collagen for isotopic analysis. *Journal of Archaeological Science*, **17**, 431–451.

Ambrose S.H. & DeNiro M.J. (1986) The isotopic ecology of East African mammals. *Oecologia*, **69**, 395–406.

Ambrose S.H. & DeNiro M.J. (1989) Climate and habitat reconstruction using stable carbon and nitrogen isotope ratios of collagen in prehistoric herbivore teeth from Kenya. *Quaternary Research*, **31**, 407–422.

Ambrose S.H. & Norr L. (1993) Carbon isotopic evidence for routing of dietary protein to bone collagen, and whole diet to bone apatite carbonate: purified diet growth experiments. In: Lambert J. & Grupe G. (eds) *Molecular Archaeology of Prehistoric Human Bone*, pp. 1–37. Springer-Verlag, Berlin.

Anderson I.C. & Levine J.S. (1986) Relative rates of nitric oxide and nitrous oxide production by nitrifiers, denitrifiers, and nitrate respirers. *Applied and Environmental Microbiology*, **51**, 938–945.

Anderson J.H. (1964) The metabolism of hydroxylamine to nitrite by *Nitrosomonas*. *Biochemical Journal*, **91**, 8–17.

Andersson K.K. & Hooper A.B. (1983) O_2 and H_2O and each the source of one O in NO_2^- produced from NH_3 by *Nitrosomonas*: ^{15}N evidence. *FEBS Letters*, **164**, 236–240.

Andersson K.K., Philson S.B. & Hooper A.B. (1982) ^{18}O isotope shift in ^{15}N NMR analysis of biological N-oxidations: $H_2O-NO_2^-$-exchange in the ammonia-oxidizing bacterium *Nitrosomonas*. *Proceedings of the National Academy of Sciences USA*, **79**, 5871–5875.

Andrews T.J. & Abel K.M. (1979) Photosynthetic carbon metabolism in seagrasses. ^{14}C-labeling evidence for the C3 pathway. *Plant Physiology*, **63**, 650–656.

Andrews T.J. & Abel K.M. (1981) Kinetics and subunit interactions of ribulose bisphospate carboxylase–oxygenase from the cyanobacterium, *Synechococcus* sp. *Journal of Biological Chemistry*, **256**, 8445–8451.

Appleby G., Colbeck J. & Holdsworth E.S. (1980) β-carboxylation enzymes in marine phytoplankton and isolation and purification of pyruvate carboxylase from *Amphidinium carterae* (Dinophyceae). *Journal of Phycology*, **16**, 290–295.

Arnelle D. & O'Leary M. (1992) Binding of carbon dioxide to phosphoenolpyruvate carboxykinase deduced from carbon kinetic isotope effects. *Biochemistry*, **31**, 4363–4368.

Arntz W.E. (1977) Results and problems of an 'unsuccessful' benthos cage predation experiment (western Baltic). In: Keegan B.F., Ceidigh P.O. & Boaden P.J.S. (eds) *Biology of Benthic Organisms*, pp. 31–44. Pergamon Press, Oxford.

Arthur M.A., Dean W.E. & Claypool G.E. (1985) Anomalous ^{13}C enrichment in modern marine organic carbon. *Nature*, **315**, 216–218.

Aselmann I. & Crutzen P.J. (1989) Global distribution of natural wetlands and rice paddies, their net primary productivity, seasonality and possible methane emissions. *Journal of Atmospheric Chemistry*, **8**, 307–358.

Ayliffe L.K. & Chivas A.R. (1990) Oxygen isotope composition of the bone phosphate of Australian kangaroos: potential as a palaeoenvironmental recorder. *Geochimica et Cosmochimica Acta*, **54**, 2603–2609.

Bada J.L., Schoeninger M.J. & Schimmelmann A. (1989) Isotopic fractionation during peptide bond hydrolysis. *Geochimica et Cosmochimica Acta*, **53**, 3310–3337.

Bada J.L., Peterson R.O., Schimmelmann A. & Hedges R.E.M. (1990) Moose teeth as monitors of environmental isotopic parameters. *Oecologia*, **82**, 102–106.

Badger M.R. (1987) The CO_2-concentrating mechanism in aquatic phototrophs. In: Hatch M.D. & Boardman N.K. (eds) *The Biochemistry of Plants*, pp. 219–274. Academic Press, San Diego.

Badger M.R. & Andrews T.J. (1982) Photosynthesis and inorganic carbon usage by the marine cyanobacterium, *Synechococcus* sp. *Plant Physiology*, **70**, 517–523.

Badger M.R. & Price G.D. (1989) Carbonic anhydrase activity associated with the cyanobacterium *Synechococcus* PCC7942. *Plant Physiology*, **89**, 51–60.

Badger M.R., Bassett M. & Comins H.N. (1985) A model for HCO_3^- accumulation and photosynthesis in the cyanobacterium *Synechococcus* sp. *Plant Physiology*, **77**, 465–471.

Balderston W.L., Sherr B. & Payne W.J. (1976) Blockage by acetylene of nitrous oxide reduction in *Pseudomonas perfectomarinus*. *Applied and Environmental Microbiology*, **31**, 504–508.

Barnola J.M., Raynaud D., Korotkevich Y.S. & Lorius C. (1987) Vostok ice core provides 160 000-year record of atmospheric CO_2. *Nature*, **329**, 408–414.

Bartsiokas A. & Middleton A.P. (1992) Characterization and dating of recent and fossil bone by X-ray diffraction. *Journal of Archaeological Science*, **19**, 63–72.

Beardall J. (1989) Photosynthesis and photorespiration in marine phytoplankton. *Aquatic Botany*, **34**, 105–130.

Beardall J., Mukerji D., Glover H.E. & Morris I. (1976) The path of carbon in photosynthesis by marine phytoplankton. *Journal of Phycology*, **12**, 409–417.

Beardall J., Griffiths H. & Raven J.A. (1982) Carbon isotope discrimination and the CO_2 accumulating mechanism in *Chlorella emersonii*. *Journal of Experimental Biology*, **33**, 729–737.

Beauchamp B., Krouse H.R., Harrison J.C., Nassichuk W.W. & Eliuk L.S. (1989) Cretaceous cold-seep communities and methane derived carbonates in the Canadian Arctic. *Science*, **244**, 53–56.

Behrens E.W. & Frishman S.A. (1971) Stable carbon isotopes in blue-green algal mats. *Journal of Geology*, **79**, 95–100.

Behrensmeyer A.K. (1978) Taphonomic and ecologic information from bone weathering. *Paleobiology*, **4**, 150–162.

Bender M.M. (1971) Variations in the $^{13}C/^{12}C$ ratios of plants in relation to the pathway of photosynthetic carbon dioxide fixation. *Phytochemistry*, **10**, 1239–1244.

Benedict C.R. (1978) Nature of obligate photoautotrophy. *Annual Reviews of Plant Physiology*, **29**, 67–93.

Benedict C.R. & Scott J.R. (1976) Photosynthetic carbon metabolism of a marine grass. *Plant Physiology*, **57**, 876–880.

Benedict C.R., Wong W.W.L. & Wong J.H.H. (1980) Fractionation of the stable isotopes of inorganic carbon by seagrasses. *Plant Physiology*, **65**, 512–517.

Benninger C.K. & Martens C.S. (1983) Sources and fates of sedimentary organic matter in the White Oak and Neuse River Estuaries. In: *Water Resources Research Institute of the University of North Carolina, Report No. 194*, 60.

Benson B.B. & Parker P.D.M. (1961) Nitrogen/argon and nitrogen isotope ratios in aerobic sea water. *Deep-Sea Research*, **7**, 237–253.

Berg C.J. & Alatalo P. (1984) Potential of chemosynthesis in molluscan mariculture. *Aquaculture*, **39**, 165–179.

Berry J.A. (1989) Studies of mechanisms affecting the fractionation of carbon isotopes in photosynthesis. In: Rundel P.W., Ehleringer J.R. & Nagy K.A. (eds) *Stable Isotopes in Ecological Research*, pp. 82–94. Springer-Verlag, New York.

Beviss-Challinor M.H. & Field J.G. (1982) Analysis of a benthic community food web using isotopically labeled potential food. *Marine Ecology Progress Series*, **9**, 223–230.

Bidigare R.R., Kennicutt M.C., Keeney-Kennicutt W.L. & Macko S.A. (1990) Isolation and purification of chlorophylls *a* and *b* for the determination of stable carbon and nitrogen isotope compositions. *Analytical Chemistry*, **63**, 130–133.

Binkley D., Sollins P. & McGill W.B. (1985) Natural abundance of nitrogen-15 as a tool for tracing alder-fixed nitrogen. *Soil Science Society of America Journal*, **49**, 444–447.

Bjorøy M., Hall K., Gillyon P. & Jumeau J. (1991) Carbon isotope variations in *n*-alkanes and isoprenoids of whole oils. *Chemistry and Geology*, **93**, 13–20.

Black A.S. & Waring S.A. (1977) The natural abundance of ^{15}N in the soil-water system of a small catchment area. *Australian Journal of Soil Research*, **15**, 51–57.

Black C.C. & Bender M.M. (1976) C-13 values in marine organisms from the Great Barrier Reef. *Australian Journal of Plant Physiology*, **3**, 25–32.

Blackmer A.M. & Bremner J.M. (1977) N-isotope discrimination in denitrification of nitrate in soils. *Soil Biology and Biochemistry*, **9**, 73–77.

Blair N., Leu A., Munoz E., Olsen J., Kwong E. & Marais D.D. (1987) Carbon isotopic fractionation in heterotrophic microbial metabolism. *Applied and Environmental Microbiology*, **50**, 996–1001.

Blake D.R. & Rowland F.S. (1988) Continuing worldwide increase in tropospheric methane. *Science*, **239**, 1129–1131.

Bocherens H., Fizet M., Cuif J.-P., Jaeger J.-J., Michard J.-G. & Mariotti A. (1988) Premières mesures d'abondances isotopiques naturelles en ^{13}C et ^{15}N de la matière organique fossile de dinosaure. Application á l'étude du régime alimentaire du genre *Anatosaurus* (Ornithischia, Hadrosauridae): *Comptes Rendus de l'Académie des Sciences de Paris*, **306**, 1521–1525.

Bocherens H., Fizet M., Mariotti A., Lange-Badre B., Vandermeersch B., Borel J.P. & Bellon G. (1991) Isotopic biogeochemistry (^{13}C, ^{15}N) of fossil vertebrate collagen: application to the study of a past food web including Neanderthal man. *Journal of Human Evolution*, **20**, 481–492.

Bolin B. (1986) How much CO_2 will remain in the atmosphere? The carbon cycle and projections for the future. In: Bolin B., Jager J., Doos B.R. & Warrick R.A. (eds) *The Greenhouse Effect, Climate Change and Ecosystems*, SCOPE 29. J. Wiley & Sons, Chichester.

Bombin M. & Muehlenbachs K. (1985) $^{13}C/^{12}C$ ratios of Pleistocene mummified remains from Beringia. *Quaternary Research*, **23**, 123–129.

Boreham C.J., Fookes C.J.R., Popp B.N. & Hayes J.M. (1989) Origins of etioporphyrins in sediments: evidence from stable carbon isotopes. *Geochimica et Cosmochimica Acta*, **53**, 2451–2455.

Boreham P.F.L. & Ohiagu C.E. (1978) The use of serology in evaluating invertebrate predator–prey relationships: a review. *Bulletin of Entomological Research*, **68**, 171–194.

Bormann F.H., Likens G.E. & Melillo J.M. (1977) Nitrogen budget for an aggrading northern hardwood ecosystem. *Science*, **196**, 981–983.

Born M., Doerr H. & Levin I. (1990) Methane consumption in aerated soils of the temperate zone. *Tellus*, **42B**, 2–8.

Botello A.V. & Macko S.A. (1980) Presencia de hidrocarburos fosiles (*n*-parafinas) en sedimentos recientes de algunas lagunas costeras en el litoral pacifico de Mexico. *Anales del Centro de Ciencias del Mar y Limnologia*, **7**, 159–168.

Botello A.V. & Macko S.A. (1982) Oil pollution and carbon isotope ratios in organisms and recent sediments of coastal lagoons in the Gulf of Mexico. *Oceanologica Acta* SP, 55–62.

Botello A.V., Mandelli E.F., Macko S.A. & Parker P.L. (1980) Organic carbon isotope ratios of recent sediments from coastal lagoons of the Gulf of Mexico. *Geochimica et Cosmochimica Acta*, **44**, 557–559.

Böttcher J., Streber O., Voerkelius S. & Schimdt H.-L. (1990) Using isotope fractionation of nitrate-nitrogen and nitrate-oxygen for evaluation of microbial denitrification in a sandy aquifer. *Journal of Hydrology*, **114**, 413–424.

Boutton T.W., Wong W.W., Hachey D.L., Lee L.S., Cabrera M.P. & Klein P.D. (1983) Comparison of quartz and pyrex tubes for combustion of organic samples for stable carbon isotope analysis. *Analytical Chemistry*, **55**, 1832–1833.

Bowden R.D., Steudler P.A., Melillo J.M. & Aber J.D. (1990) Annual nitrous oxide fluxes from temperate forest soils in the northeastern United States. *Journal of Geophysical Research*, **95**, 13997–14005.

Bowden W.B. (1986) Gaseous nitrogen emissions from undisturbed terrestrial ecosystems: an assessment of their impacts on local and global nitrogen budgets. *Biogeochemistry*, **2**, 244–279.

Bowes G. (1985) Pathways of CO_2 fixation by aquatic organisms. In: Lucas W.J. & Berry J.A. (eds) *Inorganic Carbon Uptake by Aquatic Photosynthetic Organisms*, pp. 187–210. American Society of Plant Physiology, Rockville.

Brand U. & Veizer J. (1981) Chemical diagenesis of a multicomponent carbonate system – 2: Stable isotopes. *Journal of Sedimentary Petrology*, **51**, 987–997.

Bremer E. & van Kessel C. (1990) Appraisal of the nitrogen-15 natural-abundance method for quantifying dinitrogen fixation. *Soil Science Society of America Journal*, **54**, 404–411.

Bremner J.M. & Hauck R.D. (1982) Advances in methodology for research on nitrogen transformations in soils. In: Stevenson F.J. (ed.) *Nitrogen in Agricultural Soils*, pp. 467–502. Agronomy No. 22, American Society of Agronomy, Madison.

Brenninkmeijer C.A.M., Manning M.R., Lowe D.C., Wallace G. & Sparks R.J. (1990) Carbon-14, carbon-13, oxygen-18 and concentration measurements of atmospheric carbon monoxide in the southern hemisphere. In: *Symposium on Atmospheric Chemistry and Global Pollution*, p. 33. Chamrousse, France, September 5–11, 1990.

Broadbent F.E., Rauschkolb R.S., Lewis K.A. & Chang G.Y. (1980) Spatial variability in nitrogen-15 and total nitrogen in some virgin and cultivated soils. *Soil Science Society of America Journal*, **44**, 524–527.

Broecker W.S. & Peng T.H. (1982) *Tracers in the Sea*, pp. 412–433. Lamont Doherty Geological Observatory, Palisades, New York.

Broecker W.S. & Peng T.H. (1989) The cause of the glacial to interglacial atmospheric CO_2 change: a polar alkalinity hypothesis. *Global Biogeochemical Cycles*, **3**, 215–239.

Brooks A.S., Hare P.E., Kokis J.E., Miller G.H., Ernst R.D. & Wendorf F. (1990) Dating Pleistocene archeological sites by protein diagenesis in ostrich eggshell. *Science*, **248**, 60–64.

Brooks J.K., Kennicutt M.C. II, Fay R.R., McDonald T.J. & Sassen R. (1984) Thermogenic gas hydrates in the Gulf of Mexico. *Science*, **225**, 409–411.

Brooks J.M., Kennicutt M.C. II, Fisher C.R. *et al.* (1987) Deep-sea hydrocarbon seep communities: evidence for energy and nutritional carbon sources. *Science*, **238**, 1138–1142.

Brooks P.D., Stark J.M.. McInteer B.B. & Preston T. (1989) Diffusion method to prepare soil extracts for automated nitrogen-15 analysis. *Soil Science Society of America Journal*, **53**, 1707–1711.

Brown A.B. (1974) Bone Sr as a dietary indicator in human skeletal populations. *Wyoming Contributions to Geology*, **13**, 47–48.

Brugnoli E., Hubick K.T., von Caemmorer S., Wong S.C. & Farquhar G.D. (1988). Correlation between the carbon isotope discrimination in leaf starch and sugars of C_3 plants and the ratio of intercellular and atmospheric partial pressures of carbon dioxide. *Plant Physiology*, **88**, 1418–1424.

Burke R.A., Barber T.R. & Sackett W.M. (1988) Methane flux and stable hydrogen and carbon composition of sedimentary methane from the Florida Everglades. *Global Biogeochemical Cycles*, **2**, 329–340.

Burleigh R. & Brothwell D. (1978) Studies on Amerindian dogs. 1. Carbon isotopes in relation to maize in the diet of domestic dogs from Early Peru and Ecuador. *Journal of Archaeological Science*, **5**, 355–362.

Burnett W.C. & Schaeffer O.A. (1980) Effect of ocean dumping of $^{13}C/^{12}C$ ratios in marine sediments from the New York Bight. *Estuarine and Coastal Marine Science*, **11**, 605–611.

Burns B.D. & Beardall J. (1987) Utilization of inorganic carbon by marine microalgae. *Journal of Experimental Marine Ecology*, **107**, 75–86.

Burris R.H. & Wilson P.W. (1957) Methods for the measurements of nitrogen fixation. In: Colwick S.P. & Kaplan N.O. (eds) *Methods in Enzymology*, vol. 4, pp. 355–366. Academic Press, New York.

Burris R.H., Eppling F.J., Wahlin H.B. & Wilson P.W. (1943) Detection of nitrogen fixation with isotope nitrogen. *Journal of Biological Chemistry*, **148**, 349–357.

Butler J.H., Jones R.D., Garber J.H. & Gordon L.I. (1987) Seasonal distributions and turnover of reduced trace gases and hydroxylamine in Yaquina Bay, Oregon. *Geochimica et Cosmochimica Acta*, **51**, 697–706.

Butler J.H., Elkins J.W., Thompson T.M. & Egan K.B. (1989) Tropospheric and dissolved N_2O of the west Pacific and east Indian Oceans during the El Niño southern oscillation event of 1987. *Journal of Geophysical Research*, **94**, 14865–14877.

Cai D.-L., Tan F.C. & Edmond J.M. (1988) Sources and transport of particulate organic carbon in the Amazon River and Estuary. *Estuarine Coastal and Shelf Science*, **26**, 1–14.

Calder J.A. & Parker P.L. (1968) Stable carbon isotope ratios as indices of petrochemical pollution of aquatic systems. *Environmental Science and Technology*, **2**, 535–539.

Cantrell C.A., Shetter R.E., McDaniel A.H. *et al.* (1990) Carbon kinetic isotope effect in the oxidation of methane by the hydroxy radical. *Journal of Geophysical Research*, **95**, 22455–22462.

Capone D.G. (1983) Benthic nitrogen fixation. In: Carpenter E.J. & Capone D.G. (eds) *Nitrogen in the Marine Environment*, pp. 105–137. Academic Press, New York.

Capone D.G. (1988) Benthic nitrogen fixation. In: Blackburn T.H. & Sorensen J. (eds) *Nitrogen Cycling in Coastal Marine Environments*, pp. 85–123, Scope 33. John Wiley & Sons, Chichester.

Capone D.G. & Budin J.M. (1982) Nitrogen fixation associated with rinsed roofs and rhizomes of the eelgrass *Zostera marina*. *Plant Physiology*, **70**, 1601–1604.

Capone D.G. & Carpenter E.J. (1982) Nitrogen fixation in the marine environment. *Science*, **217**, 1140–1142.

Cariolou M.A. & Morse D.E. (1988) Purification and characterization of calcium-binding conchiolin shell peptides from mollusc, *Haliotis rufescens*, as a function of development. *Journal of Comparative Physiology B*, **157**, 717–729.

Carpenter E.J. (1983) Nitrogen fixation by marine *Oscillatoria* (*Trichodesmium*) in the world's oceans. In: Carpenter E.J. & Capone D.G. (eds) *Nitrogen in the Marine Environment*, pp. 65–103. Academic Press, New York.

Carpenter E.J. & Price C.C. (1977) Nitrogen fixation, distribution and production of *Oscillatoria* (*Trichodesmium*) spp. D in the western Sargasso and Caribbean Seas. *Limnology and Oceanography*, **22**, 60–72.

Carpenter E.J. & Capone D.G. (1992) Nitrogen fixation in *Trichodesmium* blooms. In: Carpenter E.J., Capone D.G. & Reuter J.G. (eds) *Marine Pelagic Cyanobacteria: Trichodesmium and other Diazotrophs*, pp. 211–217. Kluwer Publishers, Dordrecht.

Carpenter E.J., Van Ralte C.D. & Valiela I. (1978) Nitrogen fixation by algae in Massachusetts salt marsh. *Limnology and Oceanography*, **23**, 318–327.

Carpenter S.J. & Lohmann K.C. (1989) $\delta^{18}O$ and $\delta^{13}C$ variation in late Devonian marine cements from Golden Spike and Nevis reefs, Alberta, Canada. *Journal of Sedimentary Petrology*, **59**, 792–814.

Carr G.J. & Ferguson S.J. (1990a) Nitric oxide formed by nitrite reductase of *Paracoccus denitrificans* is sufficiently stable to inhibit cytochrome oxidase activity and is reduced by its reductase under aerobic conditions. *Biochimica et Biophysica Acta*, **1017**, 57–62.

Carr G.J. & Ferguson S.J. (1990b) The nitric oxide reductase of *Paracoccus denitrificans*. *Biochemical Journal*, **269**, 423–429.

Cary S.C., Fisher C.R. & Felbeck H. (1988) Mussel growth supported by methane as sole carbon and energy source. *Science*, **240**, 78–80.

Cavanaugh C.M. (1983) Symbiosis of chemoautotrophic bacteria and marine invertebrates from sulphide-rich habitats. *Nature*, **302**, 58–61.

Cavanaugh C.M. (1985) Symbiosis of chemoautotrophic bacteria and marine invertebrates from hydrothermal vents and reducing sediments. *Biological Society of Washington Bulletin*, **6**, 373–388.

Cavanaugh C.M., Gardiner S.L., Jones M.L., Jannasch H.W. & Waterbury J.B. (1981) Prokaryotic cells in the hydrothermal vent tube worm *Riftia pachyptila* Jones: possible chemoautotrophic symbionts. *Science*, **213**, 340–342.

Cavanaugh C.M., Levering P.R., Maki J.S., Mitchell R. & Lidstrom M.E. (1987) Symbiosis of methylotrophic bacteria and deep-sea mussels. *Nature*, **325**, 346–348.

Chambers L.A. & Trudinger P.A. (1979) Microbiological fraction of stable sulfur isotopes: a review and critique. *Geomicrobiology Journal*, **1**, 249–293.

Chambers L.A., Trudinger P.A., Smith J.W., Burns M.S. (1975) Fractionation of sulfur isotopes by continuous cultures of *Desulfovibrio desulfuricans*. *Canadian Journal of Microbiology*, **21**, 1602–1607.

Chappellaz J., Barnola J.M., Raynaud D., Korotkevich Y.S. & Lorius C. (1990) Icecore record of atmospheric methane over the past 160000 years. *Nature*, **345**, 127–131.

Checkley D.M. & Entzeroth L.C. (1985) Elemental and isotopic fractionation of carbon and nitrogen by marine, planktonic copepods and implications to the marine nitrogen cycle. *Journal of Plankton Research*, **7**, 553–568.

Checkley D.M. & Miller C.A. (1989) Nitrogen isotope fractionation by oceanic zooplankton. *Deep-Sea Research*, **36**, 1449–1456.

Cheng H.H., Bremner J.M. & Edwards A.P. (1964) Variations in nitrogen-15 abundance in soils. *Science*, **146**, 1574–1575.

Chickerur B.S., Tung M.S. & Brown W.E. (1980) A mechanism for the incorporation of carbonate into apatite. *Calcified Tissue International*, **32**, 55–62.

Chien S.H., Shearer G. & Kohl D.H. (1977) The nitrogen isotope effect associated with nitrate and nitrate loss from waterlogged soils. *Soil Science Society of America Journal*, **41**, 63–69.

Childress J.J., Fisher C.R., Brooks J.M., Kennicutt M.C. II, Bidigare R. & Anderson A.E. (1986) A methanogenic marine molluscan (Bivalvia; Mytilidae) symbiosis: mussels fueled by gas. *Science*, **233**, 1306–1308.

Christensen J.P., Murray J.W., Devol A.H. & Codispoti L.A. (1987a) Denitrification in continental shelf sediments has major impact on the oceanic budget. *Global Biogeochemical Cycles*, **1**, 97–116.

Christensen J.P., Smethie W.M. Jr. & Devol A.H. (1987b) Benthic nutrient regeneration and denitrification on the Washington continental shelf. *Deep-Sea Research*, **34**, 1027–1047.

Christensen P.B., Nielsen L.P., Revsbech N.P. & Sorensen J. (1988) Microzonation of denitrification activity in stream sediments as studied with a combined oxygen and nitrous oxide micro sensor. *Applied and Environmental Microbiology*, **55**, 1234–1241.

Christensen S., Groffman P., Mosier A. & Zak D.R. (1990) Rhizosphere denitrification: a minor process but indicator of decomposition activity. In: Revsbech N.P. & Sorensen J. (eds) *Denitrification in Soil and Sediment*, pp. 199–211. Plenum Press, New York.

Cicerone R.J. (1989) Analysis of sources and sinks of atmospheric nitrous oxide (N_2O). *Journal of Geophysical Research*, **94**, 18265–18271.

Cicerone R.J. & Oremland R.S. (1988) Biogeochemical aspects of atmospheric methane. *Global Biogeochemical Cycles*, **2**, 299–327.

Cifuentes L.A., Sharp J.H. & Fogel M.L. (1988) Stable carbon and nitrogen isotope biogeochemistry in the Delaware Estuary. *Limnology and Oceanography*, **33**, 1102–1115.

Cifuentes L.A., Fogel M.L., Pennock J.R. & Sharp J.H. (1989) Biogeochemical factors that influence the stable nitrogen isotope ratio of dissolved ammonium in the Delaware

Estuary. *Geochimica et Cosmochimica Acta*, **53**, 2713–2721.

Clarke F.E. (1981) The nitrogen cycle, viewed with poetic license. In: Clarke F.E. & Rosswall T. (eds) *Terrestrial Nitrogen Cycles*, pp. 13–24. Ecological Bulletins, Stockholm.

Claypool G.E. & Kaplan I.R. (1974) The origin and distribution of methane in marine sediments. In: Kaplan I.R. (ed.) *Natural Gases in Marine Sediments*, pp. 99–139. Plenum Press, New York.

Clayton R.N. & Mayeda T.K. (1963) The use of bromine pentafluoride in the extraction of oxygen from oxides and silicates for isotopic analysis. *Geochimica et Cosmochimica Acta*, **27**, 43–52.

Cline J.D. (1975) Denitrification and isotopic fractionation in two contrasting marine environments: the eastern tropical North Pacific Ocean and the Cariaco Trench. PhD Thesis, University of California, Los Angeles.

Cline J.D. & Kaplan I.R. (1975) Isotopic fractionation of dissolved nitrate during denitrification in the eastern tropical North Pacific Ocean. *Marine Chemistry*, **3**, 271–299.

Cline J.D., Wisegarver D.P. & Kelly-Hansen K. (1987) Nitrous oxide and vertical mixing in the equatorial Pacific during the 1982–1983 El Nino. *Deep-Sea Research*, **34**, 857–873.

Coakley J.P., Carey J.H. & Eadie B.J. (submitted) Specific organic components as tracers of contaminated fine sediment dispersal in Lake Ontario near Toronto. *Hydrobiology*.

Cobabe E. (1991) Lucinid bivalve evolution and the detection of chemosymbiosis in the fossil record. PhD Thesis, Harvard University.

Codispoti L.A. (1989) Phosphorous vs. nitrogen limitation of new and export production. In: Berger W.H., Smetacek V.S. & Wefer G. (eds) *Productivity of the Oceans: Present and Past*, pp. 377–394. John Wiley & Sons, New York.

Codispoti L.A. & Christensen J.P. (1985) Nitrification, denitrification and nitrous oxide cycling in the eastern tropical South Pacific Ocean. *Marine Chemistry*, **16**, 277–300.

Codispoti L.A., Elkins J.W., Yoshinari T., Friederich G.E., Sakamoto C.M. & Packard T.T. (1992) Nitrous oxide cycling in upwelling regions underlain by low oxygen waters. In: Desai B.N. (ed.) *Oceanography of the Indian Ocean*, 271–284. Oxford, CIBH Publishing CO.

Coffin R.B., Fry B., Peterson B.J. & Wright R.T. (1989) Carbon isotopic compositions of estuarine bacteria. *Limnology and Oceanography*, **34**, 1305–1310.

Coffin R.B., Velinsky D.J., Devereux R., Price W.A. & Cifuentes L.A. (1990) Stable carbon isotope analysis of nucleic acids to trace sources of dissolved substrates used by estuarine bacteria. *Applied and Environmental Microbiology*, **56**, 2012–2020.

Cohen Y. & Gordon L.I. (1978) Nitrous oxide in the oxygen minimum of the eastern tropical North Pacific: evidence for its consumption during denitrification and possible mechanisms for its production. *Deep-Sea Research*, **25**, 509–524.

Cohen Y. & Gordon L.I. (1979) Nitrous oxide production in the ocean. *Journal of Geophysical Research*, **84**, 347–353.

Colby J., Dalton H. & Whittenbury R. (1979) Biological and biochemical aspects of microbial growth on C-1 compounds. *Annual Reviews of Microbiology*, **33**, 481–517.

Coleman B. (1991) Second International Symposium on inorganic carbon utilization by aquatic photosynthetic organisms. *Canadian Journal of Botany*, **69**, 907–1160.

Coleman D.C. & Fry B. (eds) (1991) *Carbon Isotope Techniques*. Academic Press, San Diego.

Coleman M.L., Shepherd T.J., Durham J.J., Rouse J.E. & Moore G.R. (1982) Reduction of water with zinc for hydrogen isotope analysis. *Analytical Chemistry*, **54**, 995–998.

Colman B. & Gehl K.A. (1983) Physiological characteristics of photosynthesis in *Phorphyridium cruentum*: evidence for bicarbonate transport in a unicellular red alga. *Journal of Phycology*, **19**, 216–219.

Comstock J.P. & Ehleringer J.R. (1988) Contrasting photosynthetic behavior in leaves and twigs of *Hymenoclea salsola*, a green-twigged, warm desert shrub. *American Journal of Botany*, **75**, 1360–1370.

Condon A.G., Richards R.A. & Farquhar G.D. (1987) Carbon isotope discrimination is positively correlated with grain yield and dry matter production in field-grown wheat. *Crop Science*, **27**, 996–1001.

Conkright M.E. & Sackett W.M. (1986) A stable carbon isotope evaluation of the contribution of terriginous carbon to the marine food web in Bayboro Harbor, Tampa Bay, Florida. *Contributions to Marine Science*, **29**, 131–139.

Conrad R. (1990) Flux of NO_x between soil and atmosphere: importance and soil microbial metabolism. In: Revsbech N.P. & Sorensen J. (eds) *Denitrification in Soil and Sediment*, pp. 105–128. Plenum Press, New York.

Conrad R., Seiler W., Bunse G. & Giehl H. (1982) Carbon monoxide in sea water (Atlantic Ocean). *Journal of Geophysical Research*, **87**, 8839–8852.

Conway N.M. (1990) The nutritional role of endosymbiotic bacteria in animal-bacteria symbioses: *Solemya velum*, a case study. PhD Thesis, MIT/WHOI, 390pp.

Conway N. & McDowell-Capuzzo J. (1991) Incorporation and utilization of bacterial lipids by the *Solemya velum* symbiosis. *Marine Biology*, **108**, 277–291.

Conway N. & McDowell-Capuzzo J. (1992) High taurine levels in the *Solemya velum* symbiosis. *Comparative Biochemistry and Physiology*, **102B**, 175–185.

Conway N., McDowell-Capuzzo J. & Fry B. (1989) The role of endosymbiotic bacteria in the nutrition of *Solemya velum*: evidence from a stable isotope analysis of endosymbionts and host. *Limnology and Oceanography*, **34**, 249–255.

Conway N., Howes B., McDowell-Capuzzo J., Turner R.D. & Cavanaugh C. (1992) Characterization and site description of *Solemya borealis* (Bivalvia; Solemyidae), another bivalve-bacteria symbiosis. *Marine Biology*, **112**, 601–613.

Cook F.D., Wellman R.P. & Krouse H.R. (1973) Nitrogen isotope fractionation in the nitrogen cycle. In: Ingerson E. (ed.) *IAGC Symposium, Vol. 2, Biogeochemistry*, pp. 49–64. The Clarke Co., Washington, D.C.

Cooper L.W. & McRoy C.P. (1988) Stable carbon isotope ratio variations in marine macrophytes along intertidal gradients. *Oecologia*, **77**, 238–241.

Cooper T.G. & Wood H.G. (1971) The carboxylation of phosphoenolpyruvate and pyruvate. II. The active species of CO_2 utilized by phosphoenolpyruvate carboxylase and pyruvate carboxylase. *Journal of Biological Chemistry*, **246**, 5488–5490.

Cooper T.G., Tchen T.T., Wood H.G. & Benedict C.R. (1968) The carboxylation of phosphoenolpyruvate and pyruvate. I. The active species of CO_2 utilized by phosphoenolpyruvate carboxykinase, carboxytransphosphorylase and pyruvate carboxylase. *Journal of Biological Chemistry*, **243**, 3857–3863.

Coplen T.B., Kendall C. & Hopple J. (1983) Comparison of stable isotope reference samples. *Nature*, **302**, 236–238.

Corliss J.B. & Ballard R.D. (1977) Oases of life in the cold abyss. *Nat. Geogr.*, **152**, 441–453.

Corliss J.B., Dymomd J., Gordon L.I. *et al.* (1979) Sub-marine thermal springs on the Galapagos Rift. *Science*, **203**, 1073–1083.

Corredor J.E. & Capone D.J. (1985) Studies on nitrogen diagenesis in coral reef sands. *Proceedings of the 5th International Coral Reef Congress*, **3**, 395–399.

Craig H. (1953) The geochemistry of the stable carbon isotopes. *Geochimica et Cosmochimica Acta*, **3**, 53–92.

Craig H. (1954) Carbon-13 in plants and the relationships between carbon-13 and carbon-14 variations in nature. *Journal of Geology*, **62**, 115–149.

Craig H. (1961) Isotopic variations in meteoric water. *Science*, **133**, 1702–1703.

Craig H. & Gordon L. (1965) Deuterium and oxygen-18 variation in the ocean and marine atmosphere. *Proceedings of the Conference on Stable Isotopes in Oceanography Studies and Paleotemperatures*, pp. 9–130. Laboratory of Geology and Nuclear Science, Pisa.

Craig H. & Chou C.C. (1982) Methane: the record in polar ice cores. *Geophysical Research Letters*, **9**, 1221–1224.

Craig H., Welhan J.A., Kim K., Poreda R. & Lupton J.E. (1980) Geochemical studies of the 21°N EPR hydrothermal fluids. *EOS, Transactions of the American Geophysical Union*, **61**, 992 (Abstract).

Craig H., Horibe Y. & Sowers T. (1988a) Gravitational separation of gases and isotopes in polar ice caps. *Science*, **242**, 1675–1678.

Craig H., Chou C.C., Welhan J.A., Stevens C.M. & Engelkemeir A. (1988b) The isotopic composition of methane in polar ice cores. *Science*, **242**, 1535–1538.

Crompton A.W. & Hiiemae K. (1969) How molar teeth work. *Discovery*, **5**, 23–34.

Crowson R.A., Showers W.J., Wright E.K. & Hoering T.C. (1991) Preparation of phosphate samples for oxygen isotope analysis. *Analytical Chemistry*, **63**, 2397–2400.

Crutzen P.J. (1970) The influence of nitrogen oxides on the atmospheric ozone content. *Quarterly Journal of the Royal Meteorological Society*, **96**, 320–325.

Crutzen P.J. (1981) Atmospheric chemical processes of the oxides of nitrogen, including nitrous oxides. In: Delwiche C.C. (ed.) *Denitrification, Nitrification, and Atmospheric Nitrous Oxide*, pp. 17–44. John Wiley and Sons, New York.

Dahms A.S. & Boyer P.D. (1973) Occurrence and characteristics of ^{18}O exchange reactions catalyzed by sodium- and potassium-dependent adenosine triphosphatases. *Journal of Biological Chemistry*, **248**, 3155–3162.

Dando P.R., Southward A.J. & Southward E.C. (1986) Chemoautotrophic symbionts in the gills of the bivalve molluscs *Lucinoma borealis* and the sedimentary chemistry of its habitat. *Proceedings of the Royal Society, London B*, **227**, 227–247.

Dansgaard W. (1964) Stable isotopes in precipitation. *Tellus*, **16**, 436–468.

David K.A.V. & Fay P. (1977) Effects of long term treatment with acetylene on nitrogen fixing microorganisms. *Applied and Environmental Microbiology*, **34**, 640–646.

David M.B., Fuller R.D., Fernandez I.J. *et al.* (1990) Spodosol variability and response to acidic deposition. *Soil Science Society of America Journal*, **54**, 541–548.

Davidson E.A., Stark J.M. & Firestone M.K. (1990) Microbial production and consumption of nitrate in an annual grassland. *Ecology*, **71**, 1968–1975.

Davidson J.A., Cantrell C.A., Tyler S.C., Shetter R.E., Cicerone R.J. & Calvert J.G. (1987) Carbon kinetic isotope effect in the reaction of CH_4 with HO. *Journal of Geophysical Research*, **92**, 2195–2199.

Dawson T.E. & Ehleringer J.R. (1991) Streamside trees that do not use stream water. *Nature*, **350**, 335–337.

Dean W.E., Arthur M.A. & Claypool G.E. (1986) Depletion of ^{13}C in Cretaceous marine organic matter: source, diagenetic, or environmental signal? *Marine Geology*, **70**, 119–157.

Deegan L.A., Peterson B.J. & Portier R. (1990) Stable isotopes and cellulase activity as evidence for detritus as a food source for juvenile Gulf Menhaden. *Estuaries*, **13**, 14–19.

Degens E.T. (1969) Biogeochemistry of stable carbon isotopes. In: Eglinton E. & Murphy M.J.T. (eds) *Organic Geochemistry*, pp. 304–329. Springer-Verlag, Berlin.

Degens E.T., Guillard R.R.L., Sackett W.M. & Hellebust J.A. (1968a) Metabolic fractionation of carbon isotopes in marine plankton – I. Temperature and respiration experiments. *Deep-Sea Research*, **15**, 1–9.

Degens E.T., Behrendt M., Gotthardt B. & Reppmann E. (1968b) Metabolic fractionation of carbon isotopes in marine plankton – II. Data on samples collected off the coasts of Peru and Ecuador. *Deep-Sea Research*, **15**, 11–20.

Deines P. (1980) The isotopic composition of reduced organic carbon. In: Fritz P. & Fontes J.C. (eds) *Handbook of Environmental Isotope Geochemistry*, vol. 1, pp. 329–406.

DeLaune R.D. & Lindau S.W. (1987) $\delta^{13}C$ signature of organic carbon in estuarine bottom sediment as an indicator of carbon export from adjacent marshes. *Biogeochemistry*, **4**, 225–230.

Delmas J., Ascencio J.M. & Legrand D. (1980) Polar ice evidence that atmospheric CO_2 20 000 B.P. was 50% of present. *Nature*, **284**, 155–157.

Delmas P.D., Tracy R.P., Riggs B.L. & Mann K.G. (1984) Identification of the noncollagenous proteins of bovine bone by two-dimensional gel electrophoresis. *Calcified Tissue International*, **36**, 308–316.

DeLong E.F. (1993) Single-cell identification using fluoresently labeled, ribosomal RNA-specific probes. In: Kemp P.F., Sherr B.F., Sherr E.B. & Cole J.J. (eds) *Handbook of Methods in Aquatic Microbial Ecology*, pp. 285–294. Lewis Publishers, Boca Raton.

DeLucia E.H. & Schlesinger W.H. (1991) Resource-use efficiency and drought tolerance in adjacent Great Basin and Sierran plants. *Ecology*, **71**, 51–58.

Delwiche C.C. (1970) The nitrogen cycle. *Scientific American*, **233**, 137–149.

Delwiche C.C. (1981) The nitrogen cycle and nitrous oxide. In: Delwiche C.C. (ed.) *Denitrification, Nitrification, and Atmospheric Nitrous Oxide*, pp. 1–15. John Wiley & Sons, New York.

Delwiche C.C. & Steyn P.L. (1970) Nitrogen isotope fractionation in soils and microbial reactions. *Environmental Science and Technology*, **4**, 929–935.

DeNiro M.J. (1985) Postmortem preservation and alteration of *in vivo* bone collagen isotope ratios in relation to paleodietary reconstruction. *Nature*, **317**, 806–809.

DeNiro M.J. (1987) Stable isotopes and archaeology. *American Scientist*, **75**, 182–191.

DeNiro M.J. & Epstein S. (1977) Mechanism of carbon isotope fractionation associated with lipid synthesis. *Science*, **197**, 261–263.

DeNiro M.J. & Epstein S. (1978) Influence of diet on the distribution of carbon isotopes in animals. *Geochimica et Cosmochimica Acta*, **42**, 495–506.

DeNiro M.J. & Epstein S. (1981a) Influence of diet on the distribution of nitrogen isotopes in animals. *Geochimica et Cosmochimica Acta*, **42**, 495–506.

DeNiro M.J. & Epstein S. (1981b) Hydrogen isotope ratios of mouse tissues are influenced by a variety of factors other than diet. *Science*, **214**, 1374–1376.

DeNiro M.J. & Weiner S. (1988a) Chemical, enzymatic and spectroscopic characterization of "collagen" and other organic fractions from prehistoric bones. *Geochimica et Cosmochimica Acta*, **52**, 2197–2206.

DeNiro M.J. & Weiner S. (1988b) Organic matter within crystalline aggregates of hydroxyapatite: a new substrate for stable isotope and possibly other biogeochemical analyses of bone. *Geochimica et Cosmochimica Acta*, **52**, 2415–2423.

DeNiro M.J. & Weiner S. (1988c) Use of collagenase to purify collagen from prehistoric bones for stable isotopic analysis. *Geochimica et Cosmochimica Acta*, **52**, 2425–2431.

Descolas-Gros C. & Fontugne M.R. (1985) Carbon fixation in marine phytoplankton: carboxylase activities and stable carbon-isotope ratios; physiological and paleoclimatological aspects. *Marine Biology*, **87**, 1–6.

Descolas-Gros C. & Fontugne M.R. (1988) Carboxylase activities and carbon isotope ratios of Mediterranean phytoplankton. *Oceanologica Acta*, Special Issue, 245–250.

Descolas-Gros C. & Fontugne M.R. (1990) Stable carbon isotope fractionation by marine phytoplankton during photosynthesis. *Plant, Cell & Environment*, **13**, 217–218.

Des Marais D.J., Mitchell J.M., Meinschein W.G. & Hayes J.M. (1980) The carbon isotopic biogeochemistry of individual hydrocarbons in bat guano and the ecology of insectivorous bats in the region of Carlsbad, New Mexico. *Geochimica et Cosmochimica Acta*, **44**, 2075–2086.

Deuser W.G., Degens E.T. & Guillard R.R.L. (1968) Carbon isotope relationships between plankton and sea water. *Geochimica et Cosmochimica Acta*, **32**, 657–660.

Dickinson R.E. & Cicerone R.J. (1986) Future global warming from atmospheric trace gases. *Nature*, **319**, 109–115.

Dickman M.D. & Thode H.G. (1985) The rate of lake acidification in four lakes north of Lake Superior and its relationship to downcore sulphur isotope ratios. *Water, Air, and Soil Pollution*, **26**, 233–253.

Dinçer T., Al-Mugrin A. & Zimmermann U. (1974) Study of the infiltration and re-charge through the sand dunes in arid zones with special reference to the stable isotopes and thermonuclear tritium. *Journal of Hydrology*, **23**, 79–109.

Dixon G.K. & Merrett M.J. (1988) Bicarbonate utilization by the marine diatom *Phaeo-dactylum tricornutum* (Bohlin). *New Phytology*, **109**, 47–51.

Dixon G.K., Patel B.N. & Merrett M.J. (1987) Role of intracellular carbonic anhydrase in inorganic-carbon assimilation by *Phorphyridium purpureum*. *Planta*, **172**, 508–513.

Dixon G.K., Brownlee C. & Merrett M.J. (1989) Measurement of internal pH in the coccolithophore *Emiliania huxleyi* using 2′,7′-bis-(2-carboxyethyl)-5(and-6)carboxyfluorescein acetoxymethylester and digital imaging microscopy. *Planta*, **178**, 443–449.

Dole M. (1935) The relative atomic weight of oxygen in water and air. *Journal of the American Chemical Society*, **57**, 2731.

Dongmann G., Nurnberg H.W., Forstel H. & Wagener K. (1974) On the enrichment of $H_2^{18}O$ in the leaves of transpiring plants. *Radiation and Environmental Biophysics*, **11**, 41–52.

Doohan M.E. & Newcomb E.H. (1976) Leaf ultrastructure and $\delta^{13}C$ values of three seagrasses from the great barrier reef. *Australian Journal of Plant Physiology*, **3**, 9–23.

Dreimanis A. (1968) Mastodons, their geologic age and extinction in Ontario, Canada. *Canadian Journal of Earth Sciences*, **4**, 663–675.

Dugdale R.C. & Goering J.J. (1967) Uptake of new and regenerated forms of nitrogen in primary productivity. *Limnology and Oceanography*, **12**, 196–206.

Dugdale R.C., Menzel D.W. & Ryther J.H. (1961) Nitrogen fixation in the Sargasso Sea. *Deep-Sea Research*, **7**, 293–300.

Dugdale R.C., Goering J.J. & Ryther J.H. (1964) High nitrogen fixation rates in the Sargasso Sea and the Arabian Sea. *Limnology and Oceanography*, **9**, 507–510.

Dunton K.H. & Schell D.M. (1987) Dependence of consumers on macroalgal (*Laminaria solidungula*) carbon in an arctic kelp community: $\delta^{13}C$ evidence. *Marine Biology*, **93**, 615–625.

Eadie B.J. & Jeffrey L.M. (1973) $\delta^{13}C$ analyses of oceanic particulate organic matter. *Marine Chemistry*, **1**, 199–209.

Eadie B.J., Jeffrey L.M. & Sackett W.M. (1978) Some observations on the stable carbon isotope composition of dissolved and particulate organic carbon in the marine environment. *Geochimica et Cosmochimica Acta*, **42**, 1265–1269.

Eanes E.D. & Posner A.S. (1970) A note on the crystal growth of hydroxyapatite precipitated from aqueous solutions. *Material Research Bulletin*, **5**, 377–394.

Edwards A.P. (1973) Isotopic tracer techniques for identification of sources of nitrate pollution. *Journal of Environmental Quality*, **2**, 382–387.

Edwards G.E. & Huber S.C. (1981) The C4 pathway. In: Hatch M.D. & Boardman N.K. (eds) *The Biochemistry of Plants*, pp. 237–281. Academic Press, London.

Edwards T.W.D. & Fritz P. (1986) Assessing meteoric water composition and relative humidity from ^{18}O and ^{2}H in wood cellulose: paleoclimatic implications for southern Ontario, Canada. *Applied Geochemistry*, **1**, 715–723.

Ehdaie B., Hall A.E., Farquhar G.D., Nguyen H.T. & Waines J.G. (1991) Water-use efficiency and carbon isotope discrimination in wheat. *Crop Science*, **31**, 1282–1288.

Ehhalt D.H. (1979) Der atmosphärische Kreislauf von Methan. *Naturwissenschaften*, **66**, 307–311.

Ehleringer J.R. (1989) Carbon isotope ratios and physiological processes in arid land plants. In: Rundel P.W., Ehleringer J.R. & Nagy K.A. (eds) *Stable Isotopes in Ecological Research*, pp. 41–54. Springer-Verlag, New York.

Ehleringer J.R. (1991) $^{13}C/^{12}C$ fractionation and its utility in terrestrial plant studies. In: Coleman D.C. & Fry B. (eds) *Carbon Isotope Techniques*, p. 274. Academic Press, New York.

Ehleringer J.R. & Pearcy R.W. (1983) Variation in quantum yields for CO_2 uptake among C_3 and C_4 plants. *Plant Physiology*, **73**, 555–559.

Ehleringer J.R. & Cooper T.A. (1988) Correlations between carbon isotope ratio and microhabitat in desert plants. *Oecologia*, **76**, 562–566.

Ehleringer J.R. & Osmond C.B. (1989) Stable isotopes. In: Pearcy R.W., Ehleringer J.R., Mooney H.A. & Rundel P.W. (eds) *Plant Physiological Ecology*, p. 457. Chapman & Hall, New York.

Ehleringer J.R., Shulze E.-D., Ziegler H., Lange O.L., Farquhar G.D. & Covoan I.R. (1985) Xylem-tapping mistletoes: water or nutrient parasites? *Science*, **227**, 1479–1481.

Ehleringer J.R., Field C.B., Lin Z.F. & Kuo C.Y. (1986) Leaf carbon isotope and mineral composition in subtropical plants along an irradiance cline. *Oecologia*, **70**, 520–526.

Ehleringer J.R., Rundel P.W. & Nagy K.A. (1986) Stable isotopes in physiological ecology and food web research. *Trends in Evolution and Ecology*, **1**, 42–45.

Ehleringer J.R., Comstock J.P. & Cooper T.A. (1987a) Leaf-twig carbon isotope ratio differences in photosynthetic-twig desert shrubs. *Oecologia*, **71**, 318–320.

Ehleringer J.R., Lin Z.F., Field C.B., Sun G.C. & Kuo C.Y. (1987b) Leaf carbon isotope ratios of plants from a subtropical monsoon forest. *Oecologia*, **72**, 109–114.

Ehleringer J.R., White J.W., Johnson D.A. & Brick M. (1990) Carbon isotope discrimination, photosynthetic gas exchange, and transpiration efficiency in beans and range grasses. *Acta Oecologia*, **11**, 611–625.

Ehleringer J.R., Phillips S.L., Schuster W. & Sandquist D.R. (1991) Differential utilization of summer rains by desert plants. *Oecologia*, **88**, 430–434.

Elkins J.W. (1978) Aquatic sources and sinks for nitrous oxide. PhD Thesis, Harvard University.

Elkins J.W., Wofsy S.C., McElroy M.B., Kolb C.E. & Kaplan W.A. (1978) Aquatic sources and sinks for nitrous oxide. *Nature*, **275**, 602–606.

Elrifi I.R. & Turpin D.H. (1986) Nitrate and ammonium induced photosynthetic suppression in N-limited *Selenastrum minutum*. *Plant Physiology*, **81**, 273–279.

Elrifi I.R., Holmes J.J., Weger H.G., Mayo W.P. & Turpin D.H. (1988) RuBP limitation of photosynthetic carbon fixation during NH_3 assimilation: interaction between photosynthesis, respiration and ammonium assimilation in N-limited green algae. *Plant Physiology*, **87**, 395–406.

Elster H., Gil-Av E. & Weiner S. (1991) Amino acid racemization in fossil bone. *Journal of Archaeological Science*, **18**, 605–617.

Emrich K., Emhalt E.H. & Vogel J.C. (1970) Carbon isotope fractionation during the precipitation of calcium carbonate. *Earth and Planetary Science Letters*, **8**, 363–371.

Engel M.H. & Macko S.A. (1984) Separation of amino acid enantiomers for stable nitrogen and carbon isotopic analyses. *Analytical Chemistry*, **56**, 2598–2600.

Engel M.H. & Macko S.A. (1986) Stable isotope evaluation of the origins of amino acids in fossils. *Nature*, **323**, 531–533.

Engel M.H., Macko S.A. & Silfer J.A. (1990) Carbon isotope composition of individual amino acids in the Murchison meteorite. *Nature*, **348**, 47–49.

Enright J.T., Newman W.A., Hessler R.R. & McGowan J.A. (1981) Deep-ocean hydrothermal vent communities. *Nature*, **280**, 219–221.

Eppley R.W. & Peterson B.J. (1979) Particulate organic matter flux and planktonic new production in the deep ocean. *Nature*, **282**, 671–680.

Eppley R.W., Renger E.H., Venrick E.L. & Mullin M.M. (1973) A study of plankton dynamics and nutrient cycling in the central gyre of the North Pacific Ocean. *Limnology and Oceanography*, **18**, 534–551.

Epstein S., Buchsbaum R., Lowenstam H.A. & Urey H.C. (1953) Revised carbonate–water isotopic temperature scale. *Bulletin of the Geological Society of America*, **64**, 1315–1326.

Epstein S., Yapp C.J. & Hall J.H. (1976) The determination of D/H ratio of non-exchangeable hydrogen in cellulose extracted from aquatic and land plants. *Earth & Planetary Science Letters*, **30**, 241–251.

Epstein S., Thompson P. & Yapp C.J. (1977) Oxygen and hydrogen isotopic ratios in plant cellulose. *Science*, **218**, 1209–1215.

Erben H.K., Hoefs J. & Wedepohl K.H. (1979) Paleobiological and isotopic studies of eggshells from a declining dinosaur species. *Paleobiology*, **5**, 380–414.

Ericson J.E., Sullivan C.H. & Boaz N.T. (1981) Diets of Pliocene mammals from Omo, Ethiopia, deduced from carbon isotope ratios in tooth apatite. *Palaeogeography, Palaeoclimatology, and Palaeoecology*, **36**, 69–73.

Estep M.F. & Dabrowski H. (1980) Tracing food webs with stable hydrogen isotopes. *Science*, **209**, 1537–1538.

Estep M.F. & Hoering T.C. (1980) Biogeochemistry of the stable hydrogen isotopes. *Geochimica et Cosmochimica Acta*, **44**, 1197–1206.

Estep M.L.F. & Vigg S. (1985) Stable carbon and nitrogen isotope tracers of trophic dynamics in natural and fisheries of the Lahontan Lake system, Nevada. *Canadian Journal of Fisheries and Aquatic Science*, **42**, 1712–1719.

Estep M.F., Tabita F.R. & Baalen C.V. (1978a) Purification of ribulose bisphosphate carboxylase and carbon isotope fractionation by whole cells and carboxylase from *Cylindrotheca* sp. (Bacillariophyceae). *Journal of Phycology*, **14**, 183–188.

Estep M.F., Tabita F.R., Parker P.L. & Baalen C.V. (1978b) Carbon isotope fractionation by ribulose bisphospate carboxylase from various organisms. *Plant Physiology*, **61**, 680–687.

Evans J.R., Sharkey T.D., Berry J.A. & Farquhar G.D. (1986) Carbon isotope discrimination measured concurrently with gas exchange to investigate CO_2 diffusion in leaves of higher plants. *Australian Journal of Plant Physiology*, **13**, 281–292.

Faganeli J., Vukovie A., Saleh F.I. & Pezdic J. (1986) C:N:P ratios and stable carbon and hydrogen isotopes in the benthic marine algae *Ulva rigida* C. Ag. and *Fucus virsoides* J. Ag. *Journal of Experimental Marine Biology and Ecology*, **102**, 153–166.

Falcone A.B., Shug A.L. & Nicholas D.J.D. (1963) Some properties of a hydroxylamine oxidase from *Nitrosomonas europaea*. *Biochimica et Biophysica Acta*, **77**, 199–208.

Falkowski P.G. (1991) Species variability in the fractionation of ^{13}C and ^{12}C by marine phytoplankton. *Journal of Plankton Research*, **13**(suppl), 21–28.

Faller L.D. & Elgavish G.A. (1984) Catalysis of oxygen-18 exchange between inorganic phosphate and water by gastric H, K-ATPase. *Biochemistry*, **23**, 6584–6590.

Farquhar G.D. (1983) On the nature of carbon isotope discrimination in C_4 species. *Australian Journal of Plant Physiology*, **10**, 205–226.

Farquhar G.D. & Richards R.A. (1984) Isotopic composition of plant carbon correlates with water-use efficiency of wheat genotypes. *Australian Journal of Plant Physiology*, **11**, 539–552.

Farquhar G.D., O'Leary M.H. & Berry J.A. (1982) On the relationship between carbon isotope discrimination and the intercellular carbon dioxide concentration in leaves. *Australian Journal of Plant Physiology*, **9**, 121–137.

Farquhar G.D., Ehleringer J.R. & Hubick K.T. (1989a) Carbon isotope discrimination and photosynthesis. *Annual Review of Plant Physiology and Plant Molecular Biology*, **40**, 503–537.

Farquhar G.D., Hubick K.T., Condon A.G. & Richards R.A. (1989b) Carbon isotope fractionation and plant water-use efficiency. In: Rundel P.W., Ehleringer J.R. & Nagy K.A. (eds) *Stable Isotopes in Ecological Research*, pp. 21–40. Springer-Verlag, New York.

Farran A., Grimalt J., Albaigés J., Botello A.V. & Macko S.A. (1987) Assessment of petroleum pollution in a Mexican river by molecular markers and carbon isotope ratios. *Marine Pollution Bulletin*, **18**, 284–289.

Feigin A., Kohl D.H., Shearer G. & Commoner B. (1974a) Variation in the natural nitrogen-15 abundance in nitrate mineralized during incubation of several Illinois soils. *Soil Science Society of America Proceedings*, **38**, 90–95.

Feigin A., Shearer G., Kohl D.H. & Commoner B. (1974b) The amount and nitrogen-15 content of nitrate in soil profiles from two central Illinois fields in a corn–soybean rotation. *Soil Science Society of America Proceedings*, **38**, 465–471.

Felbeck H. (1983) Sulfide oxidation and carbon fixation by the gutless clam *Solemya reidi*: an animal–bacterial symbiosis. *Journal of Comparative Physiology*, **152**, 3–11.

Felbeck H., Childress J.J. & Somero G.N. (1981) Calvin–Benson cycle and sulfide oxidation enzymes in animals from sulfide-rich habitats. *Nature*, **293**, 291–293.

Feller R.J., Taghon G.L., Gallagher E.D., Kenny G.E. & Jumars P.A. (1979) Immunological methods for food web analysis in a soft-bottom benthic community. *Marine Biology*, **54**, 61–74.

Feller R.J., Zargursky G. & Day E.A. (1985) Deep-sea food web analysis using cross-reacting antisera. *Deep-Sea Research*, **4**, 485–497.

Fiala-Médioni A., Alayse A.M. & Cahet G. (1986) Evidence of *in situ* uptake and incorporation of bicarbonate and amino acids by a hydrothermal vent mussel. *Journal of Experimental Marine Biology and Ecology*, **96**, 191–198.

Field C. & Mooney H.A. (1983) Leaf age and seasonal effects on light, water, and nitrogen use efficiency in a California shrub. *Oecologia*, **56**, 348–355.

Field C., Merino J. & Mooney H.A. (1983) Compromises between water-use efficiency and nitrogen-use efficiency in five species of California evergreens. *Oecologia*, **60**, 384–389.

Field C.B., Ball J.T. & Berry J.A. (1989) Photosynthesis: principles and field techniques. In: Pearcy R.W., Ehleringer J.R., Mooney H.A. & Rundel P.W. (eds) *Plant Physiological Ecology*, pp. 209–253. Chapman & Hall, New York.

Fielder R. & Proksch G. (1975) The determination of nitrogen-15 by emission and mass spectrometry in biochemical analysis: a review. *Analytica Chimica Acta*, **78**, 1–62.

Firestone M.K. & Davidson E.A. (1989) Microbiological basis of NO and N_2O production and consumption in soil. In: Andreae M.O. & Schimel D.S. (eds) *Exchange of Trace Gases between Terrestrial Ecosystems and the Atmosphere*, pp. 7–21. Dahlem Konferenzen. John Wiley & Sons, Chichester.

Fisher C.R. (1990) Chemoautotrophic and methanotrophic symbioses in marine invertebrates. *Reviews in Aquatic Sciences*, **2**, 399–436.

Fisher C.R., Childress J.J., Oremland R.S. & Bidigare R.R. (1987) The importance of methane and thiosulfate in the metabolism of the symbionts of two deep-sea mussels. *Marine Biology*, **96**, 59–71.

Fisher C.R., Childress J.J., Arp A.J. *et al.* (1988a) Microhabitat variation in the hydrothermal vent mussel, *Bathymodiolus thermophilus*, at the Rose Garden vent on the Galapagos Rift. *Deep-Sea Research*, **35**, 1769–1791.

Fisher C.R., Childress J.J. & Arp A.J. *et al.* (1988b) Physiology, morphology, and biochemical composition of *Riftia pachyptila* at Rose Garden in 1985. *Deep-Sea Research*, **35**, 1745–1758.

Fisher C.R., Childress J.J. & Arp A.J. *et al.* (1988c) Variation in hydrothermal-vent clam, *Calyptogena magnifica*, at the Rose Garden vent on the Galapagos spreading center. *Deep-Sea Research*, **35**, 1811–1832.

Fisher C.R., Kennicutt M.C. II & Brooks J.M. (1990) Stable carbon isotopic evidence for carbon limitation in hydrothermal vent vestimentiferans. *Science*, **247**, 1094–1096.

Flanagan L.B. & Ehleringer J.R. (1991a) Stable isotope composition of stem and leaf water: applications to the study of plant water use. *Functional Ecology*, **5**, 270–277.

Flanagan L.B. & Ehleringer J.R. (1991b) Effects of mild water stress and diurnal changes in temperature and humidity on the stable oxygen and hydrogen isotopic composition of leaf water in *Cornus stolonifera L.*. *Plant Physiology*, **97**, 298–305.

Flanagan L.B., Bain J.F. & Ehleringer J.R. (1991) Stable oxygen and hydrogen isotope composition of leaf water in C_3 and C_4 plant species under field conditions. *Oecologia*, **88**, 394–400.

Flanagan L.B., Ehleringer J.R. & Marshall J.D. (1992) Differential uptake of summer precipitation among co-occurring trees and shrubs in a pinyon-juniper woodland. *Plant, Cell and Environment*, **15**, 831–836.

Flipse W.J. & Bonner F.T. (1985) Nitrogen-isotope ratios of nitrate in ground water under fertilized fields, Long Island, New York. *Ground Water*, **23**, 59–67.

Focht D.D. (1973) Isotope fractionation of ^{15}N and ^{14}N in microbiological nitrogen transformations: a theoretical model. *Journal of Environmental Quality*, **3**, 247–252.

Fogel M.L. & Cifuentes L.A. (1993) Isotope fractionation during primary production. In: Engel M.H. & Macko S.A. (eds) *Organic Geochemistry*, pp. 73–98, Plenum Press, New York.

Fogel M.L., Cifuentes L.A., Velinsky D.J. & Sharp J.H. (1992) Relationship of carbon availability in estuarine phytoplankton to isotopic composition. *Marine Ecology Progress Series*, **82**, 291–300.

Folinsbee R.E., Fritz P., Krouse H.R. & Robblee A.R. (1970) Carbon-13 and oxygen-18 in dinosaur, crocodile, and bird eggshells indicate environmental conditions. *Science*, **168**, 1353–1355.

Follet R.F. & Walker D.J. (1989) Ground water quality concerns about nitrogen. In: Follet R.F. (ed.) *Nitrogen Management and Ground Water Protection*, pp. 1–22. Elsevier, New York.

Fontugne M., Descolas-Gros C. & Billy G.D. (1991) The dynamics of CO_2 fixation in the Southern Ocean as indicated by carboxylase activities and organic carbon isotopic ratios. *Marine Chemistry*, **35**, 371–380.

Fox L.E. & Wofsy S.C. (1983) Kinetics of removal of iron colloids from estuaries. *Geochimica et Cosmochimica Acta*, **47**, 211–216.

Francalacci P. (1989) Dietary reconstruction at Arene Candide Cave (Liguria, Italy) by means of trace element analysis. *Journal of Archaeological Science*, **16**, 109–124.

Francey R.J. & Tans P.P. (1987) Latitudinal variation in oxygen-18 of atmospheric CO_2. *Nature*, **327**, 495–497.

Francey R.J., Gifford R.M., Sharkey T.D. & Weis B. (1985) Physiological influences on carbon isotope discrimination in heron pine (*Lagarostrobos Franklinii*). *Oecologia*, **66**, 211–218.

Franchi I.A., Exley R.A., Gilmour I. & Pillinger C.T. (1989) Stable isotope and abundance measurements of solvent extractable compounds in Murchison. 14th Symposium on Antarctic meteorites. June, 1989. National Institute of Polar Research, Tokyo (extended abstract).

Freedman P.A., Gillyon E.C.P. & Jumeau E.J. (1988) Design and application of a new instrument for GC-isotope ratio MS. *American Laboratory*, June, 114–119.

Freeman K.H., Hayes J.M., Trendel J.-M. & Albrecht P. (1990) Evidence from carbon isotope measurements for diverse origins of sedimentary hydrocarbons. *Nature*, **343**, 254–256.

Freundlich J.C., Kuper R., Breunig P. & Bertram H.-G. (1989) Radiocarbon dating of ostrich eggshells. *Radiocarbon*, **31**, 1030–1034.

Freyer H.D. (1978a) Seasonal trends of NH_4^+ and NO_3^- nitrogen isotope composition in rain collected at Jülich, Germany. *Tellus*, **30**, 83–92.

Freyer H.D. (1978b) Preliminary ^{15}N studies on atmospheric nitrogenous trace gases. *Pure and Applied Geophysics*, **116**, 393–404.

Freyer H.D. (1979) On the ^{13}C record in tree rings. Part I. ^{13}C variations in northern hemispheric trees during the last 150 years. *Tellus*, **31**, 124–137.

Freyer H.D. (1986) Interpretation of the northern hemisphere record of ^{13}C/^{12}C trends of atmospheric CO_2. In: Trabalka J.R. & Reschle D.E. (eds) *The Changing Carbon Cycle: A Global Analysis*, pp. 125–150. Springer-Verlag, Berlin.

Freyer H.D. & Aly A.I.M. (1974) Nitrogen-15 variations in fertilizer nitrogen. *Journal of Environmental Quality*, **3**, 405–406.

Freyer H.D. & Aly A.I.M. (1975) Nitrogen-15 studies on identifying fertilizer excess in environmental systems. In: *Isotope Ratios as Pollutant Source and Behavior Indicators*, pp. 21–33. IAEA, Vienna.

Friedli H., Moor H., Oeschger H., Siegenthaler U. & Stauffer B. (1984) ^{13}C/^{12}C ratios in CO_2 extracted from Antarctic ice. *Geophysical Research Letters*, **11**, 1145–1148.

Friedli H., Lotscher H., Oeschger H., Siegenthaler U. & Stauffer B. (1986) Ice core record of the $^{13}C/^{12}C$ ratio of atmospheric CO_2 in the past two centuries. *Nature*, **324**, 237–238.

Friend A.D., Woodward F.I. & Switsor V.R. (1989) Field measurements of photosynthesis, stomatal conductance, leaf nitrogen, and $\delta^{13}C$ along altitudinal gradients in Scotland. *Functional Ecology*, **3**, 117–133.

Fry B. (1983) Fish and shrimp migrations in the northern Gulf of Mexico analyzed using stable C, N, and S isotope ratios. *Fisheries Bulletin*, **81**, 789–801.

Fry B. (1986) Stable sulfur isotopic distributions and sulfate reduction in lake sediments of the Adirondack Mountains, New York. *Biogeochemistry*, **2**, 329–343.

Fry B. (1988) Food web structure on Georges Bank from stable C, N, and S isotopic compositions. *Limnology and Oceanography*, **33**, 1182–1190.

Fry B. (1989) Sulfate fertilization and changes in stable sulfur isotopic compositions of lake sediments. In: Rundel P.W., Ehleringer J.R. & Nagy K.A. (eds) *Stable Isotopes in Ecological Research*, pp. 445–453. Springer-Verlag, New York.

Fry B. (1991) Stable isotope diagrams of freshwater food webs. *Ecology*, **72**, 2293–2297.

Fry B. & Parker P.L. (1979) Animal diet in Texas seagrass meadows: $\delta^{13}C$ evidence for the importance of benthic plants. *Estuarine and Coastal Marine Sciences*, **8**, 499–509.

Fry B. & Arnold C. (1982) Rapid $^{13}C/^{12}C$ turnover during growth of brown shrimp (*Penaeus aztecus*). *Oecologia*, **54**, 200–204.

Fry B. & Sherr E.B. (1984) $\delta^{13}C$ measurements as indicators of carbon flow in marine and freshwater ecosystems. *Contributions to Marine Science*, **27**, 13–47.

Fry B. & Wainright S.C. (1991) Diatom sources of ^{13}C-rich carbon in marine food webs. *Marine Ecology Progress Series*, **76**, 149–157.

Fry B., Scalan R.S. & Parker P.L. (1977) Stable carbon isotope evidence for two sources of organic matter in coastal sediments: seagrasses and plankton. *Geochimica et Cosmochimica Acta*, **41**, 1875–1877.

Fry B., Jeng W., Scalan R.S., Parker P.L. & Baccus J. (1978a) $\delta^{13}C$ food web analysis of a Texas sand dune community. *Geochimica et Cosmochimica Acta*, **42**, 1299–1302.

Fry B., Joern A. & Parker P.L. (1978b) Grasshopper food web analysis: use of carbon isotope ratios to examine feeding relationships among terrestrial herbivores. *Ecology*, **59**, 498–506.

Fry B., Scalan R.S., Winters K. & Parker P.L. (1982) Sulphur uptake by salt grasses, mangroves and seagrasses in anaerobic sediments. *Geochimica et Cosmochimica Acta*, **46**, 1121–1124.

Fry B., Gest H. & Hayes J.M. (1983a) Sulphur isotope composition of deep-sea hydrothermal vent animals. *Nature*, **306**, 51–52.

Fry B., Scalan R.S. & Parker P.L. (1983b) $^{13}C/^{12}C$ ratios in marine food webs of the Torres Strait, Queensland. *Australian Journal of Marine and Freshwater Research*, **34**, 707–716.

Fry B., Anderson R.K., Entzeroth L., Byrd J.L. & Parker P.L. (1984) C-13 enrichment and oceanic food web structure in the northwestern Gulf of Mexico. *Contributions to Marine Science*, **27**, 49–63.

Fry B., Cox J., Gest H. & Hayes J.M. (1986) Discrimination between ^{34}S and ^{32}S during bacterial metabolism of inorganic sulfur compounds. *Journal of Bacteriology*, **165**, 328–330.

Fry B., Brandt W., Mersch F.J., Tholke K. & Garritt R. (1992) Automated analysis system for coupled $\delta^{13}C$ and $\delta^{15}N$ measurements. *Analytical Chemistry*, **64**, 288–291.

Fuchs G., Thauer R., Ziegler H. & Stichler W. (1979) Carbon isotope fractionation by *Methanobacterium thermoautotrophicum*. *Archives of Microbiology*, **120**, 135–139.

Fulco A.J. (1983) Fatty acid metabolism in bacteria. *Progress in Lipid Research*, **22**, 133–160.

Fung I., John J., Lerner J. *et al.* (1991) Three-dimensional model of synthesis of the global methane cycle. *Journal of Geophysical Research*, **96**, 13033–13065.

Gaebler O.H., Choitz H.C., Vitti T.G. & Vukmirovich R. (1963) Significance of ^{15}N excess in nitrogenous compounds of biological origin. *Canadian Journal of Biochemistry*, **41**, 1089–1097.

Gaebler O.H., Vitti T.G. & Vukmirovich R. (1966) Isotope effects in metabolism of ^{14}N and ^{15}N from unlabeled dietary proteins. *Canadian Journal of Biochemistry*, **44**, 1249–1257.

Garber E.A. & Hollocher T.C. (1982) Positional isotope equivalence of nitrogen in N_2O produced by the denitrifying bacterium *Pseudomonas stutzeri*. *Journal of Biological Chemistry*, **257**, 4705–4708.

Garten C.T. Jr. (1991) Nitrogen isotope composition of ammonium and nitrate and bulk precipitation and forest throughfall. *International Journal of Environmental Analytical Chemistry*, **87**, 33–45.

Garten C.T. Jr & Taylor G.E. Jr (1992) Foliar $\delta^{13}C$ within a temperate deciduous forest: spatial, temporal, and species sources of variation. *Oecologia*, **90**, 1–7.

Gat J.R. & Gonfiantini R. (eds) (1981) *Stable isotope hydrology: deuterium and oxygen-18 in the water cycle*. International Atomic Energy Agency, Vienna, Technical Reports Series, no. 210.

Gearing P., Plucker F.E. & Parker P.L. (1977) Organic carbon stable isotope ratios of continental margin sediments. *Marine Chemistry*, **5**, 251–266.

Gearing J.N., Gearing P.J., Rudnick D.T., Requejo A.G. & Hutchins M.J. (1984) Isotopic variability of organic carbon in a phytoplankton-based temperate estuary. *Geochimica et Cosmochimica Acta*, **48**, 1089–1098.

Gebauer G. (1991) Natural nitrogen isotope ratios in different compartments of Norway spruce for a healthy and declining stand. In: *Stable Isotopes in Plant Nutrition, Soil Fertility and Environmental Studies*, pp. 131–138. Symposium Proceedings Series IAEA-SM-313/56.

Gebauer G. & Schulze E.-D. (1991) Carbon and nitrogen isotope ratios in different compartments of a healthy and declining *Picea abies* forest in the Fichtelgebirge, NE Bavaria. *Oecologia*, **87**, 198–207.

Gerlach S.A., Ekstrom D.K. & Eckhardt P.H. (1976) Filter feeding in the hermit crab *Pagurus bernhardus*. *Oecologia*, **24**, 257–265.

Giere O. (1985) Structure and position of bacterial endosymbionts in the gill filaments of Lucinidae from Bermuda (Mollusca, Bivalvia). *Zoomorphology*, **105**, 296–301.

Giere O. & Langheld C. (1987) Structural organisation, transfer, and biological fate of endosymbiotic bacteria in gutless oligochaetes. *Marine Biologist*, **93**, 641–650.

Giere O., Felbeck H., Dawson R. & Liebezeit G. (1984) The gutless marine oligochaete *Phallodrilus leukodermatus* Giere, a tubificid of structural, ecological and physiological significance. *Hydrobiologia*, **115**, 83–89.

Giere O., Conway N., Gastrock G. & Schmidt C. (1990) Regulation of the ecology of a gutless annelid by endosymbiotic bacteria. *Marine Ecology Progress Series*, **68**, 287–299.

Gilmour I., Swart P.K. & Pillinger C.T. (1984) The isotopic composition of individual petroleum lipids. *Organic Geochemistry*, **6**, 665–670.

Giovannoni S.J., Delong E.F., Olsen G.J. & Pace N.R. (1988) Phylogenetic group-specific oligodeoxynucleotide probes for identification of single microbial cells. *Journal of Bacteriology*, **170**, 720–726.

Giovannoni S.J., DeLong E.F., Schmidt T.M. & Pace N.R. (1990) Tangential flow filtration and preliminary phylogenetic analysis of marine picoplankton. *Applied and Environmental Microbiology*, **56**, 2572–2575.

Glover H.E. (1989) Ribulosebisphosphate carboxylase/oxygenase in marine organisms. *International Reviews of Cytology*, **115**, 67–138.

Goerick R. & Fry B. (1994) Variations of marine plankton $\delta^{13}C$ with latitude, temperature, and dissolved CO_2 in the world ocean. *Global Biogeochemical Cycles*, **8** (in press).

Goering J.J. & Parker P.L. (1972) Nitrogen fixation by epiphytes on sea grasses. *Limnology and Oceanography*, **17**, 320–323.

Goering J.J., Dugdale R.C. & Menzel D.W. (1966) Estimates of *in situ* rates of nitrogen uptake by *Trichodesmium* sp. in the tropical Atlantic Ocean. *Limnology and Oceanography*, **11**, 614–620.

Goering J., Alexander B. & Haubenstock N. (1990) Seasonal variability of stable carbon and nitrogen isotope ratios of organisms in a North Pacific Bay. *Estuarine, Coastal and Shelf Science*, **30**, 239–260.

Goeyens L., De Vries R.T.P., Bakker J.F. & Helder W. (1987) An experiment on the relative importance of denitrification, nitrate reduction and ammonification in coastal marine sediment. *Netherlands Journal of Sea Research*, **21**, 171–175.

Goldfine H. (1972) Comparative aspects of bacterial lipids. *Advances in Microbial Physiology*, **8**, 1–58.

Goldhaber M.B. & Kaplan I.R. (1975) Controls and consequences of sulfate reduction rates in recent marine sediments. *Soil Sciences*, **119**, 42–55.

Gonfiantini R. (1978) Standards for stable isotope measurements in natural compounds. *Nature*, **271**, 534–536.

Gonfiantini R., Gratziu S. & Tongiorgi E. (1965) Oxygen isotope composition of water in leaves. In: *Isotope and Radiation in Soil–Plant–Nutrition Studies*. International Atomic Energy Commission, Vienna, pp. 405–410.

Goodfriend G.A. (1988) Mid-Holocene rainfall in the Negev Desert from ^{13}C of land snail shell organic matter. *Nature*, **333**, 757–760.

Goodfriend G.A. (1991) Holocene trends in ^{18}O in land snail shells from the Negev Desert and their implications for changes in rainfall source areas. *Quaternary Research*, **35**, 417–426.

Goodfriend G.A. & Hood D.G. (1983) Carbon isotope analysis of land snail shells: implications for carbon sources and radiocarbon dating. *Radiocarbon*, **25**, 810–830.

Goodfriend G.A. & Magaritz M. (1987) Carbon and oxygen isotope composition of shell carbonate of desert land snails. *Earth and Planetary Science Letters*, **86**, 377–388.

Goodfriend G.A., Magaritz M. & Gat J.R. (1989) Stable isotope composition of land snail body water and its relation to environmental waters and shell carbonate. *Geochimica et Cosmochimica Acta*, **53**, 3215–3221.

Goodroad L.L. & Keeney D.R. (1984) Nitrous oxide emissions from forest, marsh and prairie ecosystems. *Journal of Environmental Quality*, **13**, 448–452.

Gordon D.C. Jr. (1977) Variability of particulate organic carbon and nitrogen along the Halifax–Bermuda section. *Deep-Sea Research*, **24**, 257–270.

Goreau T.J., Kaplan W.A., Wofsy S.C., McElroy M.B., Valois F.W. & Watson S.W. (1980) Production of NO_2^- and N_2O by nitrifying bacteria at reduced concentrations of oxygen. *Applied and Environmental Microbiology*, **40**, 526–532.

Goretski J. & Hollocher T.C. (1990) The kinetic and isotopic competence of nitric oxide as an intermediate in denitrication. *Journal of Biological Chemistry*, **265**, 889–895.

Goretski J., Zafiriou O.C. & Hollocher T.C. (1990) Steady-state nitric oxide concentrations during denitrication. *Journal of Biological Chemistry*, **265**, 11535–11538.

Gormly J.R. & Spaulding R.F. (1979) Sources and concentrations of nitrate-nitrogen in ground water of the Central Platte Region, Nebraska. *Ground Water*, **3**, 291–301.

Goyal A. & Tolbert N.E. (1989) Uptake of inorganic carbon by isolated chloroplasts from air-adapted *Dunaliella*. *Plant Physiology*, **89**, 1264–1269.

Grissom C.B. & Cleland W.W. (1988) Isotope effect studies of the chemical mechanism of pig heart NADP isocitrate dehydrogenase. *Biochemistry*, **27**, 2934–2943.

Gutknecht J., Bisson M.A. & Tosteson F.C. (1977) Diffusion of carbon dioxide through lipid bilayer membranes: effects of carbonic anhydrase, bicarbonate and unstirred layers. *Journal of General Physiology*, **69**, 779–794.

Guy R.D., Fogel M.F., Berry J.A. & Hoering T.C. (1987) Isotope fractionation during oxygen production and consumption by plants. In: Biggins J. (ed.) *Progress in Photosynthesis Research*, pp. 597–600. Martinus Nijhoff, Dordrecht.

Guy R.D., Vanlerberghe G.C. & Turpin D.H. (1989) Significance of phosphoenolpyruvate carboxylase during ammonium assimilation: carbon isotope discrimination in photosynthesis and respiration by the N-limited green algae *Selenastrum minutum*. *Plant Physiology*, **89**, 1150–1157.

Hackney C.T. & Haines E.B. (1980) Stable carbon isotope composition of fauna and organic matter collected in a Mississippi estuary. *Estuarine and Coastal Marine Science*, **10**, 703–708.

Haines E.B. (1976a) Stable carbon isotope ratios in the biota, soils and tidal water of a Georgia Salt Marsh. *Estuarine and Coastal Shelf Science*, **4**, 609–616.

Haines E.B. (1976b) Relation between the stable carbon isotope composition of fiddler crabs, plants, and soils in a salt marsh. *Limnology and Oceanography*, **21**, 880–883.

Haines E.B. & Montague C.L. (1979) Food sources of estuarine invertebrates analyzed using $^{13}C/^{12}C$ ratios. *Ecology*, **60**, 48–56.

Hall A.E., Mutters R.G., Hubick K.T. & Farquhar G.D. (1990) Genotypic differences in carbon isotope discrimination by cowpea under wet and dry field conditions. *Crop Science*, **30**, 300–305.

Hallberg G.R. (1986) From hoes to herbicides: agriculture and groundwater quality. *Journal of Soil and Water Conservation*, **41**, 357–364.

Hallberg G.R. (1989) Nitrate in ground water in the United States. In: Follett R.F. (ed.) *Nitrogen Management and Ground Water Protection*, pp. 35–74. Elsevier, New York.

Hansen J., Fung I., Lacis A. *et al.* (1988) Global climate changes as forecast by Goddard Institute for Space Studies three-dimensional model. *Journal of Geophysical Research*, **93**, 9341–9364.

Hardy R.W.F., Burns R.C. & Holsten R.P. (1973) Applications of the acetylene–ethylene assay for measurement of nitrogen fixation. *Soil Biology and Biochemistry*, **5**, 47–81.

Hare P.E. (1980) Organic geochemistry of bone and its relation to the survival of bone in the natural environment. In: Behrensmeyer A.K. & Hill A.P. (eds) *Fossils in the Making*, pp. 208–219. University of Chicago Press, Chicago.

Hare P.E., Fogel M.L., Stafford T.W., Mitchell A.D. & Hoering T.C. (1991) The isotopic composition of carbon and nitrogen in individual amino acids isolated from modern and fossil proteins. *Journal of Archaeological Science*, **18**, 277–292.

Harrigan P., Zieman J.C. & Macko S.A. (1989) The base of nutritional support for the gray snapper (*Lutjanus griseus*): an evaluation based on a combined stomach content and stable isotope analysis. *Bulletin of Marine Science*, **44**, 65–77.

Harrison A.G. & Thode H.G. (1958) Mechanism of bacterial reduction of sulphate from isotope fractionation studies. *Transactions of the Faraday Society*, **54**, 84–92.

Harrison W.G. (1983) Nitrogen in marine environment. Use of isotopes. In: Carpenter E.J. & Capone D.G. (eds) *Nitrogen in the Marine Environment*, pp. 763–807. Academic Press, New York.

Hartmann U.M. & Nielson H. (1969) $\delta^{34}S$ Werte in retzenten Meeressedimenten und ihre Deutung am Beispiel einiger Sedimentprofile aus der westlichen Östsee. *Geologische Rundschau*, **58**, 621–655.

Hashimoto L.K., Kaplan W.A., Wofsy S.C. & McElroy M.B. (1983) Transformations of fixed nitrogen and N_2O in the Cariaco Trench. *Deep-Sea Research*, **30**, 575–590.

Hassan A.A., Termine J.D. & Haynes C.V. Jr. (1977) Mineralogical studies on bone apatite and their implications for radiocarbon dating. *Radiocarbon*, **19**, 364–374.

Hattersley P.W. (1982) ^{13}C values of C_4 types in grasses. *Australian Journal of Plant Physiology*, **9**, 139–154.

Hattersley P.W. (1983) The distribution of C_3 and C_4 grasses in Australia in relation to climate. *Oecologia*, **57**, 113–128.

Hattori A. (1983) Denitrification and dissimilatory nitrate reduction. In: Carpenter E.J. & Capone D.G. (eds) *Nitrogen in the Marine Environment*, pp. 191–232. Academic Press, New York.

Hauck R.D. (1973) Nitrogen tracers in nitrogen cycle studies – past and future needs. *Journal of Environmental Quality*, **2**, 317–327.

Hauck R.D. & Bremner J.M. (1976) Use of tracers for soil and fertilizer nitrogen research. *Advances in Agronomy*, **28**, 219–266.

Hauck R.D., Melsted S.W. & Yankwich P.E. (1958) Use of N-isotope distribution in nitrogen gas during denitrification. *Soil Science*, **86**, 287–291.

Hauck R.D., Bartholomew W.V., Bremner J.M. *et al.* (1972) Use of variations in natural nitrogen isotope abundance for environmental studies: a questionable approach. *Science*, **177**, 453–454.

Hayes J.M. (1982) Fractionation *et al.*: an introduction to isotopic measurements and terminology. *Spectra*, **8**, 3–8.

Hayes J.M., Takigiku R., Ocampo R., Callot H.J. & Albrecht P. (1987) Isotopic compositions and probable origins of organic molecules in the Eocene Messel shale. *Nature*, **329**, 48–51.

Hayes J.M., Popp B.N., Takigiku R. & Johnson M.W. (1989) An isotopic study of biogeochemical relationships between carbonates and organic carbon in the Greenhorn Formation. *Geochimica et Cosmochimica Acta*, **53**, 2961–2972.

Hayes J.M., Freeman K.H., Popp B.N. & Hoham C.H. (1990) Compound-specific isotope analysis: a novel tool for reconstruction of ancient biogeochemical processes. *Organic Geochemistry*, **16**, 1115–1128.

Heaton T.H.E. (1984) Sources of nitrate in phreatic groundwater in the western Kalihari. *Journal of Hydrology*, **67**, 249–259.

Heaton T.H.E. (1985) Isotopic and chemical aspects of nitrate in the ground-water of the Springbok Flats. *Water South Africa*, **11**, 199–208.

Heaton T.H.E. (1986) Isotopic studies of nitrogen pollution in the hydrosphere and atmosphere: a review. *Chemical Geology*, **59**, 87–102.

Heaton T.H.E. (1987a) $^{15}N/^{14}N$ ratios of nitrate and ammonium in rain at Pretoria, South Africa. *Atmospheric Environment*, **21**, 843–852.

Heaton T.H.E. (1987b) The $^{15}N/^{14}N$ ratios of plants in South Africa and Namibia: relationship to climate and coastal/saline environments. *Oecologia*, **74**, 236–246.

Heaton T.H.E. & Collett G.M. (1985). The analysis of $^{15}N/^{14}N$ ratios in natural samples, with emphasis on nitrate and ammonium in precipitation. CSIR Research Report 624, Council of Scientific and Industrial Research, Pretoria.

Heaton T.H.E., Talma A.S. & Vogel J.C. (1983) Origin and history of nitrate in confined groundwater in the western Kalihari. *Journal of Hydrology*, **62**, 243–262.

Heaton T.H.E., Vogel J.C., von la Chevallerie G. & Collett G. (1986) Climatic influence on the isotopic composition of bone nitrogen. *Nature*, **322**, 822–823.

Hedges J.I., Clark W.A. & Cowie G.L. (1988) Organic matter sources to the water column and surficial sediments of a marine bay. *Limnology and Oceanography*, **33**, 1116–1136.

Heimann M. & Keeling C.D. (1989) A three dimensional model of atmospheric CO_2 transport based on observed winds: 2. Model description and simulated tracer elements. In: Peterson D.H. (ed.) *Aspects of Climate Variability in the Pacific and the Western Americas*, pp. 237–275. Geophysical Monograph 55, American Geophysical Union, Washington, DC.

Heimann M., Keeling C.D. & Compton J.T. (1989) A three-dimensional model of atmospheric CO_2 transport based on observed winds: 3. Seasonal cycle and synoptic time scale variations. In: Peterson D.H. (ed.) *Aspects of Climate Variability in the Pacific and the Western Americas*, pp. 277–303. Geophysical Monograph 55, American Geophysical Union, Washington, DC.

Herman D.J. & Rundel P.W. (1989) Nitrogen isotope fractionation in burned and unburned chaparral soils. *Soil Science Society of America Journal*, **53**, 1229–1236.

Hermes J.D., Roeske C.A., O'Leary M.H. & Cleland W.W. (1982) Use of multiple isotope effects to determine enzyme mechanisms and intrinsic isotope effects. Malic enzyme and glucose-6-phosphate dehydrogenase. *Biochemistry*, **21**, 5106–5114.

Herring G.M. (1972) The organic matrix in bone. In: Bourne G.H. (ed.) *The Biochemistry and Physiology of Bone*, pp. 128–190. Academic Press, New York.

Herry A. & LePennec M. (1987) Endosymbiotic bacteria in the gills of the littoral bivalve molluscs, *Thyasira flexuosa* (Thyasiridae) and *Lucinella divaricata* (Lucinidae). *Symbiosis*, **4**, 25–36.

Hessen D.O., Andersen T. & Lyche A. (1990) Carbon metabolism in a humic lake: pool sizes and cycling through zooplankton. *Limnology and Oceanography*, **35**, 84–99.

Hessler R.R., Smithey W.M., Boudrias M.A., Keller C.H., Lutz R.A. & Childress J.J. (1988) Temporal change in megafauna at the Rose Garden hydrothermal vent (Galapagos Rift, eastern tropical Pacific). *Deep-Sea Research*, **35**, 1681–1709.

Hobson K.A. (1991) Use of stable carbon and nitrogen isotope analysis in seabird dietary studies. PhD Dissertation, University of Saskatchewan, Saskatoon.

Hobson K.A. & Clark R.G. (1992) Assessing avian diets using stable isotopes II: factors influencing the diet-tissue fractionation. *The Condor*, **94**, 189–197.

Hobson K.A. & Welch H.E. (1992) Determination of trophic relationships within a high arctic marine food web using $\delta^{13}C$ and $\delta^{15}N$ analysis. *Marine Ecology Progress Series*, **84**, 9–18.

Hoch M.P., Kirchman D.L. & Fogel M.L. (1989) Nitrogen isotope fractionation in the uptake of ammonium by a marine bacterium. In: *Annual Report of the Director, Geophysical Laboratory*, pp. 117–123. Carnegie Institution of Washington Number 2150, Washington, DC.

Hoefs J. (1987) *Stable Isotope Geochemistry*. (Minerals and rocks; 9). Springer-Verlag, Berlin.

Hoering T. (1957) The isotopic composition of ammonia and the nitrate ion in rain. *Geochimica et Cosmochimica Acta*, **12**, 97–102.

Hoering T. & Ford H.T. (1960) The isotope effect in the fixation of nitrogen by Azotobacter. *Journal of the American Chemical Society*, **82**, 376–378.

Högberg P. (1990) Forests losing large quantities of nitrogen have elevated $^{15}N : {}^{14}N$ ratios. *Oecologia*, **84**, 229–231.

Hollocher T.C. (1984) Source of the oxygen atoms of nitrate in the oxidation of nitrite by *Nitrocacter agilis* and evidence against a P-O-N anhydride mechanism in oxidative phosphorylation. *Archives of Biochemistry and Biophysics*, **233**, 721–727.

Hollocher T.C., Tate M.E. & Nicholas D.J.D. (1981) Oxidation of ammonia by *Nitrosomonas europaea*. Definitive ^{18}O-tracer evidence that hydroxylamine formation involves a monooxygenase. *Journal of Biological Chemistry*, **256**, 10834–10836.

Hopkins T.L. (1987) Midwater food web in McMurdo Sound, Ross Sea, Antarctica. *Marine Biology*, **96**, 93–106.

Hovland M., Talbot M.R., Qvale H., Olaussen S. & Aasberg L. (1987) Methane-related carbonate cements in pockmarks of the North Sea. *Journal of Sedimentary Petrology*, **57**, 881–892.

Howarth R.W. & Teal J.M. (1980) Energy flow in a salt marsh ecosystem: the role of reduced inorganic sulfur compounds. *American Naturalist*, **116**, 862–872.

Howarth R.W., Marino R., Lane J. & Cole J.J. (1988a) Nitrogen fixation in fresh water, estuarine and marine ecosystem. 1. Rates and importance. *Limnology and Oceanography*, **33**, 669–687.

Howarth R.W., Marino R. & Cole J.J. (1988b) Nitrogen fixation in fresh water, estuarine and marine ecosystem. 2. Biogeochemical controls. *Limnology and Oceanography*, **33**, 688–701.

Hubick K.T., Shorter R. & Farquhar G.D. (1988) Heritability and genotype X environment interactions of carbon isotope discrimination and transpiration efficiency in peanut (*Arachis hypogaea* L). *Australian Journal of Plant Physiology*, **15**, 799–813.

Hübner H. (1986) Isotope effects of nitrogen in the soil and biosphere. In: Fritz P. & Fontes J.C. (eds) *Handbook of Environmental Isotope Geochemistry*, vol. 2b, *The Terrestrial Environment*, pp. 361–425. Elsevier, Amsterdam.

Hynes A.J., Wine P.H. & Ravishankara A.R. (1986) Kinetics of the OH + CO reaction under atmospheric conditions. *Journal of Geophysical Research*, **91**, 11815–11820.

Ingerson E. (1953) Nonradiogenic isotopes in geology. *Bulletin of the Geological Society of America*, **64**, 301–374.

Jannasch H.W. (1984) Chemosynthesis: the nutritional basis for life at deep-sea vents. *Oceanus*, **27**, 73–78.

Jannasch H.W. & Mottl M.J. (1985) Geomicrobiology of deep-sea hydrothermal vents. *Science*, **229**, 717–725.

Jannasch H.W. & Wirsen C.O. (1979) Chemosynthetic primary production of East Pacific sea floor spreading centers. *Bioscience*, **29**, 592–598.

Jeffrey L.M. (1969) Development of a method for isolation of gram quantities of dissolved matter from sea water and some chemical and isotopic characteristics of the isolated material. PhD Thesis, Texas A&M University.

Jenkins M.C. & Kemp W.W. (1984) The coupling of nitrification and denitrification in two estuarine sediments. *Limnology and Oceanography*, **29**, 609–619.

Johnson D.A., Asay K.H., Tieszen L.L., Ehleringer J.R. & Jefferson P.G. (1990) Carbon isotope discrimination: potential in screening cool-season grasses for water-limited environments. *Crop Science*, **30**, 338–343.

Johnson K.S. (1982) Carbon dioxide hydration and dehydration kinetics in seawater. *Limnology and Oceanography*, **27**, 849–855.

Johnson K.S., Childress J.J., Hessler R.R., Sakamoto-Arnold C.M. & Beehler C.L. (1988) Chemical and biological interactions in the Rose Garden hydrothermal vent field. *Deep-Sea Research*, **35**, 1723–1744.

Johnston A.M. & Raven J.A. (1986) Dark fixation studies on the intertidal macroalga *Ascophyllum nodosum*. *Journal of Phycology*, **22**, 78–83.

Johnston A.M. & Raven J.A. (1987) The C_4-like characteristics of the intertidal macroalga *Ascophyllum nodosum* (L.) Le Jolis (Fucales, Pheophyta). *Phycologia*, **26**, 159–166.

Johnston A.M. & Raven J.A. (1989) Extraction, partial purification and characterization of phosphoenolpyruvate carboxykinase from *Ascophyllum nodosum* (Phaeophyceae). *Journal of Phycology*, **25**, 568–576.

Johnston A.M. & Raven J.A. (1992) Effect of aeration rates on growth rates and natural abundance $^{13}C/^{12}C$ ratio of *Phaeodactylum tricornutum*. *Marine Ecology Progress Series*, **87**, 295–300.

Johnston H. (1972) Newly recognized vital nitrogen cycle. *Proceedings of the National Academy of Science USA*, **69**, 2369–2372.

Johnston H.S. (1984) Human effects on the global atmosphere. *Annual Review of Physical Chemistry*, **35**, 481–505.

Jones M.L. (1984) The giant tube worms. *Oceanus*, **27**, 47–54.

Jordan D.B. & Ogren W.L. (1981) Species variation in the specificity of ribulose biphosphate carboxylase/oxygenase. *Nature*, **291**, 513–515.

Jordan F., Kuo D.J. & Monse E.U. (1978) Carbon kinetic isotope effects on pyruvate decarboxylation catalyzed by yeast pyruvate decarboxylase and models. *Journal of the American Chemical Society*, **100**, 2872–2878.

Jørgensen K.S., Jensen H.B. & Sorensen J. (1984) Nitrous oxide production from nitrification and denitrification in marine sediment at low oxygen concentrations. *Canadian Journal of Microbiology*, **30**, 1073–1078.

Jouzel J., Russell G.I., Suozzo R.J., Koster R.D., White J.W.C. & Broecker W.S. (1987) Simulation of the HDO and $H_2^{18}O$ atmospheric cycles for present day conditions. *Journal of Geophysical Research*, **92**, 14739–14760.

Kalisz S. & Teeri J.A. (1986) Population-level variation in photosynthetic metabolism and growth in *Sedum wrightii*. *Ecology*, **67**, 20–26.

Kaplan A., Friedberg D., Schwarz R., Ariel R., Seijffers J. & Reinhold L. (1989) The 'CO_2 concentrating mechanism' of cyanobacteria: physiological, molecular and theoretical studies. In: Briggs W.R. (ed.) *Photosynthesis*, pp. 243–255. A.R. Liss, New York.

Kaplan A., Schwarz R., Lieman-Hurwitz J. & Reinhold L. (1991) Physiological and molecular aspects of the inorganic carbon-concentrating mechanisms in cyanobacteria. *Plant Physiology*, **97**, 851–855.

Kaplan I.R. (1975) Stable isotopes as a guide to biogeochemical proceses. *Proceedings of the Royal Society London, Series B*, **189**, 183–211.

Kaplan I.R. & Rittenberg S.C. (1964) Microbial fractionation of sulfur isotopes. *Journal of General Microbiology*, **34**, 195–212.

Kaplan W. (1984) Sources and sinks of nitrous oxide. In: Tiedje J.M. & Klug M.J. (eds) *Microbial Ecology*, pp. 478–483. Proceedings of the 3rd International Symposium on Microbial Ecology, East Lansing, Michigan.

Kaplan W. & Wofsy S.C. (1985) The biochemistry of nitrous oxide: a review. In: Jannasch H.W. & Williams P.J.-B. (eds) *Advances in Aquatic Microbiology*, vol. 3, pp. 181–206. Academic Press, Florida.

Karamanos R.E. & Rennie D.A. (1978) N isotope fractionation during exchange reactions with soil clay. *Canadian Journal of Soil Science*, **58**, 53–60.

Karamanos R.E. & Rennie D.A. (1980a) Changes in natural ^{15}N abundance associated with pedogenic processes in soil. II. Changes on different slope positions. *Canadian Journal of Soil Science*, **60**, 365–372.

Karamanos R.E. & Rennie D.A. (1980b) Variations in natural nitrogen-15 abundance as an aid in tracing fertilizer nitrogen transformations. *Soil Science Society of America Journal*, **44**, 57–62.

Karl D.C., Wirsen C.O. & Jannasch H. (1980) Deep-sea primary production at the Galapagos hydrothermal vents. *Science*, **213**, 333–336.

Karl D.M. & Knauer G.A. (1984) Vertical distribution, transport, and exchange of carbon in the northeast Pacific Ocean: evidence for multiple zones of biological activity. *Deep-Sea Research*, **31**, 221–243.

Karl D.M., Knauer G.A., Martin J.H. & Ward B.B. (1984) Bacterial chemolithotrophy in the ocean is associated with sinking particles. *Nature*, **309**, 54–56.

Kasper H.F. (1982) Denitrification in marine sediments: measurements of capacity and estimate of *in situ* rate. *Applied and Environmental Microbiology*, **43**, 522–527.

Kastner M., Garrison R.E., Kolodny Y., Shemesh A. & Reimers C.E. (1990) Simultaneous changes of oxygen isotopes in PO_4^{3-} and CO_3^{2-} in apatites, with emphasis on the Monterey Formation, California. In: Burnett W.C. & Riggs S.R. (eds) *Genesis of Neogene to Modern Phosphorites*, pp. 312–324. Cambridge University Press, New York.

Katzenberg M.A. (1989) Stable isotope analysis of archaeological faunal remains from southern Ontario. *Journal of Archaeological Science*, **16**, 319–329.

Kay R.F. (1975) The functional adaptations of primate molar teeth. *American Journal of Physical Anthropology*, **43**, 195–215.

Keegan W.F. & DeNiro M.J. (1988) Stable carbon and nitrogen-isotope ratios of bone collagen used to study coral-reef and terrestrial components of prehistoric Bahamian diet. *American Antiquity*, **53**, 320–336.

Keeling C.D. (1958) The concentration and isotopic abundances of atmospheric carbon dioxide in rural areas. *Geochimica et Cosmochimica Acta*, **13**, 322–334.

Keeling C.D. (1961) The concentration and isotopic abundances of atmospheric carbon dioxide in rural and marine air. *Geochimica et Cosmochimica Acta*, **24**, 277–298.

Keeling C.D., Bacastow R.B., Carter A.F. *et al.* (1989a) A three dimensional model of atmospheric CO_2 transport based on observed winds: 1. Analysis of observational data. In: Peterson D.H. (ed.) *Aspects of Climate Variability in the Pacific and the Western Americas*, pp. 165–236. Geophysical Monograph 55, American Geophysical Union, Washington, DC.

Keeling C.D., Piper S.C. & Heimann M. (1989b) A three-dimensional model of atmospheric CO_2 transport based on observed winds: 4. Mean annual gradients and inter-annual variations. In: Peterson D.H. (ed.) *Aspects of Climate Variability in the Pacific and the Western Americas*, pp. 305–363. Geophysical Monograph 55, American Geophysical Union, Washington, DC.

Keeney D.R. (1982) Nitrogen management for maximum efficiency and minimum pollution. In: Stevenson F.J. (ed.) *Nitrogen in Agricultural Soils*, pp. 605–649. American Society of Agronomy, Madison.

Keith M.L., Anderson G.M. & Eichler R. (1964) Carbon and oxygen isotopic composition of mollusk shells from marine and fresh-water environments. *Geochimica et Cosmochimica Acta*, **28**, 1757–1786.

Kelly D.P. (1982) Biochemistry of the chemolithotrophic oxidation of inorganic sulphur. *Philosophical Transactions of the Royal Society London B*, **298**, 499–528.

Kemp N.E. (1984) Organic matrices and mineral crystallites in vertebrate scales, teeth, and skeletons. *American Zoologist*, **24**, 965–974.

Kennicutt M.C. & Brooks J.M. (1990) Unusual normal alkane distributions in offshore New Zealand sediments. *Organic Geochemisty*, **15**, 193–197.

Kennicutt M.C. II, Brooks J.M., Bidigare R.R., Fay R.R., Wade T.L. & McDonald T.J. (1985) Vent-type taxa in a hydrocarbon seep region on the Louisiana slope. *Nature*, **317**, 351–353.

Kennicutt M.C. II, Brooks J.M. & Denoux J.R. (1988) Leakage of deep reservoired petroleum to the near surface on the Gulf of Mexico continental slope. *Marine Chemistry*, **24**, 39–59.

Kennicutt M.C. II, Brooks J.M. & Burke R.A. (1989) Hydrocarbon seepage, gas hydrates, and authigenic carbonate in the northwestern Gulf of Mexico. *Proceedings of the 21st Annual Offshore Technology Conference*, May 1–4, 1989, OTC 5954: 663–667.

Kennicutt M.C. II, Burke R.A., MacDonald I.R., Brooks J.M., Denoux G.J. & Macko S.A. (1992) Stable isotope partitioning in seep and vent organisms: chemical and ecological significance. *Chemical Geology (Isotope Geoscience Section)*, **101**, 293–310.

Kennicutt M.C., Macko S.A., Harvey H.R. & Bidigare R.R. (1992) Preservation of sargassum under anoxic conditions: isotopic and molecular evidence. In: Whelan J.K. & Farrington J.W. (eds) *Organic Matter: Productivity, Accumulation and Presentation in Recent and Ancient Sediments*, pp. 123–141. Columbia University Press, New York.

Kerby N.W. & Raven J.A. (1985) Transport and fixation of inorganic carbon by marine algae. *Advances in Botanical Research*, **11**, 71–123.

Kerr R.A. & Quinn J.G. (1980) Chemical comparison of dissolved organic matter isolated from different oceanic environments. *Marine Chemistry*, **8**, 217–229.

Keunen J.G. & Beudeker R.F. (1982) Microbiology of thiobacilli and other sulphur-oxidizing autotrophs, mixotrophs and heterotrophs. *Philosophical Transactions of the Royal Society London B*, **298**, 473–497.

Khalil M.A.K. & Rasmussen R.A. (1988) Carbon monoxide in Earth's atmosphere: indications of a global increase. *Nature*, **332**, 242–245.

Kim K.-R. & Craig H. (1990) Two isotope characterizations of N_2O in the Pacific Ocean and constraints on its origin in deep water. *Nature*, **347**, 58–60.

Kimber R.W.L. & Hare P.E. (1992) Wide range of racemization of amino acids in peptides from human fossil bone and its implications for amino acid racemization dating. *Geochimica et Cosmochimica Acta*, **56**, 739–743.

King S.L., Quay P.D. & Lansdown J.M. (1989) $^{13}C/^{12}C$ kinetic isotope effect for soil oxidation of methane at ambient atmospheric concentration. *Journal of Geophysical Research*, **94**, 18373–18377.

Kioboe T., Kaas H., Kruse B., Mohlenberg F., Tiselius P. & Aertebjerg G. (1990) The structure of the pelagic food web in relation to water column structure in the Skagerrak. *Marine Ecology Progress Series*, **59**, 19–32.

Kluge M., Brulfert J., Ravelomanana D., Lipp J. & Ziegler H. (1991) Crassulacean acid metabolism in *Kalanchoë* species collected in various climatic zones of Madagascar: a survey by $\delta^{13}C$ analysis. *Oecologia*, **88**, 407–414.

Knowles R. (1990) Acetylene inhibition technique: development, advantages, and potential problems. In: Revsbech N.P. & Sorensen J. (eds) *Denitrification in Soil and Sediment*, pp. 151–166. Plenum Press, New York.

Knowles R. & Blackburn T.H. (eds) (1993) *Nitrogen Isotope Techniques*. Academic Press, San Diego.

Koch P.L. (1989) Paleobiology of late Pleistocene mastodonts and mammoths from southern Michigan, and western New York, PhD Thesis, University of Michigan.

Koch P.L. (1991) The isotopic ecology of Pleistocene proboscideans. *Journal of Vertebrate Paleontology*, **11**(suppl), 40A.

Koch P.L., Fisher D.C. & Dettman D.L. (1989) Oxygen isotopic variation in the tusks of extinct proboscideans: a measure of season of death and seasonality. *Geology*, **17**, 515–519.

Koch P.L., Behrensmeyer A.K., Tuross N. & Fogel M.L. (1990) Isotopic fidelity during bone weathering and burial. In: *Annual Report of the Director of the Geophysical Laboratory*, pp. 105–110. Carnegie Institution of Washington, Washington, DC.

Koch P.L., Halliday A.N., Walter L.M., Stearley R.F., Huston T.J. & Smith G.R. (1992) Sr isotopic composition of hydroxyapatite from recent and fossil salmon: the record of lifetime migration and diagenesis. *Earth and Planetary Science Letters*, **108**, 277–287.

Kohl D.H. & Shearer G. (1980) Isotopic fractionation associated with symbiotic N_2 fixation and uptake of NO_3^- by plants. *Plant Physiology*, **66**, 51–56.

Kohl D.H., Shearer G.B. & Commoner B. (1971) Fertilizer nitrogen: contribution to nitrate in surface water in a corn belt watershed. *Science*, **174**, 1331–1334.

Kohl D.H., Shearer G.B. & Commoner B. (1973) Variation of ^{15}N in corn and soil following application of fertilizer nitrogen. *Soil Science Society of America Proceedings*, **37**, 888–892.

Koike I. (1990) Measurement of sediment denitrification using ^{15}N tracer method. In: Revsbech N.P. & Sorensen J. (eds) *Denitrification in Soil and Sediment*, pp. 291–300. Plenum Press, New York.

Koike I. & Hattori A. (1978a) Denitrification and ammonium formation in anaerobic coastal sediments. *Applied and Environmental Microbiology*, **35**, 278–282.

Koike I. & Hattori A. (1978b) Simultaneous determination of nitrification and nitrate reduction in coastal sediments by a ^{15}N dilution technique. *Applied and Environmental Microbiology*, **35**, 853–857.

Koike I. & Sørensen J. (1988) Nitrate reduction and denitrification in marine sediments. In: Blackburn T.H. & Sorensen J. (eds) *Nitrogen Cycling in Coastal Marine Environments*, pp. 251–273. John Wiley & Sons, Chichester.

Koike I., Wada E., Tsuji T. & Hattori A. (1972) Studies on denitrification in a brackish lake. *Archives of Hydrobiology*, **69**, 508–520.

Kolodny Y. & Kaplan I.R. (1970) Carbon and oxygen isotopes in apatite CO_2 and co-existing calcite from sedimentary phosphorite: *Journal of Sedimentary Petrology*, **40**, 954–959.

Kolodny Y., Luz B. & Navon O. (1983) Oxygen isotope variation in phosphate of biogenic apatites, I. Fish bone apatite – rechecking the rules of the game. *Earth and Planetary Science Letters*, **64**, 398–404.

Körner Ch. & Diemer M. (1987) *In situ* photosynthetic responses to light, temperature and carbon dioxide in herbaceous plants from low and high altitude. *Functional Ecology*, **1**, 179–194.

Körner Ch., Farquhar G.D. & Roksandic Z. (1988) A global survey of carbon isotope discrimination in plants from high altitude. *Oecologia*, **74**, 623–632.

Körner Ch., Farquhar G.D. & Wong S.C. (1991) Carbon isotope discrimination by plants follows latitudinal and altitudinal trends. *Oecologia*, **88**, 30–40.

Kreitler C.W. (1975) Determining the source of nitrate in ground water by nitrogen isotope studies. Report 83, Bureau of Economic Geology, University of Texas, Austin.

Kreitler C.W. (1979) Nitrogen-isotope ratio studies of soils and groundwater nitrate from alluvial fan aquifers in Texas. *Journal of Hydrology*, **42**, 147–170.

Kreitler C.W. & Jones D.C. (1975) Natural soil nitrate: the cause of the nitrate contamination of ground water in Runnels County, Texas. *Ground Water*, **13**, 53–61.

Kreitler C.W. & Browning L.A. (1983) Nitrogen-isotope analysis of groundwater nitrate in carbonate aquifers: natural sources versus human pollution. *Journal of Hydrology*, **61**, 285–301.

Kreitler C.W., Ragone S.E. & Katz B.G. (1978) N^{15}/N^{14} ratios of ground-water nitrate, Long Island, New York. *Ground Water*, **16**, 404–409.

Kremer B.P. (1981) Dark reactions in photosynthesis. In: Platt T. (ed.) *Physiological Bases of Phytoplankton Ecology*, pp. 44–54. *Canadian Bulletin of Fisheries and Aquatic Sciences*, Ottawa.

Krishnamurthy R.V. & DeNiro M.J. (1982) Sulfur interference in the determination of hydrogen concentration and stable isotopic composition in organic matter. *Analytical Chemistry*, **54**, 153–154.

Krishnamurthy R.V. & Epstein S. (1985) Tree ring D/H ratio from Kenya, East Africa and its palaeoclimatic significance. *Nature*, **317**, 160–162.

Kroopnick P.M. (1985) The distribution of ^{13}C of TCO_2 in the world oceans. *Deep-Sea Research*, **32**, 57–84.

Kroopnick P. & Craig H. (1972) Atmospheric oxygen: isotopic composition and solubility fractionation. *Science*, **175**, 54–55.

Kroopnick P. & Craig H. (1976) Oxygen isotope fractionation in dissolved oxygen in the deep sea. *Earth Planetary Science Letters*, **32**, 375–388.

Krouse H.R. (1980) Sulphur isotopes in our environment. In: Fritz P. & Fontes J.C. (eds) *Handbook of Environmental Isotope Geochemistry, Vol. 1. The Terrestrial Environment*, pp. 435–471. Elsevier, Amsterdam.

Krouse H.R. (1988) Sulfur isotope studies of the pedosphere and biosphere. In: Rundel P.W., Ehleringer J.R. & Nagy K.A. (eds) *Stable Isotopes in Ecological Research*, pp. 424–444. Springer-Verlag, New York.

Krouse H.R. & Tabatabai M.A. (1986) Stable sulfur isotopes. In: Tabatabai M.A. (ed.) *Sulfur in Agriculture*, pp. 169–201. Academic Press, San Diego.

Krouse H.R., McCready R.G.L., Husain S.A. & Campbell J.N. (1967) Sulphur isotope fractionation by *Salmonella* sp. *Canadian Journal of Microbiology*, **13**, 21–25.

Krueger H.W. (1991) Exchange of carbon with biological apatite. *Journal of Archaeological Science*, **18**, 355–361.

Krueger H.W. & Sullivan C.H. (1984) Models for carbon isotope fractionation between diet and bone. In: Turnland J.R. & Johnson P.E. (eds) *Stable Isotopes in Nutrition*, pp. 205–220. American Chemical Society, Symposium Series, no. 258. Washington, DC.

Kulm L.D., Seuss E., Moore J.C., *et al.* (1986) Oregon subduction zone: venting fauna and carbonates. *Science*, **231**, 561–566.

Kumar S., Nicholas D.J.D. & Williams E.H. (1983) Definitive ^{15}N NMR evidence that water serves as a source of 'O' during nitrite oxidation by *Nitrobacter agilis FEBS Letters*, **152**, 71–74.

Kunz C. (1985) Carbon-14 discharges at three light-water reactors. *Health Physics*, **49**, 25–35.

Laane R.W.P.M., Turkstra E. & Mook W.G. (1990) Stable carbon isotope composition of pelagic and benthic organic matter in the North Sea and adjacent estuaries. In: Ittekkot V., Kempe S., Michaelis W. & Spitcy A. (eds) *Facets of Modern Biogeochemistry*. Springer-Verlag, Berlin.

Lacis A., Hansen J., Lee P., Mitchell T. & Lebedeff S. (1981) Greenhouse effect of trace gases, 1970–1980. *Geophysical Research Letters*, **8**, 1035–1038.

Lajtha K. & Schlesinger W.H. (1986) Plant response to variations in nitrogen availability in a desert shrubland community. *Biogeochemistry*, **2**, 29–37.

Lajtha K. & Whitford W.G. (1989) The effect of water and nitrogen amendments on photosynthesis, leaf demography, and resource-use efficiency in *Larrea tridentata*, a desert evergreen shrub. *Oecologia*, **80**, 341–348.

Lajtha K. & Barnes F.J. (1991) Carbon gain and water use in pinyon pine–juniper woodlands of northern New Mexico: field versus phytotron chamber measurements. *Tree Physiology*, **9**, 59–67.

Lajtha K. & Getz J.L. (1993) Photosynthesis and water-use efficiency in pinyon–juniper communities along an elevation gradient in northern New Mexico. *Oecologia*, **94**, 95–101.

Land L.S., Lundelius E.L. Jr. & Valastro S. Jr. (1980) Isotopic ecology of deer bones. *Palaeogeography, Palaeoclimatology, and Palaeoecology*, **32**, 143–151.

Larkum A.W.D., Kennedy I.R. & Muller W.J. (1988) Nitrogen fixation on a coral reef. *Marine Biology*, **98**, 143–155.

Law C.S. & Owens N.J.P. (1990) Significant flux of atmospheric nitrous oxide from the northwest Indian Ocean. *Nature*, **346**, 826–829.

Lawrence J.R. & White J.W.C. (1984) Growing season precipitation from D/H ratios of eastern white pine. *Nature*, **311**, 558–560.

Leavitt S.W. & Long A. (1986) Stable-carbon isotope variability in tree foliage and wood. *Ecology*, **67**, 1002–1010.

LeBlanc C.G., Bourbonniere R.A., Schwarcz H.P. & Risk M.J. (1989) Carbon isotopes and fatty acids analysis of the sediments of Negro Harbour, Nova Scotia, Canada. *Estuarine and Coastal Shelf Science*, **28**, 261–276.

Lécolle P. (1985) The oxygen isotope composition of landsnail shells as a climatic indicator: applications to hydrogeology and paleoclimatology. *Chemical Geology (Isotope Geoscience Section)*, **58**, 157–181.

Ledgard S.A.F., Freney J.R. & Simpson J.R. (1984) Variations in the natural enrichment of ^{15}N in the profiles of some Australian pasture soils. *Australian Journal of Soil Research*, **22**, 155–164.

Lee C. & Cronin C. (1982) The vertical flux of particulate organic nitrogen in the sea: decomposition of amino acids in Peru upwelling area and the equatorial Atlantic. *Journal of Marine Research*, **40**, 227–252.

Lee-Thorp J.A. (1989) Stable carbon isotopes in deep time: the diets of fossil fauna and hominids. PhD Thesis, University of Cape Town.

Lee-Thorp J.A. & van der Merwe N.J. (1987) Carbon isotope analysis of fossil bone apatite. *South African Journal of Science*, **83**, 71–74.

Lee-Thorp J.A. & van der Merwe N.J. (1991) Aspects of the chemistry of modern and fossil biological apatites. *Journal of Archaeological Science*, **18**, 343–354.

Lee-Thorp J.A., Sealey J.C. & van der Merwe N.J. (1989a) Stable carbon isotope ratio differences between bone collagen and bone apatite, and their relationship to diet. *Journal of Archaeological Science*, **16**, 585–599.

Lee-Thorp J.A., van der Merwe N.J. & Brain C.K. (1989b) Isotopic evidence for dietary differences between two extinct baboon species from Swartkrans. *Journal of Human Evolution*, **18**, 183–190.

Legendre L. & Gosselin M. (1989) New production and export of organic matter to the deep ocean: consequence of some recent discoveries. *Limnology and Oceanography*, **34**, 1374–1380.

LeGeros R.Z. (1981) Apatites in biological systems. In: Pamplin B. (ed.) *Inorganic Biological Crystal Growth, Progress in Crystal Growth and Characterization*, vol. 4, pp. 1–45. Pergamon Press, New York.

Le Pennec M. & Hily A. (1984) Anatomie, structure et ultrastructure de la branchie d'un Mytilidae des sites hydrothermaux du Pacifique oriental. *Oceanologica Acta*, **7**, 517–523.

Le Pennec M. & Prieur D. (1984) Observations sur la nutrition d'un Mytilidae d'un site hydrothermal actif de la dorsale du Pacifique oriental. *C.r. hebd. Séanc. Acad. Sci. Paris*, **298**, 493–498.

Létolle R. (1980) Nitrogen-15 in the natural environment. In: Fritz P. & Fontes J.C. (eds) *Handbook of Environmental Isotope Geochemistry*, vol. 1A, pp. 407–433. Elsevier, Amsterdam.

Levin I., Kromer B., Schoch-Fischer H., Bruns M. & Munnich K.O. (1985) 25 years of tropospheric ^{14}C observations in Central Europe. *Radiocarbon*, **27**, 1–19.

Levin I., Bosinger R., Bonani G. *et al.* (1991) Radiocarbon in atmospheric carbon dioxide and methane: global distributions and trends. *Radiocarbon*.

Libby W.F. (1955) *Radiocarbon Dating*. University of Chicago Press, Chicago.

Libby W.F., Berger R., Mead J., Alexander G. & Ross J. (1964) Replacement rates for human tissue from atmospheric radiocarbon. *Science*, **146**, 1170–1172.

Libby L.M., Pandolfi L.J., Payton P.H., Marshall J. III, Becker B. & Giertz-Sienbenlist V. (1976) Isotopic tree thermometers. *Nature*, **261**, 284–288.

Lipschultz F., Zafiriou O.C., Wofsy S.C., McElroy M.B., Valois F.W. & Watson S.W. (1981) Production of NO and N_2O by soil nitrifying bacteria. *Nature*, **294**, 641–643.

Lipschultz F., Wofsy S.C., Ward B.B., Codispoti L.A., Friederich G. & Elkins J.W. (1990) Bacterial transformations of inorganic nitrogen in the oxygen-deficient waters of the eastern tropical South Pacific Ocean. *Deep-Sea Research*, **37**, 1513–1541.

Liu K.-K. & Kaplan I.R. (1989) The eastern tropical Pacific as a source of ^{15}N-enriched nitrate in seawater off southern California. *Limnology and Oceanography*, **34**, 820–830.

Liu K.-K., Shaw P.-T. & Kaplan I.R. (1987) Modeling of nitrogen isotopic variation of nitrate within the denitrifying zone in the eastern tropical South Pacific. *EOS*, **68**, 1714.

Longin R. (1971) New method of collagen extraction for radiocarbon dating. *Nature*, **230**, 241–242.

Longinelli A. (1973) Preliminary oxygen-isotope measurements of phosphate from mammal teeth and bones. In: Labeyrie J. (ed.) *Les Méthodes Quantitative d'Étude des Variations du Climat au Cours du Pléistocene: Colloques Internationaux du C.N.R.S.*, no. 219, pp. 267–271.

Longinelli A. (1984) Oxygen isotopes in mammal bone phosphate: a new tool for paleo-hydrological and paleoclimatological research? *Geochimica et Cosmochimica Acta*, **48**, 385–390.

Longinelli A. & Nuti S. (1973) Revised phosphate–water isotopic temperature scale. *Earth and Planetary Science Letters*, **19**, 373–376.

Lonsdale P. (1977) Clustering of suspension feeding macrobenthos near abyssal hydrothermal vents at oceanic spreading centers. *Deep-Sea Research*, **24**, 857–863.

Lorius C., Jouzel J., Raynaud D., Hansen J. & Le Treut H. (1990) The ice core record: climate sensitivity and future greenhouse warming. *Nature*, **347**, 139–145.

Lovley D.R. & Klug M.J. (1986) Model for the distribution of sulfate reduction and methanogenesis in freshwater sediments. *Geochimica et Cosmochimica Acta*, **50**, 11–18.

Lowrance R.R. & Pionke H.B. (1989) Transformations and movement of nitrate in aquifer systems. In: Follett R.F. (ed.) *Nitrogen Management and Ground Water Protection*, pp. 374–392. Elsevier, New York.

Lucotte M. (1989) Organic carbon isotope ratios and implications for the maximum turbidity zone of the St. Lawrence Upper Estuary. *Estuarine and Coastal Shelf Science*, **29**, 293–304.

Luo Y. & Sternberg L. (1992) Hydrogen and oxygen isotopic fractionation during heterotrophic cellulose synthesis. *Journal of Experimental Botany*, **43**, 47–50.

Lusis M.A. & Wiebe H.A. (1976) The rate of oxidation of sulfur dioxide in the plume of a nickel smelter stack. *Atmospheric Environment*, **10**, 793–798.

Luz B. & Kolodny Y. (1985) Oxygen isotope variations in phosphate of biogenic apatites, IV. Mammal teeth and bones. *Earth and Planetary Science Letters*, **75**, 29–36.

Luz B., Kolodny Y. & Horowitz M. (1984) Fractionation of oxygen isotopes between mammalian bone-phosphate and environmental water. *Geochimica et Cosmochimica Acta*, **48**, 1689–1693.

Luz B., Cormie A.B. & Schwarcz H.P. (1990) Oxygen isotope variations in phosphate of deer bones. *Geochimica et Cosmochimica Acta*, **54**, 1723–1728.

Maass I. & Weise G. (1981) Untersuchungen zur Regionalen Ausbreitung von Schadstoffen mit Hilfe technogener Isotopenvariationen. *Isotopenpraxis*, **17**, 156–159.

MacKay C., Pandow M. & Wolfgang R. (1963) On the chemistry of natural radiocarbon. *Journal of Geophysical Research*, **86**, 7210–7254.

Macko S.A. (1981) Stable nitrogen isotope ratios as tracers of organic geochemical processes. PhD Thesis, University of Texas, Austin.

Macko S.A. & Parker P.L. (1983) Stable nitrogen and carbon isotope ratios of beach tars on south Texas barrier islands. *Marine Environmental Research*, **10**, 93–103.

Macko S.A. & Engel M.H. (1991) Assessment of indigeneity in fossil organic matter: amino acids and stable isotopes. *Philosophical Transactions of the Royal Society London*, **333**, 367–374.

Macko S.A., Parker P.L. & Botello A.V. (1981) Persistence of spilled oil in a Texas salt marsh. *Environmental Pollution Bulletin*, **2**, 119–128.

Macko S.A., Lee W.Y. & Parker P.L. (1982) Nitrogen and carbon isotope fractionation by two species of marine amphipods: laboratory and field studies. *Journal of Experimental Marine Biology and Ecology*, **63**, 145–149.

Macko S.A., Estep M.L.F. & Lee W.Y. (1983) Stable hydrogen isotope analysis of food webs on laboratory and field populations of marine amphipods. *Journal of Experimental Marine Biology and Ecology*, **72**, 243–249.

Macko S.A., Enterzoth L. & Parker P.L. (1984) Regional differences in the stable isotopes of nitrogen and carbon on the continental shelf of the Gulf of Mexico. *Naturwissenschaften*, **71**, 374–375.

Macko S.A., Fogel-Estep M.L., Engel M.H. & Hare P.E. (1986) Kinetic fractionation of stable nitrogen isotopes during amino acid transamination. *Geochimica et Cosmochimica Acta*, **50**, 2143–2146.

Macko S.A., Fogel M.L., Hare P.E. & Hoering T.C. (1987) Isotopic fractionation of nitrogen and carbon in the synthesis of amino acids by microorganisms. *Chemical Geology (Isotope Geoscience Section)*, **65**, 79–92.

Macko S.A., Helleur R., Hartley G. & Jackman P. (1990) Diagenesis of organic matter – a study using stable isotopes of individual carbohydrates. *Organic Geochemistry*, **16**, 1129–1137.

Macko S.A., Engel M.H., Hartley G., Hatcher P., Helleur R., Jackman P. & Silfer J. (1991) Isotopic compositions of individual carbohydrates as indicators of early diagenesis of organic matter. *Chemistry and Geology*, **93**, 147–161.

Madison R.J. & Brunett J.O. (1985) Overview of the occurrence of nitrate in ground water of the United States. In: *U.S.G.S. National Water Summary 1984*, pp. 93–105. U.S. Geological Survey Water-Supply, paper 2275.

Magaritz M., Heller J. & Volokita M. (1981) Land–air boundary environment as recorded by the $^{18}O/^{16}O$ and $^{13}C/^{12}C$ isotope ratios in the shells of land snails. *Earth and Planetary Science Letters*, **52**, 101–106.

Mague T.H., Weare N.H. & Holm-Hansen O. (1974) Nitrogen fixation in the north Pacific Ocean. *Marine Biology*, **24**, 109–119.

Mague T.H., Mague F.C. & Holm-Hansen O. (1977) Physiology and chemical composition of nitrogen fixing phytoplankton in the central north Pacific Ocean. *Marine Biology*, **41**, 213–227.

Majoube M. (1971) Fractionnement en oxygène 18 et en deutérium entre l'eau et sa vapeur. *Journal de Chimie Physique*, **10**, 1473.

Manning M.R., Lowe D.C., Melhuish W.H. *et al.* (1990) The use of radiocarbon measurements in atmospheric studies. *Radiocarbon*, **32**, 37–58.

Marino B.D. & McElroy M.B. (1991) Isotopic composition of atmospheric CO_2 inferred from carbon in C_4 plant cellulose. *Nature*, **349**, 127–131.

Marino B.D., Logan J.A., McElroy M.D. & Wahlen M. (1990) Stable carbon isotope ratios of C_3 or C_4 based source terms for the isotopic composition of biomass burning emissions. Proc. of Chapman Conference on Global Biomass Burning, Williamsburg, Virginia, 1990, 34.

Marino B.D., McElroy M.B., Salawitch R.J. & Spaulding G. (1992) Glacial-to-interglacial variations in the carbon isotopic composition of atmospheric CO_2. *Nature*, **357**, 461–466.

Mariotti A. (1983) Atmospheric nitrogen is a reliable standard for natural abundance ^{15}N measurements. *Nature*, **303**, 685–687.

Mariotti A. (1984) Natural ^{15}N abundance measurements and atmospheric nitrogen standard calibration. *Nature*, **311**, 251–252.

Mariotti A., Mariotti F., Amarger N., Pizelle G., Ngambi J.-M., Champigny M.-L. & Moyse A. (1980a) Fractionnements isotopiques de l'azote lors des processus d'absorption des nitrates et de fixation de l'azote atmosphérique par les plants. *Physiologic Vegetale*, **18**, 163–181.

Mariotti A., Pierre D., Vedy J.C., Bruckert S. & Guillemot J. (1980b) The abundance of natural nitrogen-15 in the organic matter of soils along an altitudinal gradient. *Catena*, **7**, 293–300.

Mariotti A., Germon J.C., Hubert P. *et al.* (1981) Experimental determination of nitrogen kinetic isotope fractionations: some principles; illustration for the denitrification and nitrification processes. *Plant and Soil*, **62**, 413–430.

Mariotti A., Germon J.C. & Leclerc A. (1982) Nitrogen isotope fractionation associated with the $NO_2^- \rightarrow N_2O$ step of denitrification in soils. *Canadian Journal of Soil Science*, **62**, 227–241.

Mariotti A., Germon J.C., Leclerc A., Catroux G. & Letolle R. (1982) Experimental determination of kinetic isotope fractionation of nitrogen isotopes during denitrification. In: Schimdt H.-L., Förstel H. & Heinzinger K. (eds) *Stable Isotopes*, pp. 459–464. Elsevier, Amsterdam.

Mariotti A., Lancelot C. & Billen G. (1984) Natural isotopic composition of nitrogen as a tracer of origin for suspended organic matter in the Scheldt estuary. *Geochimica et Cosmochimica Acta*, **48**, 549–555.

Marlier J.F. & O'Leary M.H. (1984) Carbon kinetic isotope effects on the hydration of carbon dioxide and the dehydration of bicarbonate ion. *Journal of the American Chemical Society*, **106**, 5054–5057.

Marples T.G. (1966) A radionuclide tracer study of arthropod food chains in a *Spartina* salt march ecosystem. *Ecology*, **47**, 270–277.

Marshall J.D. & Ehleringer J.R. (1990) Are xylem-tapping mistletoes partially heterotrophic. *Oecologia*, **84**, 244–248.

Martel Y.A. & Paul E.A. (1974) The use of radiocarbon dating of organic matter in the study of soil genesis. *Soil Science Society of America Proceedings*, **38**, 501–506.

Martinelli L.A., Devol A.H., Victoria R.L. & Richey J.E. (1991) Stable carbon isotope variation in C_3 and C_4 plants along the Amazon River. *Nature*, **353**, 57–59.

Martinez L., Silver M.W., King J.M. & Alldredge A.L. (1983) Nitrogen fixation by floating diatom mats: a source of new nitrogen to oligotrophic ocean waters. *Science*, **221**, 152–154.

Masters P.M. (1987) Preferential preservation of noncollagenous protein during bone diagenesis: implications for chronometric and stable isotope measurements. *Geochimica et Cosmochimica Acta*, **51**, 3209–3214.

McArthur J.M., Coleman M.L. & Bremner J.M. (1980) Carbon and oxygen isotopic composition of structural carbonate in sedimentary francolite. *Geological Society of London Journal*, **137**, 669–673.

McCarthy J.J. & Goldman J.C. (1979) Nitrogenous nutrition of marine phytoplankton in nutrient depleted waters. *Science*, **203**, 670–672.

McCarthy J.J. & Carpenter E.J. (1983) Nitrogen cycling in near-surface waters of the open ocean. In: Carpenter E.J. & Capone D.G. (eds) *Nitrogen in the Marine Environment*, pp. 487–512. Academic Press, London and New York.

McConnaughey T. & McRoy C.P. (1979a) Food web structure and the fractionation of carbon isotopes in the Bering Sea. *Marine Biology*, **53**, 257–262.

McConnaughey T. & McRoy C.P. (1979b) ^{13}C label identifies eelgrass (*Zostera marina*) carbon in an Alaskan estuarine food web. *Marine Biology*, **53**, 263–269.

McCorkle D.C. & Emerson S.R. (1988) The relationship between pore water carbon isotopic composition and bottom water oxygen concentration. *Geochimica et Cosmochimica Acta*, **52**, 1169–1178.

McCrea J.M. (1950) On the isotopic chemistry of carbonates and a paleotemperature scale. *Journal of Chemistry and Physics*, **18**, 849–857.

McElroy M.B. & McConnel J.C. (1971) Nitrous oxide: a natural source of stratospheric NO. *Journal of Atmospheric Science*, **28**, 1095–1098.

McKane R.B., Grigal D.F. & Russelle M.P. (1990) Spatiotemporal differences in ^{15}N uptake and organization of an old-field plant community. *Ecology*, **71**, 1126–1132.

McMillan C. & Smith B.N. (1982) Comparison of $\delta^{13}C$ values for seagrasses in experimental cultures and in natural habitats. *Aquatic Botany*, **14**, 381–387.

McMillan C., Parker P.L. & Fry B. (1980) $^{13}C/^{12}C$ Ratios in seagrasses. *Aquatic Botany*, **9**, 237–249.

Mead J.I., Agenbroad L.D., Davis O.K. & Martin P.S. (1986) Dung of *Mammuthus* in arid southwest, North America. *Quaternary Research*, **25**, 121–127.

Medina E. & Minchin P. (1980) Stratification of $\delta^{13}C$ values of leaves in Amazonian rain forests. *Oecologia*, **45**, 377–378.

Medina E., Montes G., Cuevas E. & Rokcsandic Z. (1986) Profiles of CO_2 concentration and $\delta^{13}C$ values in tropical rain forests of the Upper Rio Negro Basin, Venezuela. *Journal of Tropical Ecology*, **2**, 207–217.

Medina E., Sternberg L. & Cuevas E. (1991) Vertical stratification of $\delta^{13}C$ values in closed natural and plantation forests in the Luquillo mountains, Puerto Rico. *Oecologia*, **87**, 369–372.

Mekhtiyeva V.L., Pankina R.G. & Gavrilov Y.Y. (1976) Distribution and isotopic compositions of forms of sulfur in water, animals and plants. *Geokhimiya*, **9**, 1419–1426.

Melillo J.M., Aber J.D., Linkins A.E., Turner A.R., Fry B. & Nadelhoffer K.J. (1989) Carbon and nitrogen dynamics along the decay continuum: plant litter to soil organic matter. *Plant and Soil*, **115**, 189–198.

Merlivat L. (1978) Molecular diffusivities of $H_2^{16}O$, $HD^{16}O$, and $H_2^{18}O$ in gases. *Journal of Chemistry and Physics*, **69**, 2864–2871.

Merlivat L. & Jourzel J. (1979) Global climatic interpretation of the deuterium–oxygen 18 relationship for precipitation. *Journal of Geophysical Research*, **84**, 5029–5033.

Milankovitch M. (1941) Canon of insolation and the ice-age problem. Belgrade: Royal Serbian Academy. English translation, 1969. Jerusalem: Israel program for scientific translations.

Miller G.H., Wendorf F., Ernst R. *et al.* (1991) Dating lacustrine episodes in the eastern Sahara by the epimerization of isoleucine in ostrich eggshells. *Palaeogeography, Palaeoclimatology, and Palaeoecology*, **84**, 175–189.

Minagawa M. & Wada E. (1984) Stepwise enrichment of ^{15}N along food chains: further evidence and the relation between $\delta^{15}N$ and animal age. *Geochimica et Cosmochimica Acta*, **48**, 1135–1140.

Minagawa M. & Wada E. (1986) Nitrogen isotope ratios of red tide organisms in the east China Sea: a characterization of biological nitrogen fixation. *Marine Chemistry*, **19**, 245–259.

Minagawa M., Winter D.A. & Kaplan I.R. (1984) Comparison of Kjeldahl and combustion methods for measurement of nitrogen isotope ratios in organic matter. *Analytical Chemistry*, **56**, 1859–1861.

Miyake Y. & Wada E. (1967) The abundance ratio of $^{15}N/^{14}N$ in marine environments. *Records of Oceanographic Works, Japan*, **9**, 37–53.

Miyake Y. & Wada E. (1971) The isotope effect on the nitrogen in biochemical oxidation–reduction reactions. *Records of Oceanographic Works, Japan*, **9**, 37–57.

Mix A.C. (1987) The oxygen-isotope record of glaciation. In: Ruddiman W.F. & Wright H.E. Jr. (eds) *North America and Adjacent Oceans During the Last Deglaciation: The Geology of North America*, vol. K-3, pp. 111–136. Geological Society of America, Boulder.

Mizutani H. & Wada E. (1982) Effect of high atmospheric CO_2 concentration on $\delta^{13}C$ of algae. *Origins of Life*, **12**, 377–390.

Montoya J.P. (1990) *Natural Abundance of ^{15}N in Marine and Estuarine Plankton: Studies of Biological Isotopic Fractionation and Plankton Processes*. Harvard University, Cambridge.

Montoya J.P., Horrigan S.G. & McCarthy J.J. (1991) Rapid, storm-induced changes in the natural abundance of ^{15}N in a planktonic ecosystem. *Geochimica et Cosmochimica Acta*, **55**, 3627–3638.

Montoya J.P., Wiebe P.H. & McCarthy J.J. (1992) Natural abundance of ^{15}N in particulate nitrogen and zooplankton in the Gulf Stream region and Warm-Core Ring 86A. *Deep-Sea Research*, **39**, 5363–5392.

Mook W.G. & Tan F.C. (1988) Stable carbon isotopes in rivers and estuaries. *Estuarine and Coastal Shelf Science*, **26**, 1–14.

Mook W.G. & Tan F.C. (1991) Stable carbon isotopes in rivers and estuaries. In: Degens E.T., Kempe S. & Richey J.E. (eds) *Biogeochemistry of Major World Rivers, SCOPE Report No. 42*, pp. 245–264. New York.

Mook W.G., Bommerson J.C. & Staverman W.H. (1974) Carbon isotope fractionation between dissolved bicarbonate and gaseous carbon dioxide. *Earth Planetary Science Letters*, **22**, 169–176.

Mooney H.A., Bullock S.H. & Ehleringer J.R. (1989) Carbon isotope ratios of plants of a tropical dry forest in Mexico. *Functional Ecology*, **3**, 137–142.

Moore H. (1974) Isotopic measurement of atmospheric nitrogen compounds. *Tellus*, **26**, 169–174.

Moore H. (1977) The isotopic composition of ammonia, nitrogen dioxide and nitrate in the atmosphere. *Atmosphere and Environment*, **11**, 1239–1243.

Moore K.M., Murray M.L. & Schoeninger M.J. (1989) Dietary reconstruction from bones treated with preservatives. *Journal of Archaeological Science*, **16**, 437–446.

Moore P.D. (1983) Photosynthetic pathways in aquatic plants. *Nature*, **304**, 310.

Morecroft M.D. & Woodward F.I. (1990) Experimental investigations on the environmental determination of $\delta^{13}C$ at different altitudes. *Journal of Experimental Botany*, **41**, 1303–1308.

Moriarty D.J.W. & O'Donohue M.J. (1993) Nitrogen fixation in seagrass communities during summer in the gulf of Carpentaria. *Australian Journal of Freshwater and Marine Research*, **44**, 117–125.

Moroney J.V., Kitayama M., Togasaki R.K. & Tolbert N.E. (1987) Evidence for inorganic carbon transport by intact chloroplasts of *Chlamydomonas reinhardtii*. *Plant Physiology*, **83**, 460–463.

Morris I. (1980) Pathways of carbon assimilation in marine phytoplankton. In: Falkowski P.G. (ed.) *Primary Production in the Sea*, pp. 139–159. Plenum Press, New York.

Moulton-Barrett R., Triadafilopoulos G., Michener R. & Gologorsky D. (1993) Serum ^{13}C-bicarbonate in the assessment of gastric *Helicobacter pylori* urease activity. *Am J Gastroent*, **88(3)**, 369–374.

Mullin M.M., Rau G.H. & Eppley R.W. (1984) Stable nitrogen isotopes in zooplankton: some geographic and temporal variations in the North Pacific. *Limnology and Oceanography*, **29**, 1267–1273.

Munoz J. & Merrett M.J. (1989) Inorganic-carbon transport in some marine eukaryotic microalgae. *Planta*, **178**, 450–455.

Myers E.P. (1974) The concentration and isotopic composition of carbon in marine sediments affected by sewage discharge. PhD Thesis. California Institute of Technology, Pasadena.

Myrold D.D. (1988) Denitrification in ryegrass and winter wheat cropping systems of Western Oregon. *Soil Sciences Society of America Journal*, **52**, 412–416.

Myrold D.D. (1990) Measuring denitrification in soils using ^{15}N techniques. In: Revsbech N.P. & Sorensen J. (eds) *Denitrification in Soil and Sediment*, pp. 181–198. Plenum Press, New York.

Myrold D.D. & Tiedje J.M. (1986) Simultaneous estimation of several nitrogen cycle rates using ^{15}N: theory and application. *Soil Biology and Biochemistry*, **18**, 559–568.

Nadelhoffer K.J. & Fry B. (1988) Controls on natural nitrogen-15 and carbon-13 abundances in forest soil organic matter. *Soil Science Society of America Journal*, **52**, 1633–1640.

Nagashima K. & Suzuki S. (1984) Solid-state electrochemical detector for carbon monoxide at sub-ppm concentrations. *Analytica Chimica Acta*, **162**, 153–159.

Nagy K.A. (1989) Doubly-labeled water studies of vertebrate physiological ecology. In: Rundel P.W., Ehleringer J.R. & Nagy K.A. (eds) *Stable Isotopes in Ecological Research*, pp. 270–287. Springer-Verlag, New York.

Naqvi S.W.A. (1991) N_2O production in the ocean. *Nature*, **349**, 373–374.

Naqvi S.W.A. & Noronha R.J. (1991) Nitrous oxide in the Arabian Sea. *Deep-Sea Research*, **38**, 871–890.

NASA/WMO (1985) Tropospheric trace gases: sources, distributions, and trends. In: *Atmospheric Ozone 1985: Assessment of Our Understanding of the Processes Controlling Its Distribution and Change*, pp. 57–116. World Meteorological Organization Global Ozone Research and Monitoring Report No. 16. National Aeronautics and Space Administration, Earth Science and Applications Division, Code EE, Washington, DC.

Neftel A., Oeschger H., Schwander J., Stauffer B. & Zumbrunn R. (1982) Ice core sample measurements give atmospheric CO_2 content during the past 40 000 years. *Nature*, **295**, 216–219.

Neftel A., Moor E., Oeschger H. & Stauffer B. (1985) Evidence from polar ice cores for the increase in atmospheric CO_2 in the past two centuries. *Nature*, **315**, 45–47.

Neftel A., Oeschger H., Staffelbach T. & Stauffer B. (1988) CO_2 record in the Byrd ice core 50 000–5000 B.P. *Nature*, **331**, 609–611.

Nelson B.K., DeNiro M.J., Schoeninger M.J., DePaolo D.J. & Hare P.E. (1986) Effects of diagenesis on strontium, carbon, nitrogen, and oxygen concentration and isotopic composition of bone. *Geochimica et Cosmochimica Acta*, **50**, 1941–1949.

Neuberger A. & Richards F.F. (1964) Protein turnover in mammalian tissues. In: Munro H.N. & Allison J.B. (eds) *Mammalian Protein Metabolism*, pp. 243–290. Academic Press, New York.

Nevins J.L., Altabet M.A. & McCarthy J.J. (1985) Nitrogen isotope ratio analysis of small samples: sample preparation and calibration. *Analytical Chemistry*, **57**, 2143–2145.

Nichols P.D., Klumpp D.W. & Johns R.B. (1985) A study of food chains in seagrass communities III. Stable carbon isotope ratios. *Australian Journal of Marine and Freshwater Research*, **36**, 683–690.

Nishio T., Koike I. & Hattori A. (1982) Denitrification, nitrate reduction, and oxygen consumption in coastal and estuarine sediments. *Applied and Environmental Microbiology*, **43**, 648–653.

Nishio T., Koike I. & Hattori A. (1983) Estimates of denitrification and nitrification in coastal and estuarine sediments. *Applied and Environmental Microbiology*, **45**, 444–450.

Nissenbaum A. (1973) The organic geochemistry of marine and terrestrial humic substances: implications of carbon and hydrogen isotope studies. *Advances in Organic Geochemistry 1973*, pp. 39–52.

Noe-Nygaard N. (1988) δ^{13}C-values of dog bones reveal the nature of changes in man's food resources at the Mesolithic–Neolithic transition, Denmark. *Chemical Geology (Isotope Geoscience Section)*, **73**, 87–96.

Northfelt D.W., DeNiro M.J. & Epstein S. (1981) Hydrogen and carbon isotopic ratios of cellulose nitrate and saponifiable lipid fractions prepared from annual growth rings of California redwood. *Geochimica et Cosmochimica Acta*, **45**, 1895–1898.

Nriagu J. & Harvey H.H. (1978) Isotopic variation as an index of sulphur pollution in lakes around Sudbury, Ontario. *Nature*, **273**, 223–224.

Nriagu J. & Coker R.D. (1983) Sulphur in sediments chronicles past changes in lake acidification. *Nature*, **303**, 692–694.

Nriagu J. & Soon Y.K. (1985) Distribution and isotopic composition of sulfur in lake sediments of northern Ontario. *Geochimica et Cosmochimica Acta*, **49**, 823–834.

Nriagu J., Coker R.D. & Barrie L.A. (1991) Origin of sulphur in Canadian Arctic haze from isotope measurements. *Nature*, **349**, 142–145.

Nydal R. & Lovseth K. (1983) Tracing bomb ^{14}C in the atmosphere 1962–1980. *Journal of Geophysical Research*, **88**, 3621–3642.

O'Brien B.J. (1984) Soil organic carbon fluxes and turnover rates estimated from radiocarbon enrichments. *Soil Biology and Biochemistry*, **16**, 115–120.

Ocampo R., Callot H.J., Albrecht P., Popp B.N., Horowitz M.R. & Hayes J.M. (1989) Different isotopic compositions of C_{32} etioporphyrin II in oil shale. Origin of etioporphyrin II from heme? *Naturwissenschaften*, **76**, 419–421.

Ockelmann K.W. & Vahl O. (1970) On the biology of the polychaete *Glycera alba*, especially its burrowing and feeding. *Ophelia*, **8**, 275–294.

O'Leary M.H. (1976) Carbon isotope effect on the enzymatic decarboxylation of pyruvic acid. *Biochemical Biophysical Research Communications*, **73**, 614–618.

O'Leary M.H. (1981) Carbon isotope fractionations in plants. *Phytochemistry*, **20**, 553–567.

O'Leary M.H. (1984) Measurement of the isotope fractionation associated with diffusion of carbon dioxide in aqueous solution. *Journal of Physical Chemistry*, **88**, 823–825.

O'Leary M.H. (1988a) Carbon isotopes in photosynthesis. *BioScience*, **38**, 328–336.

O'Leary M.H. (1988b) Transition state structures in enzyme-catalyzed decarboxylations. *Accounts of Chemical Research*, **21**, 450–455.

O'Leary M.H., Rife J.E. & Slater J.D. (1981) Kinetic and isotope effect studies of maize phosphoenolpyruvate carboxylase. *Biochemistry*, **20**, 7308–7314.

Olson R.J. (1981) ^{15}N tracer studies of the primary nitrite maximum. *Journal of Marine Research*, **39**, 203–226.

Oren A. & Blackburn T.H. (1979) Estimation of sediment denitrification rates at *in situ* nitrate concentrations. *Applied and Environmental Microbiology*, **37**, 174–176.

Osmond C.B. & Holtum J.A.M. (1981) Crassulacean acid metabolism. In: Hatch M.D. & Boardman N.K. (eds) *The Biochemistry of Plants*, pp. 283–328. Academic Press, London.

Osmond C.B., Bender M.M. & Burris R.H. (1976) Pathways of CO_2 fixation in the CAM plant *Kalanchoe daigremontiana*. III. Correlation with $\delta^{13}C$ value during growth and water stress. *Australian Journal of Plant Physiology*, **3**, 787–799.

Osmond C.B., Valaane N., Haslam S.M., Uotila P. & Roksandic Z. (1981) Comparisons of $\delta^{13}C$ values in leaves of aquatic macrophytes from different habitats in Britain and Finland; some implications for photosynthetic processes in aquatic plants. *Oecologia*, **50**, 117–124.

Ostrom N.E. & Macko S.A. (1991a) Sources, cycling and distribution of water column particulate and sedimentary organic matter in northern Newfoundland fjords and bays: a stable isotope study. In: Whelan J.K. & Farrington J.W. (eds) *Organic Matter: Productivity, Accumulation and Preservation in Recent and Ancient Sediments*, pp. 55–81. Columbia University Press, New York.

Ostrom N.E. & Macko S.A. (1991b) Late Wisconsinan to present sedimentation of organic matter off northern Newfoundland in response to climatological events. *Continental Shelf Research*, **11**, 1285–1296.

Ostrom P.H., Macko S.A., Engel M.H., Silfer J.A. & Russell D. (1990) Geochemical characterization of high molecular weight material isolated from Late Cretaceous fossils. *Organic Geochemistry*, **16**, 1139–1144.

Oudot C., Andrie C. & Montel Y. (1990) Nitrous oxide production in the tropical Atlantic Ocean. *Deep-Sea Research*, **37**, 183–202.

Owens N.J.P. (1985) Variations in the natural abundance of ^{15}N in estuarine suspended particulate matter: a specific indicator of biological processing. *Estuarine and Coastal Shelf Science*, **20**, 505–510.

Owens N.J.P. (1987) Natural variations in ^{15}N in the marine environment. *Advances in Marine Biology*, **24**, 389–451.

Paneth P. & O'Leary M.H. (1985) Carbon isotope effect on dehydration of bicarbonate ion catalysed by carbonic anhydrase. *Biochemistry*, **24**, 5143–5147.

Park R. & Epstein S. (1961) Metabolic fractionation of ^{13}C and ^{12}C in plants. *Plant Physiology*, **36**, 133–138.

Parker P.L. (1962) The isotopic composition of the carbon of fatty acids. Carnegie Institution of Washington Annual Report 1961–1962, 187–190.

Parker P.L. (1964) The biogeochemistry of the stable isotopes of carbon in a marine bay. *Geochimica et Cosmochimica Acta*, **28**, 1155–1164.

Patel B.N. & Merrett M.J. (1986) Inorganic-carbon uptake by the marine diatom *Pheodactylum tricornutum*. *Planta*, **169**, 222–227.

Patriquin D. & Knowles R. (1972) Nitrogen fixation in the rhizosphere of marine angiosperms. *Marine Biology*, **16**, 49–58.

Paull C.K., Jull A.J.T., Toolin L.J. & Linick T. (1985) Stable isotope evidence for chemosynthesis in an abyssal seep community. *Nature*, **317**, 709–711.

Payne W.J. (1973) Reduction of nitrogenous oxide by microorganisms. *Bacteriology Reviews*, **37**, 409–452.

Pearce A.J., Stewart M.K. & Sklash M.G. (1986) Storm runoff generation in humid headwater catchments, 1. Where does the water come from? *Water Resources Research*, **22**, 1263–1272.

Pearcy R.W. & Pfitsch W.A. (1991) Influence of sunflecks on the $\delta^{13}C$ of *Adenocaulon bicolor* plants occurring in contrasting forest understory microsites. *Oecologia*, **86**, 457–462.

Pearcy W.G. & Stuiver M. (1983) Vertical transport of carbon-14 into deep-sea food webs. *Deep-Sea Research*, **30**, 427–440.

Pearman G.I., Etheridge D., de Silva F. & Fraser P.J. (1986) Evidence of changing concentrations of atmospheric CO_2, N_2O and CH_4 from air bubbles in Antarctic ice. *Nature*, **320**, 248–250.

Peterson B.J. & Howarth R.W. (1983) Sulfur and carbon isotopes as tracers of organic matter flow in salt marshes. *Estuaries*, **6**, 305.

Peterson B.J. & Fry B. (1987) Stable isotopes in ecosystem studies. *Annual Review of Ecology and Systematics*, **18**, 293–320.

Peterson B.J., Fry B., Hullar M. & Saupe S. (1994) The distribution and stable carbon isotopic composition of dissolved organic carbon in estuaries. *Estuaries* (in press).

Peterson B.J. & Howarth R.W. (1987) Sulfur, carbon and nitrogen isotopes used to trace organic matter flow in the salt-marsh estuaries of Sapelo Island, Georgia. *Limnology and Oceanography*, **32**, 1195–1213.

Peterson B.J., Howarth R.W. & Garitt R.H. (1985) Multiple stable isotopes used to trace the flow of organic matter in estuarine food webs. *Science*, **227**, 1361–1363.

Peterson D.H. (ed.) (1989) *Aspects of Climate Variability in the Pacific and the Western Americas*. Geophysical Monograph 55, American Geophysical Union, Washington, DC.

Piepenbrink H. (1989) Examples of chemical changes during fossilisation. *Applied Geochemistry*, **4**, 273–280.

Pierotti D. & Rasmussen R.A. (1980) Nitrous oxide measurements in the eastern tropical Pacific Ocean. *Tellus*, **32**, 56–72.

Pollock W., Heidt L.E., Lueb R. & Ehhalt D. (1980) Measurement of stratospheric water vapor by cryogenic collection. *Journal of Geophysical Research*, **85**, 5555–5568.

Poth M. & Focht D.D. (1985) ^{15}N kinetic analysis of N_2O production by *Nitrosomonas europaea*: an examination of nitrifier denitrification. *Applied and Environmental Microbiology*, **49**, 1134–1141.

Potts M., Krumbein W. & Metzger J. (1978) Nitrogen fixation rates in anaerobic sediments determined by acetylene reduction, a new ^{15}N field assay and simultaneous N determination. In: *Environmental Biogeochemistry and Geomicrobiology*, vol. 3, pp. 753–769. Ann Arbor Scientific Publications, Ann Arbor.

Price G.D. & Badger M.R. (1989a) Ethoxyzolamide inhibition of CO_2 uptake in the cyanobacterium *Synechococcus* PCC7942 without apparent inhibition of internal carbonic anhydrase activity. *Plant Physiology*, **89**, 37–43.

Price G.D. & Badger M.R. (1989b) Ethoxyzolamide inhibition of CO_2 dependent photosynthesis in the cyanobacterium *Synechococcus* PCC7942. *Plant Physiology*, **89**, 44–50.

Prinn, R., Cunnold D. & Rasmussen R. *et al.* (1990) Atmospheric emissions and trends of nitrous oxide deduced from 10 years of ALE-GAGE data. *Journal of Geophysical Research*, **95**(D11) 18369–18385.

Prins H.B.A. & Elzenga J.T.M. (1989) Bicaronate utilization: function and mechanism. *Aquatic Botany*, **34**, 59–83.

Quade J., Cerling T.E. & Bowman J.R. (1989) Development of Asian monsoon revealed by marked ecological shift during the latest Miocene in northern Pakistan. *Nature*, **342**, 163–166.

Quade J., Cerling T.E., Barry J.C. *et al.* (1992) A 16-Ma record of paleodiet using carbon and oxygen isotopes in fossil teeth from Pakistan. *Chemical Geology (Isotope Geoscience Section)*, **94**, 183–192.

Quay P., King S.L., Lansdown J.M. & Wilbur D.O. (1988) Isotopic composition of methane released from wetlands: implications for the increase in atmospheric methane. *Global Biogeochemical Cycles*, **2**, 385–397.

Quay P.D., King S.L., Wilbur D.O. *et al.* (1991) Carbon isotopic composition of atmospheric CH_4: fossil and biomass burning source strengths. *Global Biogeochemical Cycles*, **5**, 25–47.

Rashid G.H. (1977) The volatilization losses of nitrogen from added urea in some soils of Bangladesh. *Plant and Soil*, **48**, 549–556.

Rasmussen R.A. & Khalil M.A.K. (1981a) Increase in the concentration of atmospheric methane. *Atmosphere and Environment*, **15**, 883–886.

Rasmussen R.A. & Khalil M.A.K. (1981b) Atmospheric methane: trends and seasonal cycles. *Journal of Geophysical Research*, **86**, 9826–9832.

Rau G.H. (1980) Carbon-13/Carbon-12 variation in subalpine lake aquatic insects: food source implications. *Canadian Journal of Fisheries and Aquatic Science*, **37**, 742–745.

Rau G.H. (1981a) Hydrothermal vent clam and tube worm $^{13}C/^{12}C$: further evidence of nonphotosynthetic food sources. *Science*, **213**, 338–340.

Rau G.H. (1981b) Low $^{15}N/^{14}N$ in hydrothermal vent animals: ecological implications. *Science*, **289**, 284–285.

Rau G.H. (1985) $^{13}C/^{12}C$ and $^{15}N/^{14}N$ in hydrothermal vent organisms: ecological and biogeochemical implications. *Biological Society of Washington Bulletin*, **6**, 243–247.

Rau G.H. & Hedges J.I. (1979) Carbon-13 depletion in a hydrothermal vent mussel: suggestion of a chemosynthetic food source. *Science*, **203**, 648–649.

Rau G.H. & Anderson N.H. (1981) Use of $^{13}C/^{12}C$ to trace dissolved and particulate organic matter utilization by populations of an aquatic invertebrate. *Oecologia*, **48**, 19–21.

Rau G.H., Sweeney R.E., Kaplan I.R., Mearns A.J. & Young D.R. (1981) Differences in animal ^{13}C, ^{15}N and D abundance between a polluted and an unpolluted coastal site: likely indicators of sewage uptake by a marine food web. *Estuarine and Coastal Shelf Science*, **13**, 701–707.

Rau G.H., Sweeney R.E. & Kaplan I.R. (1982) Plankton $^{13}C:^{12}C$ ratio changes with latitude: differences between northern and southern oceans. *Deep-Sea Research*, **29**, 1035–1039.

Rau G.H., Mearns A.J., Young D.R., Olson R.J., Schafer H.A. & Kaplan I.R. (1983) Animal $^{13}C/^{12}C$ correlates with trophic level in pelagic food webs. *Ecology*, **64**, 1314–1318.

Rau G.H., Takahashi T. & Marais D.J.D. (1989) Latitudinal variations in plankton $\delta^{13}C$: implications for CO_2 and productivity in past oceans. *Nature*, **341**, 516–518.

Rau G.H., Teyssie J.L., Rassoulzadegan F. & Fowler S.W. (1990) $^{13}C/^{12}C$ and $^{15}N/^{14}N$ variations among size-fractionated marine particles: implications for their origin and trophic relationships. *Marine Ecology Progress Series*, **59**, 33–38.

Rau G.H., Takahashi T., Marais D.J.D., Repeta D.J. & Martin J. (1992) The relationship between organic matter $\delta^{13}C$ and $[CO_2(aq)]$ in ocean surface water: data from a JGOFS site in Northeast Atlantic Ocean and a Model. *Geochimica et Cosmochimica Acta*, **56**, 1413–1419.

Raven J.A. (1970) Exogenous inorganic carbon sources in plant photosynthesis. *Biological Reviews*, **45**, 167–221.

Raven J.A. (1985) The CO_2 concentrating mechanism. In: Lucas W.J. & Berry J.A. (eds) *Inorganic Carbon Uptake by Aquatic Photosynthetic Organisms*, pp. 67–82. American Society of Plant Physiologists, Rockville, Maryland.

Raven J.A. (1987) The application of mass spectrometry to biochemical and physiological studies. *The Biochemistry of Plants*, **13**, 127–180.

Raven J.A. (1990) Use of isotopes in estimating respiration and photorespiration in micro-algae. *Marine Microbial Food Webs*, **4**, 59–86.

Raven J.A., Beardall J. & Griffiths H. (1982) Inorganic C sources for *Lemanea*, *Cladophora* and *Ranunculus* in a fast-flowing stream: measurements of gas exchange and of carbon isotope ratio and their ecological implications. *Oecologia*, **53**, 68–78.

Raven J.A., McFarlane J.J. & Griffiths H. (1987) The application of carbon isotope discrimination techniques. In: Crawford R.M. (ed.) *Plant Life in Aquatic and Amphibious Habitats*, pp. 129–149. Blackwell Scientific Publications, Oxford.

Raven J.A., Johnston A.M., Handley L.L. & McInroy S.G. (1990) Transport and assimilation of inorganic carbon by *Lichina pygmaea* under emersed and submersed conditions. *New Phytology*, **114**, 407–417.

Read J.J., Johnson D.A., Asay K.H. & Tieszen L.L. (1991) Carbon isotope discrimination, gas exchange, and water-use efficiency in crested wheatgrass clones. *Crop Science*, **31**, 1203–1208.

Rees C.E., Jenkins W.J. & Monster J. (1978) The sulphur isotopic composition of ocean water sulphate. *Geochimica et Cosmochimica Acta*, **42**, 377–381.

Reiskind J.B., Seamon P.T. & Bowes G. (1988) Alternative methods of photosynthetic carbon assimilation in marine macroalgae. *Plant Physiology*, **87**, 686–692.

Reiskind J.B., Beer S. & Bowes G. (1989) Photosynthesis, photorespiration and ecophysiological interactions in marine macroalgae. *Aquatic Botany*, **34**, 131–152.

Remde A. & Conrad R. (1990) Production of nitric oxide in *Nitrosomonas europaea* by reduction of nitrite. *Archives of Microbiology*, **154**, 187–191.

Rendina A.R., Hermes J.D. & Cleland W.W. (1984) Use of multiple isotope effects to study the mechanisms of 6-phosphogluconate dehydrogenase. *Biochemistry*, **23**, 6257–6262.

Rennie D.A., Paul E.A. & Johns I.E. (1976) Natural nitrogen-15 abundance of soil and plant samples. *Canadian Journal of Soil Science*, **56**, 43–50.

Rensberger J.M. (1986) Early chewing mechanisms in mammalian herbivores. *Paleobiology*, **12**, 474–494.

Retallick G. (1984) Completeness of the rock and fossil record: some estimates using fossil soils. *Paleobiology*, **10**, 59–78.

Revsbech N.P., Nielsen L.P., Christensen P.B. & Sørensen J. (1988) A combined oxygen and nitrous oxide microsensor for studies of nitrification. *Applied and Environmental Microbiology*, **54**, 2245–2249.

Rey C., Renugopalakrishnan V., Shimizu M., Collins B. & Glimcher M. (1991) A resolution-enhanced Fourier transform infrared spectroscopic study of the environment of the CO_3^{2-} ion in the mineral phase of enamel during its formation and maturation. *Calcified Tissue International*, **49**, 259–268.

Richards F.A. & Benson B.B. (1961) Nitrogen/argon and nitrogen isotope ratios in two anaerobic environments, the Cariaco Trench in the Caribbean Sea and Dramsfjord, Norway. *Deep-Sea Research*, **7**, 254–264.

Rieley G., Collier R.J., Jones D.M., Eglinton G., Eakina P.A. & Fallick E. (1991) Sources of sedimentary lipids deduced from stable carbon isotope analyses of individual compounds. *Nature*, **352**, 425–427.

Riga A., Van Praag H.J. & Brigodi N. (1971) Rapport isotopique natural de l'azote dans quelques sols forestiers et agricoles de Belgique soumis à divers traitements culturaux. *Geoderma*, **6**, 213–222.

Ritchie G.A.F. & Nicholas D.J.D. (1972) Identification of the sources of nitrous oxide produced by oxidative and reductive processes in *Nitrosomonas europaea*. *Biochemical Journal*, **126**, 1181–1191.

Ritchie G.A.F. & Nicholas D.J.D. (1974) The partial characterization of purified nitrite reductase and hydroxylamine oxidase from *Nitrosomonas europaea*. *Biochemical Journal*, **128**, 471–480.

Rodhe H. (1990) A comparison of the contribution of various gases to the greenhouse effect. *Science*, **248**, 1217–1219.

Roeske C.A. & O'Leary M.H. (1984) Carbon isotope effects on the enzyme-catalyzed carboxylation of ribulose biphospate. *Biochemistry*, **23**, 6275–6285.

Roeske C.A. & O'Leary M.H. (1985) Carbon isotope effect on carboxylation of ribulose bisphospate catalyzed by ribulose bisphospate carboxylase from *Rhodospirillum rubrum*. *Biochemistry*, **24**, 1603–1607.

Ronner U. & Sorensson F. (1985) Denitrification rates in the low-oxygen waters of the stratified Baltic proper. *Applied and Environmental Microbiology*, **50**, 801–806.

Rosswall T. (1983) The nitrogen cycle. In: Bolin B. & Cook R.B. (eds) *The Major Biogeochemical Cycles and their Interactions*, pp. 46–50. John Wiley & Sons, New York.

Rotty R.M. & Marland G. (1984) *Production of CO_2 from Fossil Fuel Burning by Fuel Type, 1860–1982*. Report NDP-006, Carbon Dioxide Information Center, Oak Ridge National Laboratory.

Rounick J.S. & Winterbourn M.J. (1986) Stable carbon isotopes and carbon flow in ecosystems. *BioScience*, **36**, 171–177.

Rubin M., Likins R.C. & Berry E.G. (1963) On the validity of radiocarbon dates from snail shells. *Journal of Geology*, **71**, 84–89.

Ruby E.G., Wirsen C.O. & Jannasch H.W. (1981) Chemolithotrophic sulfur-oxidizing bacteria from the Galapagos rift hydrothermal vents. *Applied and Environmental Microbiology*, **42**, 317–324.

Ruby E.G., Jannasch H.W. & Deuser W.G. (1987) Fractionation of stable carbon isotopes during chemoautotrophic growth of sulfur-oxidizing bacteria. *Applied and Environmental Microbiology*, **53**, 1940–1943.

Rust F.E. (1981) $\delta^{13}C$ of ruminant methane and its relationship to atmospheric methane. *Science*, **211**, 1044–1046.

Rust F.E. & Stevens C.M. (1980) Carbon kinetic effect in the oxidation of methane by hydroxyl. *International Journal of Chemical Kinetics*, **12**, 371–377.

Rustad L.E., Fernandez I.J., Fuller R.D., David M.B., Halteman W.A. & Nodvin S.C. (1993) Soil solution response to acidic deposition in a northern hardwood forest. *Agriculture, Ecosystems and the Environment*, **47**, 117–134.

Sackett W.M. (1964) The depositional history and isotopic organic carbon composition of marine sediments. *Marine Geology*, **2**, 173–185.

Sackett W.M. (1989) Stable carbon isotope studies on organic matter in the marine environment. In: Fritz P. & Fontes J.C. (eds) *Handbook of Environmental Isotope Geochemistry*, pp. 139–169. Elsevier, Amsterdam.

Sackett W.M. & Moore W.S. (1966) Isotopic variations of dissolved inorganic carbon.

Chemistry and Geology, **1**, 323–328.

Sackett W.M., Eckelmann W.R. & Bender M.L. (1965) Temperature dependence of carbon isotope composition in marine plankton and sediments. *Science*, **148**, 235–237.

Saino T. & Hattori A. (1980) ^{15}N natural abundance in oceanic suspended particulate matter. *Nature*, **283**, 752–754.

Saino T. & Hattori A. (1985) Variation of ^{15}N natural abundance of suspended organic matter in shallow oceanic waters. In: Sigleo A.C. & Hattori A. (eds) *Marine and Estuarine Geochemistry*, pp. 1–13. Lewis Publishers, Chelsea, Michigan.

Saino T. & Hattori A. (1987) Geographical variation of the water column distribution of suspended particulate organic nitrogen and its ^{15}N natural abundance in the Pacific and its marginal seas. *Deep-Sea Research*, **34**, 807–827.

Saltzman E.S., Brass G.W. & Price D.A. (1983) The mechanism of sulfate aerosol formation: chemical and sulfur isotopic evidence. *Geophysical Research Letters*, **10**, 513–516.

Sand-Jensen K. (1987) Environmental control of bicarbonate use among freshwater and marine macrophytes. In: Crawford R.M. (ed.) *Plant Life in Aquatic and Amphibious Habitats*, pp. 99–122. Blackwell Scientific Publications, Oxford.

Sarkar A., Bhattacharya S.K. & Mohabey D.M. (1991) Stable-isotope analyses of dinosaur eggshells: paleoenvironmental implications. *Geology*, **19**, 1068–1071.

Saupe S.M., Schell D.M. & Griffiths W. (1989) Carbon isotope ratio gradients in western arctic zooplankton. *Marine Biology*, **103**, 427–433.

Schaffner F.C. & Swart P.K. (1991) Influence of diet and environmental water on the carbon and oxygen isotopic signatures of seabird eggshell carbonate. *Bulletin of Marine Science*, **48**, 23–38.

Schell D.M. (1987) Bowhead whale feeding: allocation of regional habitat importance based on stable isotope abundances. In: Richardson W.J. (ed.) *Importance of the Eastern Alaskan Beaufort Sea to Feeding Bowhead Whales, 1985–1896*. Minerals Management Service Report 87-0037. NTIS PB88-150271/AF.

Schell D.M., Saupe S.M. & Haubenstock N. (1989a) Bowhead whale (*Balaena mysticetus*) growth and feeding as estimated by ^{12}C techniques. *Marine Biology*, **103**, 433–443.

Schell D.M., Saupe S.M. & Haubenstock N. (1989b) Natural isotope abundances baleen: markers of aging and habitat usage. *Ecological Studies*, **68**, 260–269.

Schidlowski M. (1987) Application of stable carbon isotopes to early biochemical evolution on Earth. *Annual Reviews of Earth Planetary Science*, **15**, 47–72.

Schiegl W.-G. (1970) Natural deuterium in biogenic materials. Influence of environment and geophysical applications. PhD Thesis, University of South Africa, Pretoria.

Schiegl W.E. & Vogel J.C. (1970) Deuterium content of organic matter. *Earth Planetary Science Letters*, **7**, 307–313.

Schimel J.P., Jackson L.E. & Firestone M.K. (1990) Spatial and temporal effects on plant–microbial competition for inorganic nitrogen in a California annual grassland. *Soil Biology and Biochemistry*, **21**, 1059–1066.

Schindler D.W. (1985) The coupling of elemental cycles by organisms: evidence from whole-lake chemical perturbations. In: Stumm W. (ed.) *Chemical Processes in Lakes*, pp. 225–260. John Wiley & Sons, New York.

Schmaljohann R., Faber E., Whiticar M.J. & Dando P.R. (1990) Co-existence of methane- and sulfur-based endosymbioses between bacteria and invertebrates at a site in the Skagerrak. *Marine Ecology Progress Series*, **61**, 119–124.

Schmidt H.L., Winkler F.J., Latzko E. & Wirth E. (1978) ^{13}C-kinetic isotope effects in photosynthetic carboxylation reactions and $\delta^{13}C$-values of plant material. *Israel Journal of Chemistry*, **17**, 223–224.

Schmidt J., Seiler W. & Conrad R. (1988) Emission of nitrous oxide from temperate forest soils into the atmosphere. *Journal of Atmospheric Chemistry*, **6**, 95–115.

Schmitt M., Faber E., Botz R. & Stoffers P. (1991) Extraction of methane from seawater using ultrasonic vacuum degassing. *Analytical Chemistry*, **63**, 529–532.

Schoell M. (1980) The hydrogen and carbon isotopic composition from natural gases of various origin. *Geochimica et Cosmochimica Acta*, **44**, 649–661.

Schoenheimer R. & Rittenberg D. (1939) Studies in protein metabolism. *Journal of Biological Chemistry*, **127**, 285–344.

Schoeninger M.J. (1982) Diet and the evolution of modern human form in the Middle East. *American Journal of Physical Anthropology*, **58**, 37–52.

Schoeninger M.J. (1985) Trophic level effects on $^{15}N/^{14}N$ and $^{13}C/^{12}C$ ratios in bone collagen and strontium levels in bone mineral. *Journal of Human Evolution*, **14**, 515–525.

Schoeninger M.J. & DeNiro M.J. (1982) Carbon isotope ratios of apatite from fossil bone cannot be used to reconstruct diets of animals. *Nature*, **297**, 577–578.

Schoeninger M.J. & DeNiro M.J. (1984) Nitrogen and carbon isotopic composition of bone collagen from marine and terrestrial animals. *Geochimica et Cosmochimica Acta*, **48**, 625–639.

Schubert K.R. & Evans H.J. (1976) Hydrogen evolution: a major factor affecting the efficiency of nitrogen fixation in nodulated symbionts. *Proceedings of the National Academy of Sciences USA*, **73**, 1207–1211.

Schultz D.J. (1974) Stable carbon isotope variation in organic and inorganic carbon reservoirs in the Fenoholloway River estuary and the Mississippi River estuary. PhD Thesis. Florida State University.

Schulze E.-D., Gebauer G., Ziegler H. & Lange O.L. (1991) Estimates of nitrogen fixation by trees on an andity gradient in Namibia. *Oecologia*, **88**, 451–455.

Schuphan W. (1974) Significance of nitrates in food and drinking water. In: *Effects of Agricultural Production on Nitrates in Food and Water with Particular Reference to Isotope Studies*, pp. 101–116. IAEA, Vienna.

Schutze M. (1944) A new oxidant for quantitative conversion of carbon monoxide to carbon dioxide. A contribution to the chemistry of iodine pentoxide. *Berichte der Deutschen Chemischen Gesellschaft*, **77B**, 484–487.

Schwarcz H.P., Melbye J., Katzenberg M.A. & Knyf M. (1985) Stable isotopes in human skeletons of southern Ontario: reconstructing palaeodiet. *Journal of Archaeological Science*, **12**, 187–206.

Scranton M. (1983) Gaseous nitrogen compounds in the marine environment. In: Carpenter E.J. & Capone D.G. (eds) *Nitrogen in the Marine Environment*, pp. 37–64. Academic Press, New York.

Sealy J.C. & van der Merwe N.J. (1985) Isotope assessment of Holocene human diets in the southwestern Cape, South Africa. *Nature*, **315**, 138–140.

Sealy J.C., van der Merwe N.J., Lee-Thorp J.A. & Lanham J.L. (1987) Nitrogen isotopic ecology in southern Africa: implications for environmental and dietary tracing. *Geochimica et Cosmochimica Acta*, **51**, 2707–2717.

Sealy J.C., van der Merwe N.J., Sillen A., Kruger F.J. & Krueger H.W. (1991) $^{87}Sr/^{86}Sr$ as a dietary indicator in modern and archaeological bone. *Journal of Archaeological Science*, **18**, 399–416.

Seiler W. (1974) The cycle of atmospheric CO. *Tellus*, **26**, 116–135.

Seiler W. & Conrad R. (1987) Contribution of tropical ecosystems to the global budgets of trace gases, especially CH_4, H_2, CO, and N_2O. In: Dickinson R.E. (ed.) *Geophysiology of Amazonia*, pp. 133–162. John Wiley & Sons, New York.

Seitzinger S.P. (1988) Denitrification in freshwater and coastal marine ecosystems: ecological and geochemical significance. *Limnology and Oceanography*, **33**, 702–724.

Seitzinger S.P. & Garber J.H. (1987) Nitrogen fixation and $^{15}N_2$ calibration of the acetylene reduction assay in coastal marine sediment. *Marine Ecology Progress Series*, **37**, 65–73.

Seitzinger S.P., Nixon S.W. & Pilson M.E. (1984) Denitrification and nitrous oxide production in a coastal marine ecosystem. *Limnology and Oceanography*, **29**, 73–83.

Serban A., Engel M.H. & Macko S.A. (1988) The distribution, stereochemistry and stable isotopic composition of amino acid constituents of fossil and modern mollusk shells. *Advances in Organic Geochemistry 1987, Organic Geochemistry*, **13**, 1123–1129.

Sexstone A.J., Parkin T.B. & Tiedje J.M. (1985) Temporal response of soil denitrification rates to rainfall and irrigation. *Soil Science Society of America Journal*, **49**, 99–103.

Shaffer G. & Ronner U. (1984) Denitrification in the Baltic proper deep water. *Deep-Sea Research*, **31**, 197–220.

Sharkey T.D. & Berry J.A. (1985) Carbon isotope fractionation of algae as influenced by an inducible CO_2 concentraing mechanism. In: Lucas W.J. & Berry J.A. (eds) *Inorganic Carbon Uptake by Aquatic Photosynthetic Organisms*, pp. 389–401. American Society of Plant Physiology, Rockville.

Shearer G. & Kohl D. (1986) N_2 fixation in field settings, estimations based on natural ^{15}N abundance. *Australian Journal of Plant Physiology*, **13**, 699–757.

Shearer G. & Kohl D. (1989) Estimates of N_2 fixation in ecosystems: the need for and basis of the ^{15}N natural abundance method. In: Rundel R.W., Ehleringer J.R. & Nagy K.A. (eds) *Stable Isotopes in Ecological Research*, pp. 342–374. Springer-Verlag, New York.

Shearer G., Duffy J., Kohl D.H. & Commoner B. (1974a) A steady-state model of isotopic fractionation accompanying nitrogen transformations in soil. *Soil Science Society of America Proceedings*, **38**, 315–322.

Shearer G.B., Kohl D.H. & Commoner B. (1974b) The precision of determinations of the natural abundance of nitrogen-15 in soils, fertilizers, and shelf chemicals. *Soil Science*, **118**, 308–316.

Shearer G.B., Kohl D.H. & Chien S.-H. (1978) The nitrogen-15 abundance in a wide variety of soils. *Soil Science Society of America Journal*, **42**, 899–902.

Shearer G., Kohl D.H. & Virginia R.A. (1983) Estimates of N_2-fixation from variation in the natural abundance of ^{15}N in Sonoran Desert ecosystems. *Oecologia*, **56**, 365–373.

Shemesh A. (1990) Crystallinity and diagenesis of sedimentary apatites. *Geochimica et Cosmochimica Acta*, **54**, 2433–2438.

Shemesh A., Kolodny Y. & Luz B. (1983) Oxygen isotope variation in phosphate of biogenic apatites, II. Phosphorite rocks. *Earth and Planetary Science Letters*, **64**, 405–416.

Sherr E.B. (1982) Carbon isotope composition of organic seston and sediments in a Georgia salt march estuary. *Geochimica et Cosmochimica Acta*, **46**, 1227–1232.

Shimmelmann A. & DeNiro M.J. (1986) Stable isotopic studies on chitin. II. The $^{13}C/^{12}C$ and $^{15}N/^{14}N$ ratios in arthropod chitin. *Contributions to Marine Science*, **29**, 113–130.

Shimmelmann A., DeNiro M.J., Poulicek M., Voss-Foucart A., Goffinet G. & Jeuniaux C. (1986) Stable isotopic composition of chitin from arthropods recovered in archaeological contexts as palaeoenvironmental indicators. *Journal of Archaeological Science*, **13**, 553–566.

Showers W.J. & Angle D.G. (1986) Stable isotopic characterization of organic carbon accumulation on the Amazon continental shelf. *Continental Shelf Research*, **6**, 227–244.

Shuval H.I. & Gruener N. (1974) Effects on man and animals of ingesting nitrates and nitrites in water and food. In: *Effects of Agricultural Production on Nitrates in Food and Water with Particular Reference to Isotope Studies*, pp. 117–130. IAEA, Vienna.

Siegenthaler U. & Oeschger H. (1987) Biospheric CO_2 emissions during the past 200 years reconstructed by deconvolution of ice core data. *Tellus*, **39B**, 140–154.

Silfer J.A., Engel M.H., Macko S.A. & Jumeau E.J. (1991) Stable carbon isotope analysis of amino acid enantiomers by conventional isotope ratio mass spectrometry and combined gas chromatography/isotope ratio mass spectrometry. *Analytical Chemistry*, **63**, 370–374.

Silfer J.A., Engel M.H. & Macko S.A. (1992) Kinetic fractionation of stable carbon and nitrogen isotopes during peptide bond hydrolysis: experimental evidence and geochemical implications. *Chemical Geology (Isotope Geoscience Section)*, **101**, 211–221.

Sillen A. (1986) Biogenic and diagenetic Sr/Ca in Plio–Pleistocene fossils of the Omo Shungura formation. *Paleobiology*, **12**, 311–323.

Sillen A. (1989) Diagenesis of the inorganic phase of cortical bone. In: Price T.D. (ed.) *The Chemistry of Prehistoric Human Bone*, pp. 211–229. Cambridge University Press, New York.

Sillen A. & Kavanaugh M. (1982) Strontium and paleodietary research: a review. *Yearbook of Physical Anthropology*, **25**, 67–90.

Silverman M.P. & Oyama V.I. (1968) Automatic apparatus for sampling and preparing gases for mass spectral analysis in studies of carbon isotope fractionation during methane metabolism. *Analytical Chemistry*, **40**, 1988.

Silyn-Roberts H. & Sharp R.M. (1986) Crystal growth and the role of the organic matrix in eggshell biomineralization. *Proceedings of the Royal Society London B*, **227**, 303–324.

Silyn-Roberts H. & Sharp R.M. (1989) The similarity of preferred orientation development in eggshell calcite of the dinosaurs and birds. *Proceedings of the Royal Society London B*, **235**, 347–363.

Simenstad C.A. & Wissmar R.C. (1985) $\delta^{13}C$ evidence of the origins and fates of organic carbon in estuarine and nearshore food webs. *Marine Ecology Progress Series*, **22**, 141–152.

Simkiss K. (1961) Calcium metabolism and avian reproduction. *Biological Review*, **36**, 321–367.

Smedley M.P., Dawson T.E. & Comstock J.P. *et al.* (1991) Seasonal carbon isotope discrimination in a grassland community. *Oecologia*, **85**, 314–320.

Smith A.J. & Hoare D.S. (1977) Specialist phototrophs, lithotrophs, and methylotrophs: a unity among a diversity of procaryotes? *Bacteriological Reviews*, **41**, 419–448.

Smith B.N., Oliver J. & McMillian C. (1976) Influence of carbon source, oxygen concentration, light intensity, and temperature on $^{13}C/^{12}C$ ratios in plant tissues. *Bot. Gaz.*, **137**, 99–104.

Smith D.F., Bulleid N.C., Campbell R. *et al.* (1979) Marine food-web analysis: an experimental study of demersal zooplankton using isotopically labelled prey species. *Marine Biology*, **54**, 49–59.

Smith J.A.C., Griffiths H. & Lüttge V. (1986) Comparative ecophysiology of CAM and C_3 bromeliads. I. The ecology of the Bromeliaceae in Trinidad. *Plant, Cell, and Environment*, **91**, 359–376.

Smith R.A. (1987) Water-quality trends in the nation's rivers. In: *Geological Survey Yearbook; Fiscal Year 1986*, pp. 82–86. U.S. Government Printing Office, Washington, DC.

Smith R.A., Alexander R.B. & Wolman M.G. (1987) Water-quality trends in the nation's rivers. *Science*, **235**, 1607–1615.

Smith R.G. & Bidwell R.G.S. (1989) Mechanism of photosynthetic carbon dioxide uptake by the red macroalga, *Chondrus crispus*. *Plant Physiology*, **89**, 93–99.

Smith W.K. & Hollinger D.Y. (1988) Stomatal behavior. In: Hinckley T.M. & Lassoie J.P. (eds) *Techniques and Approaches in Tree Ecophysiology*, pp. 30–46. CRC Press, Boca Raton.

Socki R.A., Karlsson H.R. & Gibson E.K. Jr (1992) Extraction technique for the determination of oxygen-18 in water using preevacuated glass vials. *Analytical Chemistry*, **64**, 829–831.

Söderlund R. & Svensson B.H. (1976) The global nitrogen cycle. In: Svensson B.H. & Söderlund R. (eds) *Nitrogen, Phosphorus and Sulphur–Global Cycles, SCOPE Report No. 7*, pp. 23–73. Ecological Bulletins, Stockholm.

Sondergaard M., Riemann B., Jensen L.M. *et al.* (1988) Pelagic food web processes in an oligotrophic lake. *Hydrobiologia*, **164**, 271–286.

Sørensen J. (1978a) Denitrification rates in a marine sediment as measured by acetylene inhibition technique. *Applied and Environmental Microbiology*, **36**, 139–143.

Sørensen J. (1978b) Occurrence of nitric and nitrous oxide in a coastal marine sediment. *Applied and Environmental Microbiology*, **36**, 809–813.

Southward A.J., Southward E.C., Dando P.R., Barrett R.L. & Ling R. (1986) Chemoautotrophic function of bacterial symbionts in small pogonophora. *Journal of the Marine Biology Association UK*, **66**, 415–437.

Soyer J., Soyer-Gobillard M.O., Thiriot-Quiévreux C., Bouvy M. & Cahet G. (1987) Chemoautotrophic bacterial endosymbionts in *Spisula subtruncata* (Bivalvia, Mactridae). Ultrastructure, metabolic significance and evolutionary implications. *Symbiosis*, **3**, 301–314.

Spies R.B., Kruger H., Ireland R. & Rice D.W. (1989) Stable isotope ratios and contaminant concentrations in a sewage distorted food web. *Marine Ecology Progress Series*, **54**, 157–170.

Spiker E.C. & Schemel L.E. (1979) Distribution and stable isotope composition of carbon in San Francisco Bay. In: *San Francisco Bay: The Urbanized Estuary: Proceedings of the 58th Annual Meeting of the Pacific Division/American Association for the Advancement of Science*, San Francisco, June 12–16, 1977, pp. 195–212.

Spiro B., Greenwood P.B., Southward A.J. & Dando P.R. (1986) $^{13}C/^{12}C$ ratios in marine invertebrates from reducing sediments: confirmation of nutritional importance of chemoautotrophic endosymbiotic bacteria. *Marine Ecology Progress Series*, **28**, 233–240.

Stafford T.W. Jr., Brendel K. & Duhamel R. (1988) Radiocarbon, ^{13}C and ^{15}N analysis of fossil bone: removal of humates with XAD-2 resin. *Geochimica et Cosmochimica Acta*, **52**, 2257–2267.

Stafford T.W. Jr., Fogel M.L., Brendel K. & Hare P.E. (1994) Late Quaternary paleoecology of the southern High Plains based on stable nitrogen and carbon isotope analysis of fossil *Bison* collagen. In: Schafer H., Carlson D.L. & Sobolik K.D. (eds) *The Archaic of the southern North American deserts*: Texas A&M Press, College Station, Texas (in press)

Stanford G. & Smith S.J. (1972) Nitrogen mineralization potentials of soils. *Soil Science Society of America Proceedings*, **36**, 465–472.

Stauffer B., Fischer G., Neftel A. & Oeschger H. (1985) Increases in atmospheric methane recorded in Antarctic ice core. *Science*, **229**, 1386–1388.

Steele K.W. & Daniel R.M. (1978) Fractionation of nitrogen isotopes by animals: a further complication to the use of variations in the natural abundance of ^{15}N for tracer studies. *Journal of Agricultural Science*, **90**, 7–9.

Steele L.P., Fraser P.J. & Rasmussen R.A. *et al.* (1987) The global distribution of methane in the troposphere. *Journal of Atmospheric Chemistry*, **5**, 125–171.

Steele L.P., Duglokencky E.J., Lang P.M., Tans P.P. & Martin R.C. (1992) Slowing of the global accumulation of atmospheric methane during the 1980s. *Nature*, **358**, 316.

Stein J.L., Cary S.C., Hessler R.R. *et al.* (1988) Chemoautotrophic symbiosis in a hydrothermal vent gastropod. *Biological Bulletin*, **174**, 373–378.

Stephenson R.L., Tann F.C. & Mann K.H. (1986) Use of stable carbon isotope ratios to compare plant material and potential consumers in a seagrass bed and a kelp bed in Nova Scotia, Canada. *Marine Ecology Progress Series*, **30**, 1–7.

Sternberg L. (1989) Oxygen and hydrogen isotope ratios in plant cellulose: Mechanisms and applications. In: Rundel P.W., Ehleringer J.R. & Nagy K.A. (eds) *Stable Isotopes in Ecological Research*, pp. 124–141. Springer-Verlag, New York.

Sternberg L. & DeNiro M.J. (1983) Isotopic composition of cellulose from C_3, C_4 and CAM plants growing in the vicinity of one another. *Science*, **220**, 947–948.

Sternberg L. & Swart P.K. (1987) Utilization of freshwater and ocean water by coastal plants of southern Florida. *Ecology*, **68**, 1898–1905.

Sternberg L., DeNiro M.J. & Johnson H.B. (1984a) Isotope ratios of cellulose from plants having different photosynthetic pathways. *Plant Physiology*, **74**, 557–561.

Sternberg L.O., DeNiro M.J. & Ting I.P. (1984b) Carbon, hydrogen and oxygen isotope ratios of cellulose from plants having intermediate photosynthetic modes. *Plant Physiology*, **74**, 104–107.

Sternberg L., DeNiro M.J. & Keeley J.E. (1984c) Hydrogen, oxygen and carbon isotope ratios of cellulose from submerged aquatic Crassulacean acid metabolism and non-Crassulacean acid metabolism plants. *Plant Physiology*, **76**, 68–70.

Sternberg L.S.L., DeNiro M.J., McJunkin D., Berger R. & Keeley J.E. (1985) Carbon, oxygen and hydrogen isotope abundances in *Stylites* reflect its unique physiology. *Oecologia*, **67**, 598–600.

Sternberg L.S.L., DeNiro M.J. & Johnson H.B. (1986) Oxygen and hydrogen isotope ratios of water from photosynthetic tissues of CAM and C_3 plants. *Plant Physiology*, **82**, 428–431.

Sternberg L.S.L., Mulkey S.S. & Wright S.J. (1989) Ecological interpretation of leaf carbon isotope ratios: influence of respired carbon dioxide. *Ecology*, **70**, 1317–1324.

Stevens C.M. & Engelkemeir A. (1988) The carbon isotopic composition of methane from some natural and anthropogenic sources. *Journal of Geophysical Research*, **93**, 725–733.

Stevens C.M. & Rust F.E. (1982) The carbon isotopic composition of atmospheric methane. *Journal of Geophysical Research*, **87**, 4879–4882.

Stevens C.M., Krout L., Walling D., Venters A., Engelkemeir A. & Ross L.E. (1972) The isotopic composition of atmospheric carbon monoxide. *Earth Planetary Science Letters*, **16**, 147–165.

Stevens C.M., Kaplan L., Gorse R. *et al.* (1980) The kinetic isotope effect for carbon and oxygen in the reaction CO + OH. *International Journal of Chemistry Kinetics*, **12**, 935–948.

Stevenson F.J. (1982) *Humus Chemistry*. John Wiley & Sons, New York.

Stewart W.D.P. (1965) Nitrogen turnover in marine and brackish habitats. 1. Nitrogen fixation. *Annals of Botany*, **29**, 229–239.

Stewart W.D.P. (1967) Nitrogen turnover in marine and brackish habitats. 2. Use of ^{15}N in measuring nitrogen fixation in the field. *Annals of Botany*, **31**, 385–407.

Stott L.D. & Kennett J.P. (1989) New constraints on early Tertiary palaeoproductivity from carbon isotopes in foraminifera. *Nature*, **342**, 526–529.

Stuermer D.H., Peters K.E. & Kaplan I.R. (1978) Source indicators of humic substances and proto-kerogen. Stable isotope ratios, elemental compositions and electron spin resonance spectra. *Geochimica et Cosmochimica Acta*, **42**, 989–997.

Stuiver M. & Braziunas T.F. (1987) Tree cellulose $^{13}C/^{12}C$ isotope ratios and climate change. *Nature*, **327**, 58–60.

Stuiver M., Burk R.I. & Quay P.D. (1984) $^{13}C/^{12}C$ ratios in tree rings and the transfer of biospheric carbon in the atmosphere. *Journal of Geophysical Research*, **89**, 11731–11748.

Stump R.K. & Frazer J.W. (1973) Simultaneous determination of carbon, hydrogen and nitrogen in organic compounds. Report 1973, UCID-16198, University of California, Livermore.

Suess H.E. (1970) Bristlecone-pine calibration of the radiocarbon time-scale 5200 B.C. to the present. In: Olsson I.U. (ed.) *Radiocarbon Variations and Absolute Chronology, Twelfth Nobel Symposium, Uppsala, August 11–15, 1969*, pp. 303–312. Almqvist & Wiksell, Stockholm and John Wiley & Sons, New York.

Sullivan C.H. & Krueger H.W. (1981) Carbon isotope analysis of separate chemical phases in modern and fossil bone. *Nature*, **292**, 333–335.

Sumi T. & Koike I. (1990) Estimation of ammonification and ammonium assimilation in surficial coastal and estuarine sediments. *Limnology and Oceanography*, **35**, 270–286.

Summons R. & Powell T.G. (1986) Chlorobiaceae in Paleozoic seas revealed by biological markers, isotopes and geology. *Nature*, **319**, 763–765.

Sweeney R.E. & Kaplan I.R. (1980a) Tracing flocculent industrial and domestic sewage transport on San Pedro shelf, southern California by nitrogen and sulfur isotope ratios. *Marine Environmental Research*, **3**, 214–224.

Sweeney R.E. & Kaplan I.R. (1980b) Natural abundances of ^{15}N as a source indicator for near-shore marine sedimentary and dissolved nitrogen. *Marine Chemistry*, **9**, 81–94.

Sweeney R.E., Liu K.-K. & Kaplan I.R. (1978) Oceanic nitrogen isotopes and their uses in determining the source of sedimentary nitrogen. In: Robinson B.W. (ed.) *Stable Isotopes in the Earth Sciences*, pp. 9–26. New Zealand Department of Science Industrial Research Bulletin 220, Wellington, New Zealand.

Sweeney R.E., Kalil E.K. & Kaplan I.R. (1980) Characterization of domestic and industrial sewage in southern California coastal sediments using nitrogen, carbon, sulfur and uranium tracers. *Marine Environmental Research*, **3**, 225–243.

Tamers M.A. & Pearson F.J. (1965) Validity of radiocarbon dates on bone. *Nature*, **208**, 1053–1055.

Tan F.C. & Strain P.M. (1979) Organic carbon isotope ratios in recent sediments in the St. Lawrence Estuary and the Gulf of St. Lawrence. *Estuarine and Coastal Shelf Science*, **8**, 213–225.

Tan F.C. & Strain P.M. (1983) Sources, sinks and distribution of organic carbon in the St. Lawrence Estuary, Canada. *Geochimica et Cosmochimica Acta*, **47**, 125–132.

Tan F.C. & Strain P.M. (1988) Stable isotope studies in the Gulf of St. Lawrence. In: Strain P.M. (ed.) *Chemical Oceanography in the Gulf of St. Lawrence*, pp. 59–77. Department of Fisheries and Oceans, Ottawa.

Tan F.C. & Walton A. (1975) The application of stable carbon isotope ratios as water quality indicators in coastal areas of Canada. In: Kaufmann B. (ed.) *Isotope Ratios as Pollutant Source and Behavior Indicators*. Proceedings of a symposium. Vienna, Austria, Nov. 19–22, 1974. Illustrated International Atomic Energy Agency, Vienna.

Tanaka N., Rye D.M., Rye R.O. II & Yoshinari T. (unpublished manuscript) High precision mass spectrometric analysis of isotopic abundance ratios in nitrous oxide by an injection of N_2O.

Tanoue E. & Handa N. (1979) Distribution of particulate organic carbon and nitrogen in the Bering Sea and the northern North Pacific Ocean. *Journal of the Oceanographic Society of Japan*, **35**, 47–62.

Tans P. (1981) $^{13}C/^{12}C$ of industrial CO_2. In: Bolin B. (ed.) *SCOPE 16: Carbon Cycle Modeling*, pp. 127–129. John Wiley & Sons, New York.

Tans P.P., Fung I.Y. & Takahashi T. (1990) Observational constraints on the global CO_2 budget. *Science*, **247**, 1431–1438.

Tate R.L. III (1987) *Soil Organic Matter: Biological and Ecological Effects*. John Wiley & Sons, New York.

Taylor C.B. (1973) Measurement of oxygen-18 ratios in environmental waters using the Epstein-Mayeda technique. Part 1: theory and experimental details of the equilibration technique. *Inst. Nucl. Sci. Pub.*, **556**, 24. Low Hutt, New Zealand.

Taylor T.G. (1970) How an eggshell is made. *Scientific American*, **222**, 88–95.

Teaford M.F. (1988) A review of dental microwear and diet in modern mammals. *Scanning Microscopy*, **2**, 1149–1166.

Teeri J.A. & Stowe L.G. (1976) Climatic patterns and the distribution of C_4 grasses in North America. *Oecologia*, **23**, 1–12.

Teeri J.A. & Schoeller D.A. (1979) $\delta^{13}C$ values of an herbivore and the ratio of C_3 to C_4 plant carbon in its diet. *Oecologia*, **39**, 197–200.

Teeri J.A. & Gurevitch J. (1984) Environmental and genetic control of Crassulacean acid metabolism in two Crassulacean species and an F1 hybrid with differing biomass $\delta^{13}C$ values. *Plant Cell and Environment*, **7**, 589–596.

Teeri J.A., Stowe L.G. & Livingstone D.A. (1980) The distribution of C_4 species of the Cyperaceae in North America in relation to climate. *Oecologia*, **47**, 307–310.

Termine J.D., Eanes E.D., Greenfield D.J. & Nylen M.J. (1973) Hydrazine deproteinated bone mineral. *Calcified Tissue Research*, **12**, 73–90.

Thackeray J.F., van der Merwe N.J., Lee-Thorp *et al.* (1990) Changes in carbon isotope ratios in the late Permian recorded in therapsid tooth apatite. *Nature*, **347**, 751–753.

Thayer G.W., Parker P.L., LaCroix M.W. & Fry B. (1978a) The stable carbon isotope ratio of some components of an eelgrass, *Zostera marina*, bed. *Oecologia*, **35**, 1–12.

Thayer G.W., Parker P.L., LaCroix M.W. & Fry B. (1978b) The stable carbon contaminant concentrations in a sewage-distorted food web. *Marine Environmental Progress Series*, **54**, 157–170.

Thayer G.W., Govoni J.J. & Connally D.W. (1983) Stable carbon isotope ratios of the planktonic food web in the northern Gulf of Mexico. *Bulletin of Marine Science*, **33**, 247–256.

Thielmann J., Tolbert N.E., Goyal A. & Senger H. (1990) Two systems for concentrating CO_2 and bicarbonate during photosynthesis by *Scenedesmus*. *Plant Physiology*, **92**, 622–629.

Thiemens M.H. & Trogler W.C. (1991) Nylon production: an unknown source of atmospheric nitrous oxide. *Science*, **251**, 932–934.

Thode H.G., Kleerekoper H. & McElcheran D.E. (1951) Isotopic fractionation in the bacterial reduction of sulphate. *Research*, **4**, 581–582.

Thode H.G., Dickman M.D. & Rao S.S. (1987) Effects of acid precipitation on sediment downcore profiles of diatoms, bacterial densities and sulphur isotope ratios in lakes north of Lake Superior. *Archives of Hydrobiology*, **4**, 397–422.

Thompson A.M. & Cicerone R.J. (1986) Possible perturbations to atmospheric CO_2, CH_4, and OH. *Journal of Geophysical Research*, **91**, 10853–10864.

Tiessen H. & Stewart J.W.B. (1983) Particle size fractions and their use in studies of soil organic matter: II. Cultivation effects on organic matter composition in size fractions. *Soil Science Society of America Journal*, **47**, 509–514.

Tiessen H., Karamanos R.E., Stewart J.W.B. & Selles F. (1984) Natural nitrogen-15 abundance as an indicator of soil organic matter transformations in native and cultivated soils. *Soil Science Society of America Journal*, **48**, 312–315.

Tieszen L.L. & Boutton T.W. (1989) Stable carbon isotopes in terrestrial ecosystem research. In: Rundel P.W., Ehleringer J.R. & Nagy K.A. (eds) *Stable Isotopes in Ecological Research, Ecological Studies*, vol. 68, pp. 167–195. Springer-Verlag, New York.

Tieszen L.L., Boutton T.W., Tesdahl K.G. & Slade N.A. (1983) Fractionation and turnover of stable carbon isotopes in animal tissues: implications for $\delta^{13}C$ analysis of diet. *Oecologia*, **57**, 32–37.

Toft N.L., Anderson J.E. & Nowak R.S. (1989) Water use efficiency and carbon isotope composition of plants in a cold desert environment. *Oecologia*, **80**, 11–18.

Toots H. & Voorhies M.R. (1965) Strontium in fossil bones and the reconstruction of food chains. *Science*, **149**, 854–855.

Trivett N.B.A. & Worthy D.E.J. (1989) Analysis and interpretation of trace gas measurements at Alert, N.W.T., with emphasis on CO_2 and CH_4. Proceedings of the Third International Conference on analysis and evaluation of atmospheric CO_2 data, past and present. *World Meteorological Organisation Report No. 59*.

Trumbore S.E., Vogel J.S. & Southon J.R. (1989) AMS ^{15}C measurements of fractionated soil organic matter: an approach to deciphering the soil carbon cycle. *Radiocarbon*, **31**, 644–654.

Tu C.K., Spiller H., Wynns G.C. & Silverman D.N. (1987) Carbonic anhydrase and the uptake of inorganic carbon by *Synechococcus* sp. (UTEX 2380). *Plant Physiology*, **85**, 72–77.

Tudge A.P. (1960) A method of analysis of oxygen isotopes in orthophosphates – its use in measurements of paleotemperatures. *Geochimica et Cosmochimica Acta*, **18**, 81–93.

Tunnicliffe V. (1991) The biology of hydrothermal vents: ecology and evolution. *Oceanography and Marine Biology Reviews*, **29**, 319–407.

Turner G.L., Bergersen F.J. & Tantala H. (1983) Natural enrichment of ^{15}N during decomposition of plant material in soil. *Soil Biology and Biochemistry*, **19**, 39–42.

Tuross N. (1989) Albumin preservation in the Taima-Taima mastodon skeleton: *Applied Geochemistry*, **4**, 255–259.

Tuross N. & Fisher L.W. (1989) The proteins in the shell of *Lingula*. In: Crick R. (ed.) *Origin, Evolution and Aspects of Biomineralization in Plants and Animals*, pp. 325–333. Plenum Press, New York.

Tuross N. & Stathoplos L. (1993) Ancient proteins in fossil bones. *Methods in Enzymology*, **224**, 121–129.

Tuross N., Eyre D.R., Holtrop M.E., Glimcher M.J. & Hare P.E. (1980) Collagen in fossil bones. In: Hare P.E., Hoering T.C. & King K. (eds) *Biogeochemistry of Amino Acids*, pp. 53–63. John Wiley & Sons, New York.

Tuross N., Fogel M.L. & Hare P.E. (1988) Variability in the preservation of the isotopic composition of collagen from fossil bone. *Geochimica et Cosmochimica Acta*, 52, 929–935.

Tuross N., Behrensmeyer A.K. & Eanes E.D. (1989a) Strontium increases and crystallinity changes in taphonomic and archaeological bone. *Journal of Archaeological Science*, 16, 661–672.

Tuross N., Behrensmeyer A.K., Eanes E.D., Fisher L.W. & Hare P.E. (1989b) Molecular preservation and crystallographic alterations in a weathering sequence of wildebeest bones. *Applied Geochemistry*, 4, 261–270.

Tuross N., Fogel M.L., Newson L. & Doran G.H. (1994) Subsistence in the Florida Archaic: the stable isotope and archaeobotanical evidence from the Windover Site. *American Antiquity* (in press).

Tyler S.C. (1986) Stable carbon isotope ratios in atmospheric methane and some of its sources. *Journal of Geophysical Research*, 91, 13232–13238.

Ulrich M.M.W., Perizonius W.R.K., Spoor C.F., Sandberg P. & Vermeer C. (1987) Extraction of osteocalcin from fossil bones and teeth. *Biochemical and Biophysical Research*, 149, 712–719.

Urey H.C. (1947) Thermodynamic properties of isotopic substances. *Journal of the Chemical Society*, 1947, 562.

US EPA (1976) *National Interim Primary Drinking Water Regulations*. Environmental Protection Agency, Washington, DC.

Vaghjiani G.L. & Ravishankara A.R. (1991) New measurement of the rate coefficient for the reaction of OH with methane. *Nature*, 350, 406–409.

van der Merwe N.J. (1982) Carbon isotopes, photosynthesis, and archaeology. *American Scientist*, 70, 596–606.

van der Merwe N.J. & Medina E. (1989) Photosynthesis and $^{13}C/^{12}C$ ratios in Amazonian rain forests. *Geochimica et Cosmochimica Acta*, 53, 1091–1094.

van der Merwe N.J. & Medina E. (1991) The canopy effect, carbon isotope ratios and foodwebs in Amazonia. *Journal of Archaeological Science*, 18, 249–259.

van der Merwe N.J., Lee-Thorp J.A. & Bell R.H.V. (1988) Carbon isotopes as indicators of elephant diets and African environments. *African Journal of Ecology*, 26, 163–172.

Van Dover C.L. (1989) Carbon and nitrogen isotopic compositions of vent and seep symbionts: a review. PhD Dissertation, MIT/WHOI Joint Program in Oceanography.

Van Dover C.L. & Fry B. (1989) Stable isotope compositions of hydrothermal vent organisms. *Marine Biology*, 102, 257–263.

Van Dover C.L., Fry B., Grassle J.F., Humphris S. & Rona P. (1988) Feeding biology of the shrimp *Rimicaris exoculata* at hydrothermal vents on the Mid-Atlantic Ridge. *Marine Biology*, 98, 209–216.

Vanlerberghe G.C., Schuller K.A., Smith R.G., Feil R., Plaxton W.C. & Turpin D.H. (1990) Relationship between NH_4^+ assimilation rate and *in vivo* phosphoenolpyruvate carboxylase activity. *Plant Physiology*, 94, 284–290.

Velinsky D.J., Pennock J.R., Sharp J.H., Cifuentes L.A. & Fogel M.L. (1989) Determination of the isotopic composition of ammonium-nitrogen at the natural abundance level from estuarine waters. *Marine Chemistry*, 26, 351–361.

Virginia R.A., Jarrell W.M., Rundel P.W., Shearer G. & Kohl D.H. (1989) The use of variation in the natural abundance of ^{15}N to assess symbiotic nitrogen fixation by woody plants. In: Rundel P.W., Ehleringer J.R. & Nagy K.A. (eds) *Stable Isotopes in Ecological Research*, pp. 345–394. Springer-Verlag, New York.

Vitousek P.M., Shearer G. & Kohl D.H. (1989) Foliar ^{15}N natural abundance in Hawaiian rainforest: patterns and possible mechanisms. *Oecologia*, 78, 383–388.

Vitousek P.M., Field C.B. & Matson P.A. (1990) Variation in foliar $\delta^{13}C$ in Hawaiian *Metrosideros polymorpha*: a case of internal resistance? *Oecologia*, 84, 362–370.

Vogel J.C. (1978a) Recycling of carbon in a forest environment. *Oecologia Planetarum*, 13, 89–94.

Vogel J.C. (1978b) Isotopic assessment of the dietary habits of ungulates. *South African*

Journal of Science, **74**, 298–301.

Vogel J.C., Talma A.S. & Heaton T.H.E. (1981) Gaseous nitrogen as evidence for denitrification in groundwater. *Journal of Hydrology*, **50**, 191–200.

Vogler E.A. & Hayes J.M. (1980) Carbon isotopic compositions of carboxyl groups of biosynthesized fatty acids. In: Douglas A.G. & Maxwell J.R. (eds) *Advances in Organic Geochemistry 1979*, pp. 697–704. Pergamon Press, Oxford.

Volz A., Ehhalt D.H. & Derwent R.G. (1981) Seasonal and latitudinal variation of ^{14}CO and the tropospheric concentration of OH radicals. *Journal of Geophysical Research*, **86**, 5163–5171.

von Caemmerer S. & Evans J.R. (1991) Determination of the average partial pressure of CO_2 in chloroplasts from leaves of several C_3 plants. *Australian Journal of Plant Physiology*, **18**, 287–305.

von Schirnding Y., van der Merwe N.J. & Vogel J.C. (1982) Influence of diet and age on carbon isotope ratios in ostrich eggshell. *Archaeometry*, **24**, 3–20.

Wada E. (1980) Nitrogen isotope fractionation and its significance in biogeochemical processes occurring in marine environments. In: Goldberg E.D., Horibe Y. & Saruhashi K. (eds) *Isotope Marine Chemistry*, pp. 375–398. Uchida Rokakuho, Tokyo.

Wada E. (1987) ^{15}N and ^{13}C abundances in marine environments with emphasis on biogeochemical structure of food web. *Isotopenpraxis*, **23**, 639–646.

Wada E. & Hattori A. (1976) Natural abundance of ^{15}N in particulate organic matter in the North Pacific Ocean. *Geochimica et Cosmochimica Acta*, **40**, 249–256.

Wada E. & Hattori A. (1978) Nitrogen isotope effects in the assimilation of inorganic nitrogenous compounds by marine diatoms. *Geomicrobiological Journal*, **1**, 85–101.

Wada E. & Hattori A. (1991) *Nitrogen in the Sea: Forms, Abundances and Rate Processes*, p. 208. CRC Press, Boca Raton.

Wada E., Kadonaga T. & Matsuo S. (1975) ^{15}N abundance in nitrogen of naturally occurring substances and global assessment of denitrification from isotopic viewpoint. *Geochemical Journal*, **9**, 139–148.

Wada E., Imaizumi R. & Takai Y. (1984) Natural abundance of ^{15}N in soil organic matter with special reference to paddy soils in Japan: biogeochemical implications on the nitrogen cycle. *Geochemical Journal*, **18**, 109–123.

Wada E., Terazaki M., Kabaya Y. & Nemoto T. (1987a) ^{15}N and ^{13}C abundances in the Antarctic Ocean with emphasis on the biogeochemical structure of the food web. *Deep-Sea Research*, **34**, 829–841.

Wada E., Minagawa M., Mizutani H., Tsuji T., Imaizumi R. & Karasawa K. (1987b) Biogeochemical studies on the transport of organic matter along the Otsuchi River Watershed, Japan. *Estuarine and Coastal Shelf Science*, **25**, 321–336.

Wada E., Mizutani H. & Minagawa M. (1991) The use of stable isotopes for food web analysis. *Critical Reviews of Food Science and Nutrition*, **30**, 361–371.

Wahlen M. & Yoshinari T. (1985) Oxygen isotope ratios in N_2O from different environments. *Nature*, **313**, 780–782.

Wahlen M., Tanaka N., Henry R. *et al.* (1988) ^{13}C, D and ^{14}C in methane. In: *Report to Congress and Environmental Protection Agency on NASA Upper Atmosphere Research Program*, pp. 315–316. NASA, Washington, DC.

Wahlen M., Tanaka N., Henry R. *et al.* (1989) Carbon-14 in methane sources and in atmospheric methane: the contribution from fossil carbon. *Science*, **249**, 286–290.

Wahlen M., Deck B., Henry R. *et al.* (1990a) Profiles of δ^{13}C and δD in CH_4 from the lower stratosphere. *Transactions of the American Geophysical Union*, 70/43, 1017.

Wahlen M., Tanaka N., Henry R. *et al.* (1990b) ^{13}C, D and ^{14}C in methane. In: *Report to Congress and Environmental Protection Agency on NASA Upper Atmosphere Research Program*, pp. 324–325. NASA, Washington, DC.

Wahlen M., Tanaka N., Deck B. *et al.* (1990c) δD in CH_4: additional constraints for a global CH_4 budget. *Transactions of the American Geophysical Union*, 71/43, 1249.

Wahlen M., Allen D., Deck B. & Herchenroder A. (1991) Initial measurements of CO_2

concentrations (1530 to 1940 AD) in air occluded in the GISP 2 ice core from Central Greenland. *Geophysical Research Letters*, **18**, 1457–1460.

Walker A.C., Hoeck H.N. & Perez L.M. (1978) Microwear of mammalian teeth as an indicator of diet. *Science*, **201**, 908–910.

Walker C.D. & Lance R.C.M. (1991) The fractionation of ^2H and ^{18}O in leaf water of barley. *Australian Journal of Plant Physiology*, **18**, 411–425.

Walker N.A. (1985) The carbon species taken up by *Chara*: a question of unstirred layers. In: Lucas W.J. & Berry J.A. (eds) *Inorganic Carbon Uptake by Aquatic Photosynthetic Organisms*, pp. 31–37. American Society of Plant Physiologists, Washington.

Wanninkhof R. (1985) Kinetic fractionation of the carbon isotopes ^{13}C and ^{12}C during transfer of CO_2 from air to seawater. *Tellus*, **37B**, 128–135.

Ward B.B. & Zafiriou O.C. (1988) Nitrification and nitric oxide in the oxygen minimum of the eastern tropical North Pacific. *Deep-Sea Research*, **35**, 1127–1142.

Ward B.B., Olson R.J. & Perry M.J. (1982) Microbial nitrification rates in the primary nitrite maximum off southern California. *Deep-Sea Research*, **29**, 247–255.

Warren P.H. (1989) Spatial and temporal variation in the structure of a freshwater food web. *Oikos*, **55**, 299–311.

Wefer W. & Killingley H.S. (1986) Carbon isotopes in organic matter from a benthic alga *Halimeda incrassata* (Bermuda): effects of light intensity. *Chemical Geology (Isotope Geoscience Section)*, **59**, 321–326.

Weiner S., Lowenstam H.A. & Hood L. (1976) Characterization of 80-million-year-old mollusk shell proteins. *Proceedings of the National Academy of Science USA*, **73**, 2541–2545.

Weiner S., Lowenstam H.A. & Hood L. (1977) Discrete molecular weight components of the organic matrices of mollusc shells. *Journal of Experimental Biology and Ecology*, **30**, 45–51.

Weiner S., Traub W. & Lowenstam H.A. (1980) Organic matrix in calcified exoskeletons. In: Westbroek P. & de Jong E.W. (eds) *Biomineralization and Biological Metal Accumulation*, pp. 205–224. D. Reidel, Publishers, New York.

Wellman R.P., Cook F.D. & Krouse H.R. (1968) Nitrogen-15: microbial alteration of abundance. *Science*, **161**, 269–270.

Welte D.H. (1969) Determination of C^{13}/C^{12} isotope ratios of individual higher *n*-paraffins from different petroleums. In: Schenck P.A. & Havenaar I. (eds) *Advances in Organic Geochemistry 1968*, pp. 269–277. Pergamon Press, Oxford.

Wheeler P.A. (1983) Phytoplankton nitrogen metabolism. In: Carpenter E.J. & Capone D.G. (eds) *Nitrogen in the Marine Environment*, pp. 309–346. Academic Press, New York.

White J.W.C. (1989) Stable hydrogen isotope ratios in plants: a review of current theory and some potential applications. In: Rundel P.W., Ehleringer J.R. & Nagy K.A. (eds) *Stable Isotopes in Ecological Research*, pp. 142–162. Springer-Verlag, New York.

White J.W.C., Cook E.R., Lawrence J.R. & Broecker W.S. (1985) The D/H ratios of sap in trees: implications for water sources and tree ring D/H ratios. *Geochimica et Cosmochimica Acta*, **49**, 237–246.

Whitfield M. & Turner D.R. (1986) The carbon dioxide system in estuaries – an inorganic perspective. *The Science of the Total Environment*, **49**, 235–255.

Whiticar M.J., Faber E. & Schoell M. (1986) Biogenic methane formation in marine and freshwater environments: CO_2 reduction vs. acetate fermentation isotope evidence. *Geochimica et Cosmochimica Acta*, **50**, 693–709.

Williams P.M. & Gordon L.I. (1970) Carbon-13:Carbon-12 ratios in dissolved and particulate organic matter in the sea. *Deep-Sea Research*, **17**, 19–27.

Williams P.M., Smith K.L., Druffel E.M. & Linick T.W. (1981) Dietary carbon sources of mussels and tubeworms from Galápagos hydrothermal vents determined from tissue ^{14}C activity. *Nature*, **292**, 448–449.

Winogradsky S. (1889) Recherches physiologiques sur les sulfobacteries. *Annals de l'Institut*

Pasteur, **3**, 49–60.

Winogradsky S. (1890) Sur les organismes de la nitrification. *Comptes rendus de l'Academie des Sciences*, **110**, 1013–1016.

Winteringham F.P.W. (1984) *Soil and Fertilizer Nitrogen*. IAEA, Vienna.

Wolin M.J. & Miller T.L. (1987) Bioconversion of organic carbon to CH_4 and CO_2. *Geomicrobiology Journal*, **55**, 239–259.

Wong W. & Sackett W.M. (1978) Fractionation of stable carbon isotopes by marine phytoplankton. *Geochimica et Cosmochimica Acta*, **42**, 1809–1815.

Wong W.W., Lee L.L. & Klein P.D. (1987a) Deuterium and oxygen-18 measurements on microliter samples of urine, plasma, saliva and human milk. *American Journal of Clinical Nutrition*, **45**, 905–913.

Wong W.W., Lee L.L. & Klein P.D. (1987b) Oxygen isotope ratio measurements on carbon dioxide generated by reaction of microliter quantities of biological fluids with guanidine hydrochloride. *Analytical Chemistry*, **59**, 690–693.

World Health Organization (1984) *Guidelines for Drinking-Water Quality*. World Health Organization, Geneva.

Wright E.K. & Hoering T.C. (1989) Separation and purification of phosphate for oxygen isotope analysis. Annual Report of the Director of the Geophysical Laboratory, Carnegie Institution of Washington, 1988–1989, pp. 137–141.

Wycoff R.W.G. (1972) *The Biogeochemistry of Amino Fossils*. Williams & Wilkins, Baltimore.

Yakir D., DeNiro M.J. & Gat J.R. (1990) Natural deuterium and oxygen-18 enrichment in leaf water of cotton plants grown under wet and dry conditions: evidence for water compartmentation and its dynamics. *Plant, Cell and Environment*, **13**, 49–56.

Yakir D. & DeNiro M.J. (1990) Oxygen and hydrogen isotope fractionation during cellulose metabolism in *Lemna gibba* L. *Plant Physiology*, **93**, 325–332.

Yakir D., DeNiro M.J. & Rundel P.W. (1989) Isotopic inhomogeneity of leaf water: evidence and implications for the use of isotopic signals transduced by plants. *Geochimica et Cosmochimica Acta*, **53**, 2769–2773.

Yapp C.J. (1979) Oxygen and carbon isotope measurements of land snail shell carbonate. *Geochimica et Cosmochimica Acta*, **43**, 629–635.

Yapp C.J. & Epstein S. (1982a) Climatic significance of the hydrogen isotope ratios in tree cellulose. *Nature*, **297**, 636–639.

Yapp C.J. & Epstein S. (1982b) A reexamination of cellulose carbon-bound hydrogen δD measurements and some factors affecting plant-water D/H relationships. *Geochimica et Cosmochimica Acta*, **46**, 955–965.

Yoshida N. (1988) ^{15}N-depleted N_2O as a product of nitrification. *Nature*, **335**, 528–529.

Yoshida N. & Matsuo S. (1983) Nitrogen isotope ratio of atmospheric N_2O as a key to the global cycle of N_2O. *Geochemical Journal*, **17**, 231–239.

Yoshida N., Hattori A., Saino T., Matsuo S. & Wada E. (1984) $^{15}N/^{14}N$ ratio of dissolved N_2O in the eastern tropical Pacific Ocean. *Nature*, **307**, 442–444.

Yoshida N., Morimoto H., Hirano M. *et al.* (1989) Nitrification rates and ^{15}N abundances of N_2O and NO_3^- in the western North Pacific. *Nature*, **342**, 895–897.

Yoshinari T. (1976) N_2O in the sea. *Marine Chemistry*, **4**, 189–202.

Yoshinari T. (1990) Emissions of N_2O from various environments – the use of stable isotope composition of N_2O as tracer for the studies of N_2O biogeochemical cycling. In: Revsbech N.P. & Sorensen J. (eds) *Denitrification in Soil and Sediment*, pp. 129–150. Plenum Press, New York.

Yoshinari T. & Knowles R. (1976) Acetylene inhibition of nitrous oxide reduction by denitrifying bacteria. *Biochemical and Biophysical Research Communications*, **69**, 705–710.

Yoshinari T. & Wahlen M. (1985) Oxygen isotope ratios in N_2O from nitrification at a wastewater treatment facility. *Nature*, **317**, 349–350.

Yoshinari T., Ueda S., Codispoti L.A. & Friederich G.E. (1990) Nitrogen and oxygen isotope composition of N_2O in the water column of Monterey Bay. *EOS*, **71**, 1143.

Young C.P. (1983) Data acquisition and evaluation of groundwater pollution by nitrates, pesticides and disease-producing bacteria. *Environmental Geology*, **5**, 11–18.

Zafiriou O.C. & McFarland M. (1981) Nitric oxide from nitrite photolysis in the central equatorial Pacific. *Journal of Geophysical Research*, **86**, 3173–3182.

Zafiriou O.C., McFarland M. & Bromund R.H. (1980) Nitric oxide in seawater. *Science*, **207**, 637–639.

Zafiriou O.C., Hanley Q.S. & Snyder G. (1989) Nitric oxide and nitrous oxide production and cycling during dissimilatory nitrite reduction by *Pseudomonas perfectomarina*. *Journal of Biological Chemistry*, **264**, 5694–5699.

Zenvirth D. & Kaplan A. (1981) Uptake and efflux of inorganic carbon in *Dunaliella salina*. *Planta*, **152**, 8–12.

Zhang J., Marshall J.D. & Jaquish B.C. (1993) Genetic differentiation in carbon isotope discrimination and gas exchange in *Pseudotsuga menziesii*: a common garden experiment. *Oecologia*, **93**, 80–87.

Ziegler H. (1989) Hydrogen isotope fractionation in plant tissues. In: Rundel P.W., Ehleringer J.R. & Nagy K.A. (eds) *Stable Isotopes in Ecological Research*, pp. 105–123. Springer-Verlag, New York.

Ziegler H., Osmond C.B., Stickler W. & Trimborn D. (1976) Hydrogen isotope discrimination in higher plants: correlation with photosynthetic pathway and environment. *Planta*, **128**, 85–92.

Zieman J.C., Macko S.A. & Mills A.L. (1984) Role of seagrasses and mangroves in estuarine food webs: temporal and spatial changes in stable isotope composition and amino acid content during decomposition. *Bulletin of Marine Science*, **35**, 380–392.

Zimba P., Sullivan M.J. & Glover H.E. (1990) Carbon fixation in cultured marine benthic diatoms. *Journal of Phycology*, **26**, 306–311.

Zimmerman J.K. & Ehleringer J.R. (1990) Carbon isotope ratios are correlated with irradiance levels in the Panamanian orchid *Catasetum viridiflavum*. *Oecologia*, **83**, 247–249.

Zumft W.G. & Kroneck P.M.H. (1990) Metabolism of nitrous oxide. In: Revsbech N.P. & Sorensen J (eds) *Denitrification in Soil and Sediment*, pp. 37–55. Plenum Press, New York.

Zyakun A.M., Bondar V.A. & Namsarev B.B. (1991) Fractionation of methane carbon isotopes by methane-oxidizing bacteria. In: *Forschungsheft C360, Reaktor der Bergakademie Freiberg*, pp. 19–27. VEB Deutscher Verlag fur Grundstoff Industrie, Leipzig.

Index

Page numbers in *italic* indicate figures and tables.